Opioid Peptides

Volume II

Pharmacology

Authors

József I. Székely, M.D., Ph.D.
Deputy Head, Medical Department
Institute for Drug Research
Budapest, Hungary

András Z. Rónai, M.D.
Staff Member
Institute of Experimental Medicine
Hungarian Academy of Sciences
Budapest, Hungary

CRC Press, Inc.
Boca Raton, Florida

Library of Congress Cataloging in Publication Data

Szekely, Jozsef I. (Jozsef Ivan)
Opioid peptides.

Includes Bibliographies and indexes.
Contents: v. 1. Research methods --
v. 2. Pharmacology -- v. 3. Opiate receptors
and their ligands.
1. Endorphins. I. Ronai, Andras Z.
II. Title. [DNLM: 1. Endorphins. QU 68 S9970]
QP552..E53S93 1983 615'.78 82-12860
ISBN 0-8493-6235-0 (v. 1)
ISBN 0-8493-6236-9 (v. 2)
ISBN 0-8493-6237-7 (v. 3)

Direct all inquiries to CRC Press, Inc., 2000 Corporate Blvd., N.W., Boca Raton, Florida, 33431.

International Standard Book Number 0-8493-6235-0 (Vol. 1)
International Standard Book Number 0-8493-6236-9 (Vol. 2)
International Standard Book Number 0-8493-6237-7 (Vol. 3)

Library of Congress Card Number 82-12860
Printed in the United States

FOREWORD

The neurosciences are among the most rapidly and spectacularly expanding branches of contemporary knowledge. This most impressing spreading of these important areas of sciences in the last two decades is documented by a series of important discoveries whereupon our general picture on the organization and on the functions of the brain is dramatically changing year by year. No wonder that on the occasion of the last International Congress of physiological sciences, organized 2 years ago this time by our scientific community here in Budapest, Hungary, about one third of all papers presented dealt with nervous regulatory phenomena.

The sudden expansion of our knowledge of the activities of the nervous system is due to new, mostly unexpected findings among which the topic of the present monograph is a typical example: two decades ago nobody could even guess the bursting career of the opioid peptides.

The authors of this book were fortunate to join the research endeavor in this fascinating field in its very early period. Thus they both have personal experience in this domain in addition to their thorough and competent expertness. I am proud of the fact that the senior author, Dr. József Székely, was my former student and co-worker for many years working with me in several fields of psychophysiology and behavioral sciences. His outstanding erudition and creativity constitute a solid guarantee for the value of the present work. Similar qualification is valid for the junior author, Dr. András Rónai, a talented and devoted young pharmacologist. They both represent the most active and productive generation of Hungarian scientists, owing to whom Hungarian neuroscience became an important component of the mainstream of world brain research.

I highly recommend this book to all those physicians, psychologists, biologists, and other related specialists, who are driven by intellectual curiosity in order to get acquainted with the important and exciting field of neuromodulators of the brain.

Budapest, June 1982

Gyorgy Adam M.D., D.Sc.
Professor of Physiology
Member of the Hungarian Academy of Sciences

PREFACE

Since the late 1960s it is rapidly becoming clear that our classical conception on neuroregulation needs profound revision. It is no more tenable that the functioning of the excitable tissues (nerve and muscle cells) is regulated exclusively by locally, i.e., transsynaptically acting neurotransmitters, and only the activity of nonexcitable tissues is controlled by hormones (produced in remotely located endocrine glands and transported via the blood flow). Furthermore, it is already unequivocally proven that the dendrites and somata of nerve cells located on postsynaptic membranes or on the effectors are not the only sites of neuroregulation. As a matter of fact, an important locus of subtle regulation of signal transmission is located on the presynaptic portion of nerve cells which produce the neurotransmitters. A new group of endogenous neuroregulators has appeared on the horizon, which cannot be termed either classical neurotransmitters or hormones. These substances, mostly peptides, resembling in certain aspects the classical neurotransmitters, in other respects the peripheral hormones, are called nowadays neuromodulators. From year to year new ones are discovered or some previously isolated substances are proven to function as neuromodulators. With understanding the mode of their action our conception on chemical neuroregulation in the organism is getting more and more complex. Simultaneously, the development of relevant methods for studying receptor binding and biochemical events linked to receptors gave a closer insight into receptor-mediated processes.

The discovery of opiate receptors and their endogenous ligands (endorphins, enkephalins, etc.) represents a typical example of these recently burgeoning investigations. The authors of this book shared the enthusiasm of many fellow researchmen in this field since in a very early phase of these studies, they had the advantage of a fruitful cooperation with highly qualified preparative biochemist (headed by L. Gráf) and peptide synthetising (headed by S. Bajusz) groups. Moreover, they had already been working in the field of narcotic analgesics for some years before the most critical period of endorphin research (1976 to 1977). Consequently, this review cannot be regarded as a completely unbiassed description of certain events, as they represented personal experience for the authors themselves. Nevertheless, great care has been taken to remain as objective as possible.

The main purpose of this book is to give a detailed survey on the pharmacology of recently discovered endogenous opioid peptides and their synthetic analogues. The data compiled in these volumes will probably enable the academic scientists and those engaged in industrial research to become better acquainted with the results of endorphin research. It is hoped that the material will be useful also for those clinicians who are investigating the role of endogenous opioids in the pathogenesis of various diseases and the contingent therapeutic application of opioid peptides or their antagonists. Since the detection of formerly unknown neurohormones may always lead to the development of new drugs, special emphasis is given to the data on the structure-activity relationships. For those who are closely involved in the research of endogenous opioids, it is believed that the first-hand information is superior than any comprehensive treatise; furthermore it is hoped that they could offer some new aspects and also certain new data not accessible to them.

Reviewing the pharmacology of opioid peptides, primarily the biological aspects of endorphin research are discussed. Especially the data available on the in vitro and in vivo pharmacological properties of opioids and their clinical significance are reviewed in detail. As for the detailed description of biochemical isolation of natural opioids, the synthesis of their analogues, the physical and chemical properties of these compounds, the necessary references can be found in the corresponding chapters. In general, it was tried to quote all the relevant papers, but those published during the last one and a half years are not included due to the time-consuming nature of processing lengthy manuscripts like this one.

The first volume of the book is devoted to the description of in vitro and in vivo methods used in the opiate research. In this part the most characteristic pharmacological properties of classical morphine congeners are discussed in view of our present knowledge on the existence and multiplicity of opiate receptors. The pharmacological properties of opioid peptides are outlined in the second volume. The third volume contains the data on the structure-activity relationships and those on the clinical significance of endorphin research are discussed also there.

Thanks are due to several outstanding scientists for permission to reprint some very instructive illustrations from their original works.

The authors want to express their gratitude to Dr. Tibor Láng, Director of the Institute for Drug Research, Budapest, Hungary for his moral support on writing this book. Dr. Erzsebet Miglecz and Dr. Ilona Berzétei have rendered valuable help with collecting, arranging, and revising the material. Ms. Teréz Sági, Ms. Gizella Szegvári, and Ms. Tünde Tóth took part in the secretarial work.

Lastly the authors wish to thank their wives, Maria Székely and Dr. Ilona Berzétei for their tolerance, encouragement, and constant support to complete these volumes.

József I. Székely
András Z. Rónai

THE AUTHORS

József I. Székely, M.D., Ph.D., is Deputy Head of the Medical Department at the Institute for Drug Research in Budapest, Hungary.

Dr. Székely, born in 1937, received his medical training at the Semmelweis University Medical School in Budapest. After having obtained his first degree (M.D.) in 1961, he spent eight years at the Physiological Department of the same university (in the late 1960s as assistant professor). In the meantime Dr. Székely acquired the M.Sc. degree in psychology at the Eötvös Loránd University in Budapest. He defended his thesis on the conditioning of evoked potentials in 1970. Dr. Székely joined the Institute for Drug Research in 1969.

He is co-author of about 60 papers more than half of them published abroad, i.e., in Western European and American journals. His recent research interests are related to the psychopharmacology, more closely to behavioural analysis of opioid peptides and other centrally acting pharmaca.

András Z. Rónai, M.D., born in 1947, received his first degree (M.D.) in 1971 at the Semmelweis University Medical School.

He started his pharmacological studies at the Department of Pharmacology of Semmelweis University Medical School in 1968 as an undergraduate working in Professor E. S. Vizi's team. After graduating he continued his activities at the same Department as junior lecturer. As a part of his postgraduate training, he spent a one-year fellowship at the Montefiore Hospital and Medical Center and Albert Einstein College (1974-1975) in the laboratory of Professor F. F. Foldes.

Late in 1975 he joined the Department of Pharmacology at the Institute for Drug Research. At present he is working as staff member in Professor Vizi's group at the Institute of Experimental Medicine of Hungarian Academy of Sciences.

Dr. Rónai, in addition to Hungarian scientific societies, is the member of European Neuroscience Association, International Brain Research Organization, and International Narcotic Research Club.

Dr. Rónai has published over 40 papers, with three exceptions in Western periodicals; besides, he is the co-editor of book entitled *Endorphins'78.* His major research interests are the pharmacology of narcotics, the pharmacological and physiological aspects of endogenous opioids, and the mechanisms of presynaptic regulation.

ABBREVIATIONS

β-LPH	Beta-lipotropin
α-EP	Alpha-endorphin
β-EP	Beta-endorphin
γ-Ep	Gamma-endorphin
E	Enkephalin
Met-E	Methionine-enkephalin
Leu-E	Leucine-enkephalin
DALA	
D-Met, Pro-EA	(2-D-Met,5-Pro) -enkephalinamide
FK 33-824	(2-D-Ala,4-MePhe,5-Met) 0 (-ol) -enkephalin

To our wives

Maria Székely and Dr. Ilona Berzétei

Opioid Peptides

Jozsef I. Szekely and Andras Z. Ronai

Volume I: Research Methods

In Vitro Analysis
The Most Characteristic In Vivo Effects of Opiates

Volume II: Pharmacology

The Mode Of Action of Opioids
Electrographic Analysis of Opioid Mechanisms
Analgesic and Behavioral Effects of Opioids
The Role of Endogenous Opioids in the Vegetative Regulation
Opioid Peptides and the Neuroendocrine Functions

Volume III: Opiate Receptors and their Ligands: Experimental and Clinical Aspects

Opiate Receptors
Detecion of Endogenous Ligands of Opiate Receptors
Regional Distribution of Enkephalins and Endorphins in the Brain and in the Peripheral Tissues
Biosynthesis, Release, and Biodegradation of Enkephalins and Endorphins
Structure-Activity Relationships
The Question of Metabolic Stability and Altered Binding Properties
The Pharmacological Profile of the Most Active Enkephalin Analogues
The Possible Role of Endogenous Opioids in Stimulation and Acupuncture-Induced Antinociception
Endogenous Opioids and Opiate Tolerance, Dependence, and Addiction
Clinical Investigation of the Opioid Peptides
Conclusions: Perspectives of Opioid Peptide Research

TABLE OF CONTENTS

Volume II

Chapter 1
The Mode of Action of Opioids 1
A. Z. Ronai and J. I. Szekely

Chapter 2
Electrographic Analysis of Opioid Mechanisms 33
J. I. Szekely

Chapter 3
Analgesic and Behavioral Effects of Opioids 109
J. I. Szekely

Chapter 4
The Role of Endogenous Opioids in the Vegetative Regulation 155
J. I. Szekely

Chapter 5
Opioid Peptides and the Neuroendocrine Functions 207
J. I. Szekely

Index 229

Chapter 1

THE MODE OF ACTION OF OPIOIDS

A. Z. Rónai and J. I. Székely

TABLE OF CONTENTS

I. The Mode of Action at Cellular Level 2
A. Modulation of the Adenylate Cyclase Activity by Opioids 2
B. The Connection of Opioid Effects with Ca^{2+} Ions 4
C. The Changes in the Composition of Neural Membranes Elicited by Opioids 5

II. Opioids in the Neural Regulation 5
A. Modulation by Opioids of the Release of Acetylcholine (Ach) 6
B. The Effect of Opioids on the Release of Norepinephrine (NE) 6
C. The Effect of Opioids on Dopamine (DA) Release 7
D. The Effect of Opioids on Brain Serotonin (5-HT) In Vitro 8
E. The Effect of Opioids on Histamine (HA) Release In Vitro 8
F. The Effect of Opioids on the Release of Substance P (SP) In Vitro 8
G. The Effect of Opioids on the Release of γ-Aminobutyric Acid (GABA) 9
H. The Effect of Opioids on the Release of Vasopressin and Oxytocin 9
I. The Effect of Opioids on the Release of Purine Nucleotides 9

III. In Vivo Metabolic Effects of Opioid Peptides 10
A. Effects on Serotonin Metabolism 10
B. Effects on Catecholamine Turnover 16
C. Effects on Acetylcholine Turnover 18
D. Other Central Metabolic Effects 19

References 21

I. THE MODE OF ACTION AT CELLULAR LEVEL

As it has been demonstrated in Chapter 1, Volume III, specific opiate receptors are confined almost exclusively to neuronal elements;[1] thus the specific opiate actions are thought to be mediated primarily if not exclusively via neural mechanisms.

The direct, specific opiate effect on cell function appears to be mostly of inhibitory character. Of the very few stimulatory actions observed under certain experimental conditions, i.e., facilitation of the Renshaw cells in the spinal cord[2] or hippocampal pyramidal neurons[3] (see also Chapter 2, Volume II) or the enhancement of acetylcholine release from isolated striatal slices in vitro[4] the latter two have been shown to be exerted by "disinhibitory" (i.e., basically inhibitory) mechanism (see below).

The propositions that endogenous opioid substances may function as neurotransmitters/modulators or (neuro)-hormones (the definitions of terms see in McLennan's,[5] Florey's,[6] and Stewart's and Channabasavaiah's[7] work) appear to be equally justified.[7-11] The site of opiate effect on the neuronal membrane can be either "presynaptic" or "postsynaptic"[8,12,13] (for the morphological-functional definition of "pre"- and "post"-synaptic neuronal membrane surfaces see in Reference 14). The potential functional significance of specific opiate binding sites on the axonal membrane[15] is not understood yet, since opioid substances do not affect specifically the axonal conduction,[16] though in high doses a nonopiate-specific impairment of impulse conduction can be demonstrated for several morphine congeners[17] (see also Chapter 2, Volume II). Primarily or exclusively presynaptic site of action can be supposed in relation to the depression of neuroeffector transmission at certain peripheral autonomic junctions[18-24] (see also below and Chapter 1, Volume I) and at various synapses in the CNS[15,25-27] (see also below). On the other hand, in certain cases one has to conclude on postsynaptic site of action to explain satisfactorily the results obtained by measuring single unit activity in nerve cells[12] or by detecting opiate-induced changes in the conductance of postsynaptic neuronal membrane[28,29] (see also Chapter 2, Volume II).

Whatever be the site of opiate action, very little is known about the chain of biochemical events initiated by the binding of an opioid agonist to its specific binding site(s) on the neuronal membrane. Three mechanisms which are likely to be involved in eliciting the opiate-specific cellular response upon the binding of opiate agonist to the receptor, shall be discussed below. These are as follows:

- Modulation of adenylate cyclase activity by opioids
- Changes in the composition of neural membranes
- The connection of opioid effects with Ca^{2+} ions (see also Chapter 2, Volume I)

To establish the possible causal interrelationships between these mechanisms or to clarify exactly their contribution to the ultimate cellular response much experimental work remains to be done.

A. Modulation of the Adenylate Cyclase Activity by Opioids

The interrelationships between 3′:5′-cyclic monophosphate (cAMP) and morphine action was first suggested by Way et al.[30,31] The proposition was based on the assumption that the interactions of opioids with biogenic amine systems (see Chapter 2, Volume I) may be connected with cAMP as a possible common denominator.[30-33] Indeed, as it has been demonstrated, exogenously administered cAMP or the enzyme-protected analogue thereof[34] and also some pharmacological manipulations enhancing the endogenous cAMP levels in the brain were capable of attenuating opiate analgesia[31,35,36] or to enhance morphine tolerance.

Measuring the changes in brain cAMP levels or the activity of the cAMP forming enzyme (adenylate cyclase) following opiate administration, rather controversial results have been reported[37-43] (see also Chapter 2, Volume I).

Upon examining the influence of opiates on the adenylate cyclase activity in vitro in brain striatal slices, homogenates or hyposmotically shocked mitochondrial-synaptosomal fraction, the data obtained by different groups were rather conflicting.[33] There is a good agreement though not without exception,[42] that the basal enzyme activity is not altered significantly by opiate agonists.[39,40,43-46] Furthermore, the majority of authors found the PGE_1[44-47] or dopamine-[44-46,48,49] induced activation of adenylate cyclase specifically depressed by opioids in vitro in different brain preparations. Others, however, could not confirm these observations.[33,40,43]

To eliminate controversies resulting from methodological problems, a new model for studying in vitro the specific opiate actions has been sought. The neuroblastoma x glioma hybrid cells NG 108-15 and the homogenates thereof provide an easily accessible source of opiate receptors[50] which remain coupled to adenylate cyclase[51-54] more tenuously than those in brain homogenates seem to be. Thus, the results obtained in this cultured cell line appeared to be more reproducible[33,54,55] than those obtained in brain preparations. Opiates[33,51,52] as well as opioid peptides[54,56-61] have been found to inhibit basal and PGE_1-stimulated adenylate cyclase activity in NG 108-15 hybrid cell cultures in a specific, opiate receptor-mediated process. The opiate receptor binding in this cell line is influenced in similar manner by cations[55,62-64] as it has been demonstrated for brain opiate receptors[65-67] (see also Chapter 1, Volume III). The coupling of opiate receptors to adenylate cyclase is modulated specifically by guanine nucleotides in NG 108-15 hybrid cells.[55,68] This cell line appears to contain a homogeneous population of opiate receptors; Met-E and the peptides of similar character are more potent opioid agents than β-EP and related longer peptides or the morphine congerens in this preparation[54,59] (see Chapter 1, Volume III and also Table 4 of Chapter 5, Volume III). The relevance of this assay system in modeling opioid effects exerted in the CNS was strongly supported by the demonstration that opiate tolerance and dependence can be developed and satisfactorily interpreted in NG 108-15 hybrid cells.[52]

Otherwise, opioid peptides are capable of inhibiting adenylate cyclase activity also in brain (striatal) homogenates.[69-71]

Using the regulation of opiate receptor binding by GTP as an indicator (see Chapter 1, Volume I), it has been shown[72-74] that some opiate receptors in the organism may, while others may not be coupled to adenylate cyclase. When they are coupled, the coupling occurs in negative fashion, i.e., the agonists when bind to these receptors, exert inhibitory effect on the cyclase activity. The functional analogy between these opiate and α_2-adrenergic and muscarinic cholinergic receptors,[61,75-78] which are also negatively coupled to cyclase, is attractive. The mechanism by which these negatively coupled receptors affect cyclase activity is much less clarified than in the case of the positively coupled ones.[78,79] One possible mechanism has emerged, namely that the actions of opioids on the activity of adenylate cyclase were mediated via Ca^{2+} ions. The activity of adenylate cyclase in brain and NG 108-15 hybrid cell line is dependent on the concentration of Ca^{2+}. Low concentrations of Ca^{2+} activate, while higher concentrations inhibit the enzyme.[61,80-82] Considering the mutual functional antagonism between the opiates and Ca^{2+} ions[83] (see below and Chapter 2, Volume I), it has been forwarded[84] that the opioids may inhibit the activity of adenylate cyclase by hindering the production of Ca^{2+} levels in the critical cellular compartment, necessary for cyclase activation.

However, in NG 108-15 hybrid cells the action of opioids was found independent of the concentration of Ca^{2+} in the incubation medium and it was not found influenced by the presence of ionophore A 23187, which is capable of enhancing the cytosolic Ca^{2+} concentration.[85]

Since in these experiments "Ca^{2+} free" medium was reported to be produced by the simple omission of Ca,$^{2+}$ and EGTA was not applied to secure Ca^{2+}-deficient environment, furthermore, the experimental design covered the demonstration of opioid effect only at a single dose level, no unanimous conclusions can be drawn from the results. In a recent study, however, using also EGTA, Wilkening et al.[61] have demonstrated that the inhibition of adenylate cyclase by opioids in the very same hybrid cell line is a function of Ca^{2+} concentration.

Thus, the possibility that the action of opioids on adenylate cyclase is mediated via Ca^{2+} ions, has to be considered, but further experimental confirmation is needed.

Considering all these experimental data, the view seems to be well founded that the regulation of adenylate cyclase activity by opioids in certain systems where such a functional connection exists may represent one of the very first biochemical events in initiating opiate-specific cellular response upon the binding of opiate agonists to the receptors. However, the "next" link in the chain, i.e., cAMP is involved in a plethora of cellular functions and nothing is known that which of them might be particularly relevant to opiate action.

B. The Connection of Opioid Effects with Ca^{2+} Ions

It has long been known that Ca^{2+} but not Ba^{2+}, Sr^{2+}, or Mg^{2+} counteracts the analgesic effect of morphine[83,86,87] (see also Chapter 2, Volume I) and attenuates the development of tolerance if applied during the chronic treatment with morphine.[88-90] This antianalgesic property of Ca^{2+} was found to be shared by Mn^{2+} and, in contrast to the finding of Hano et al.[86] also by Mg^{2+}.[87] Furthermore, Ca^{2+} ionophores seem to potentiate the morphine-antagonizing action of Ca^{2+}.[87]

Ca^{2+} antagonizes the effect of opiate agonists also in vitro in electrically stimulated longitudinal muscle-myenteric plexus preparation of guinea pig ileum, taken either from nontolerant or tolerant animals[91,92] and also in mouse vas deferens preparatum.[93] The agonist dose-response curves in guinea pig ileum were shifted parallel to the right by Ca^{2+} in a dose-dependent manner;[91,92] the kinetics of antagonism fulfilled several criteria of competitive interaction. There are some indications, however,[92] that the site of Ca^{2+} antagonism must be "behind" the opiate binding site of the receptor.

On the other hand, agents like La^{3+} or Ce^{3+} that block the Ca^{2+} influx into the cells[94] mimic some actions of morphine.[95,96] The antinociceptive action of La^{3+} could be partially antagonized by naloxone; furthermore, in morphine-tolerant mice there is a cross-tolerance to La^{3+} and in dependent state La^{3+} can substitute for morphine[95,96] (see also Chapter 2, Volume I). Absence of extracellular Ca^{2+} for a prolonged period produces phenomena in NG 108-15 hybrid cell line that are analogous with the ones seen after chronic opiate administration.[85]

Acute administration of opiate agonists to rodents after a transient enhancement,[97] specifically decreases the Ca^{2+} content in the hypothalamus, hippocampus, striatum, and cerebral cortex.[98-100]

Upon the development of opiate tolerance, brain Ca^{2+} levels are increased.[99,101] The opiate effect on brain Ca^{2+} content could be localized to the synaptic region of neurones; opiate agonists have been shown to decrease specifically and selectively the Ca^{2+} levels in the synaptosomal fraction[102] (see also Chapter 2, Volume I). Opioid substances, including the peptides, inhibit specifically the uptake of Ca^{2+} into the synaptosomes in vitro.[103,104]

This uptake-inhibitory action of opioids in vitro is even more pronounced if the Ca^{2+} uptake is stimulated by K^+-depolarization.[105,106] The presence of ATP modifies this opiate action.[106] It has been suggested[106] that opioid agonists may inhibit two components of synaptosomal Ca^{2+} uptake:

1. An ATP-dependent, K^+-independent process

2. An ATP-independent, K^+-sensitive process

The inhibitory action of opiates on depolarization-stimulated Ca^{2+} uptake into the synaptosomes, appears to be particularly relevant to their inhibitory effect on the release of certain neurotransmitters (see below).

C. The Changes in the Composition of Neural Membranes Elicited by Opioids

Chronic opiate administration has been reported to result in decreased phosphorylation of protein in synaptic plasma membranes.[107-112] The action was opiate specific, and restricted to certain protein bands.[109-112]

Under properly chosen experimental conditions[110] the decreased phosphorylation of synaptic membrane proteins by opioids including peptides, could also be demonstrated in vitro, only inadequate experimental design[113] might create confusion. The same holds for the experiments in which the acute in vivo administration of opioids was followed by the in vitro demonstration of altered synaptic membrane phosphorylation.[112]

The time course and the character of the above-discussed opiate effect suggested that it might be relevant primarily to long-term opiate actions or to the action(s) of "neuromodulator"-type endogenous opioid substances. Whether this effect is related to the opiate action on adenylate cyclase activity or involves an independent mechanism (i.e., a dual mechanism exists) remains to be established.[110]

In another "opiate positive" clonal cell line (with specific opiate receptors and opiate receptor-modulated adenylate cyclase) in N4TG1 mouse neuroblastoma cells, opioid substances specifically inhibited the synthesis of gangliosides and membrane glycoproteins.[114-116] This opiate receptor mediated process results in an alteration of the cell membrane structure with reduced content of surface-associated negatively charged components.[115] The changes in the composition of membrane would result in altered responsiveness of the exposed cells; it has been suggested,[115] that a similar mechanism may explain the altered responsiveness of opiate-sensitive neurons in vivo following chronic exposure to narcotics.

Since both ganglioside and glycoprotein biosynthesis are controlled by the intracellular concentrations of cAMP,[115,117,118] it is possible that also this opiate effect is mediated via the adenylate cyclase system.[115] The activity of glycosyl transferases present in brain synaptic membrane fractions may be regulated through cAMP mediated phosphorylation.[115,119] It is worth mentioning that membrane sialoglycosphingolipids, among other functions, may take part in the control of cellular calcium regulation and also of the ion transport.[114]

II. OPIOIDS IN THE NEURAL REGULATION

The possible connections of acute and long-term opiate effects with various neuronal systems working with different neurotransmitter substances had been discussed in Chapter 2, Volume II. In the following sections the influence of opioid substances on the functions of different neurotransmitter systems will be discussed. In Section II we shall restrict our attention on the release process of individual neurotransmitters and in more general sense, of active substances of neural origin, in vitro. In Section III the in vivo (*in situ*) effects will be reviewed, which can be regarded as integrated result of the actions exerted by opioids on mutually interdependent individual neuronal systems. Thus, the in vitro and in vivo (*in situ*) opiate effects on the release of a given neurotransmitter should not be necessarily identical unless the in vivo experimental design provides a clearcut parallelism with the in vitro studies. It is worth mentioning in advance that opioids appear to act primarily on the release process of various neural substances; in general, the opiate-induced changes in their biosynthesis and metabolism appear to be secondary phenomena.

Characteristic opioid effects can be demonstrated in isolated preparations taken from the peripheral autonomic nervous system as well as from the CNS. The peripheral preparations have the advantage that in general, fewer neuro-neuronal interactions have to be taken into consideration on interpreting the opiate effects than in the case of CNS preparations.

The great majority of effects to be discussed below are opiate-specific, i.e., they can be antagonized by the opiate antagonists naloxone or naltrexone; the few exceptions will be clearly indicated.

A. Modulation by Opioids of the Release of Acetylcholine (Ach)

Opiates depress the transmission from the myenteric plexus to the longitudinal muscle of guinea pig ileum by decreasing the amount of Ach released by nerve stimulation.[120-124] The effect is due to the impairment of Ach release process,[120,125] since neither the nerve conduction nor the Ach synthesis is hindered by the opiates in the concentration range where they are capable of reducing specifically the Ach eliberation. The effects are more prominent at low than at high-frequency stimulation, similarly to the inhibitory actions of α-adrenergic stimulants in this preparation.[126] This depressant effect of morphine congeners is shared also by opioid peptides[127-129] (see also Chapter 1, Volume I). It is not fully clarified whether the opioid substances act primarily at the nerve terminals releasing the transmitter[130,131] or by hyperpolarizing the somata of myenteric neurons;[132-134] by all probabilities, both sites of action have to be taken into consideration.

The cholinergic transmission in rabbit or rat intestine is unaffected by morphine,[124] while the transmission from the vagus nerve to the sino-auricular node has been found to be morphine-sensitive in the rat and rabbit, but not in the guinea pig.[135] In the light of recent finding that in some isolated organs irresponsive to morphine certain opioid peptides may still exert specific opioid effect,[136,137] these questions must be reexamined.

In the *in situ* experiments resembling the in vitro conditions, enkephalins similarly to morphine[138] have been shown to specifically inhibit the Ach release from cat cerebral cortex.[139,140] Met-E reduced also the K^+-stimulated efflux of Ach from rat hippocampal slices.[141]

In rat striatal slices morphine, β-EP and synthetic enkephalin analogues enhanced the ouabain-induced Ach release.[142,143] However, after the nigrostriatal dopaminergic neurons had been previously destroyed by i.c.v. 6-hydroxy-dopamine injection, the same opioid agents reduced instead of enhancing the ouabain-evoked release of Ach.[143] Since, under the same experimental conditions, exogenous dopamine has been shown previously to inhibit the Ach eliberation elicited by ouabain,[144,145] the stimulatory effect of opioids on Ach-release in nondenervated preparation must be regarded as a disinhibitory phenomenon.[143] The direct effect of opioids on the release of Ach appears to be of inhibitory character also in the striatum.

B. The Effect of Opioids on the Release of Norepinephrine (NE)

The specific effect of opioids exerted on catecholamine-release appears to be due to an action on the release process itself and not the secondary consequence of altered synthesis or uptake processes. Morphine was found to be ineffective in influencing tyrosine hydroxylase activity in vitro.[146-149] and only high doses of morphine congeners were capable of producing a nonopiate specific inhibition of catecholamine uptake[150-152] or were without effect.[150]

Of the peripheral autonomic nervous system several noradrenergic neuroeffector junctions have been shown to be affected specifically by morphine. At these junctions opioids depress the synaptic transmission by inhibiting the release of NE from the nerve terminals. Morphine-sensitive peripheral noradrenergic junctions can be demonstrated in the cat nictitating membrane[154-162] and in mouse vas deferens preparation[163,164] (see also Chapter 1, Volume I). On the other hand, the noradrenergic transmission is not affected by morphine in rabbit, guinea pig and rat vas deferens,[136,137,163] in rabbit heart,[165] pulmonary[166] and ear artery,[161,162,167] portal vein[157] and in guinea pig ileum myenteric plexus.[157] As a rule, also in these preparations, opiates are more effective in reducing the release of NE elicited in vitro by low than by high-frequency electrical stimulation.

At all the peripheral synapses where opiates act, opioid peptides are also effective.[127-129,159-162,168,169] Actually, the mouse vas deferens has a high relative sensitivity to enkephalins.[127,168,169] Furthermore, in rat vas deferens, where morphine had been found to be ineffective, β-EP[136,137] and some other opioid peptides[170,171] were reported to exert specific opiate agonist action, i.e., to depress the neuroeffector transmission by acting presynaptically, in a naloxone-reversible manner. This effect of opioid peptides could not be demonstrated in vasa of cats and guinea pigs.[136] Moreover, Met-E,[167] some enkephalin analogues and to a lesser extent β-EP[161,162] are capable of suppressing the noradrenergic transmission by a presynaptic mechanism in electrically stimulated rabbit ear artery. This effect, however, could be counteracted only by high doses of opiate antagonists.

At these peripheral autonomic junctions the opioid substances act directly upon the transmitter release sites[157,172,173] probably by reducing the availability of Ca^{2+} for the release process.[173]

As for the CNS preparations, morphine was found to be ineffective in influencing the spontaneous efflux of ^{14}C-NE from rat hypothalamus synaptosomes,[150] while it reduced the outflow of ^{3}H-NE from rat cerebrocortical slices, elicited by low-frequency field electrical stimulation.[174] Also enkephalin reduced the stimulation-(K^{+}- or field electrical stimulation)-induced release of ^{3}H-NE from cerebral (occipital) and cerebellar cortical slices, acting on opiate receptors present on noradrenergic nerve terminals.[175-177]

Otherwise, opioid substances are capable of affecting the activity of noradrenergic neurones by acting on opiate receptors on the nerve terminals and on the somata located in the locus coeruleus as well. Opioid substances have been shown to inhibit the firing of locus coeruleus neurons,[178,179] acting on the rich opiate receptor supply of noradrenergic nerve cells[179,180] (see also Chapter 1, Volume III). Furthermore, enkephalin-containing nerve terminals appear to be in close contact with noradrenergic nerve elements in this area.[181]

C. The Effect of Opioids on Dopamine (DA) Release

Clouet and Williams[150] have reported on the ineffectiveness of morphine in altering the spontaneous outflow of DA from rat striatal synaptosomes. Others, however, have found that morphine,[182,183] β-EP[184] and Met-E[141] inhibited effectively and specifically the K^{+}-stimulated DA release from striatal preparations. Using an indirect experimental approach, Vizi et al.[143] and Schwartz et al.[185] have come to the same conclusions. Methadone has been reported to be ineffective in reducing the striatal DA outflow in slice preparations, upon electrical stimulation.[186]

β-EP has also been shown to reduce the ouabain-induced DA output from isolated rat neurohypophysis, in naloxone reversible manner.[187]

Furthermore, the prolactin-releasing effect of opioids has been attributed mainly to their inhibitory action on dopaminergic neurons at hypothalamic level (see Chapter 5, Volume II).

To resolve certain contradictions between the in vitro and in vivo measurements, (see Chapter 2, Volume I and Chapter 2, Volume II), Kuschinsky and Hornykiewitz[188] have suggested that morphine may inhibit the functional dopaminergic activity by diverting the newly synthetized DA to catabolic sites; furthermore, the effect of morphine on the striatal DA-efflux in vivo may depend on the actual functional state of nigrostriatal dopaminergic neurons.[189]

D. The Effect of Opioids on Brain Serotonin (5-HT) In Vitro

Several authors have found that morphine increased the brain 5-HT metabolism and synthesis,[190-192] which might indicate accelerated release. However, morphine is a very poor releaser of 5-HT in vitro in brain synaptosomes;[193] etorphine, which significantly enhances the 5-HT metabolism in the brain,[194] was found virtually devoid of 5-HT releasing activity in vitro.[193] Met-E was also reported to be ineffective in altering the K^+-stimulated release of 5-HT from isolated hypothalamic slices.[141] It has been suggested that the increase of 5-HT metabolism induced by morphine in vivo might be due to the changes in the bioavailability of brain tryptophan and 5-hydroxytryptophan, and also to increased tryptophan hydroxylase activity.[194-198]

On the other hand, methadone and to a certain extent pethidine inhibited the uptake of 5-HT into synaptosomes and slices prepared either from whole rat brain, or from different brain regions, including the periaqueductal grey and hypothalamus.[193,199-204] Morphine had no such activity;[193,204] furthermore, there was no correlation between the analgesic potencies of opiates and their capabilities to inhibit 5-HT uptake.[193]

Though Ciofalo[199] has reported on the partial antagonism of methadone-induced blockade of 5-HT uptake by naloxone, Donzanti and Warwick[204] found that naloxone enhanced rather than reduced the inhibitory action of methadone in rat periaqueductal grey slices.

Thus, the inhibitory effect on 5-HT uptake does not appear to be a specific, opiate receptor-mediated effect; in addition, it is unlikely that this action could be related to the analgesic action of opiates.

E. The Effect of Opioids on Histamine (HA) Release In Vitro

Studying the K^+-stimulated Ca^{2+}-dependent release of histamine, the process showed some characteristics that were consistent with the eliberation of a transmitter substance.[141,205,206] However, Met-E (10^{-5} *M*) was found to be without effect on the potassium-evoked release of HA from rat hypothalamus slices.[141]

F. The Effect of Opioids on the Release of Substance P (SP) In Vitro

Substance P and Met-E appear to be mutually interrelated neuropeptides in several neural functions. SP has been reported to be present in processes of primary sensory neurones;[207,208] it was assumed to be involved in conveying sensory information.[209] One site of antinociceptive action of opioid substances may be the suppression of the transmission of noxious sensory input; the effect at these sites might be brought about by interfering with SP-mediated processes[208,210] (see also Chapter 2, Volume II). SP appears to be involved in nociceptive mechanisms not only at spinal,[211] but also at hypothalamic level.[212] On the other hand, SP has been reported to exert naloxone-reversible analgesic effect,[213-216] though this finding could not be corroborated unambiguously.[217] Furthermore, SP has been shown to suppress nalorphine-induced abstinence syndrome in mice.[218] SP bears some structural similarity to enkephalins; however, in the experiments of Terenius,[219] Chipkin and Stewart[220] SP did not show any opioid activity in the receptor binding assay and in guinea pig ileum, respectively. However, a pentapeptide fragment of SP (Phe-Phe-Gly-Leu-Met-NH_2) has been shown to possess opioid agonist properties in rat vas deferens preparation.[221]

SP and enkephalin-containing nerve terminals have been demonstrated to exhibit parallel distribution in several CNS areas[181,208] as well as in the intestine.[222]

Substance P can be released from the spinal cord preparation from newborn rats,[223] rat hypothalamic slices,[224] and synaptosomal preparation[225] in a manner similar to that of other suspected neurotransmitters.

The release of SP induced by K^+ from slices containing the nucleus spinalis nervi trigemini, could be specifically inhibited by opioid agonists including opioid peptides.[226] Using indirect experimental tools, Cuello[208] came to the very same conclusion.

G. The Effect of Opioids on the Release of γ-Aminobutyric Acid (GABA)

In an *in situ* experiment where the experimental design made the system rather similar to in vitro conditions, morphine was found to inhibit specifically the labeled GABA efflux from *in situ* perfused rat substantia nigra, preloaded with ^{3}H-GABA, elicited either by high potassium level or electrical stimulation of caudate nucleus or globus pallidus, in the presence of GABA uptake inhibitors.[227]

In vitro in rat hippocampal slices under conditions where GABA is released in neurotransmitter-like manner[228] Met-E (10^{-5} *M*) was ineffective in depressing K^+-stimulated GABA-efflux.[141]

H. The Effect of Opioids on the Release of Vasopressin and Oxytocin

The possible involvement of vasopressin in the antidiuretic effect of opiates[229,230] is a frequently investigated topic, although it seems likely now that the antidiuretic action of morphine may not be mediated via vasopressin eliberation[231] (see also Chapter 4, Volume II).

In isolated rat neurohypophysis preparation β-EP, Leu-E and synthetic enkephalin analogues have been shown to specifically enhance the ouabain-induced release of oxytocin and vasopressin as well, probably by disinhibitory mechanism, i.e., by removing an inhibitory dopaminergic control.[187,232]

However, the opposite though similarly specific effects have been reported for opioids if the oxytocin[233,234]/vasopressin[233,235] release was elicited, e.g., by electrical stimulation of the pituitary stalk.[235]

I. The Effect of Opioids on the Release of Purine Nucleotides

A transmitter role for adenosine and adenine nucleotides has been proposed in certain noncholinergic-nonadrenergic nerves in intestinal and other smooth muscles[236,237] as well as in the CNS.[238] Adenosine and adenine nucleotides have been shown to exert presynaptic inhibitory action among other tissues[238-240] also in guinea pig ileum.[241-243]

Morphine has been shown to enhance specifically the veratridine-induced purine release from rat cerebro-cortical slices in vitro[244] as well as the rate of spontaneous efflux from rat cerebro-cortical surfaces *in situ.*[245]

Whether the inhibitory effect of morphine on the release of Ach in guinea pig ileum or its suppressing action exerted upon the release of other transmitter substances involves a purinergic step, as it has been suggested by Phillis et al.,[245] still remains dubious.

In guinea pig taenia coli morphine reduced the release of a smooth muscle relaxing substance from nonadrenergic inhibitory nerves, elicited by transmural electrical stimulation in naloxone-reversible manner.[246]

III. IN VIVO METABOLIC EFFECTS OF OPIOID PEPTIDES

A. Effects on Serotonin Metabolism

Opioid peptides increase the serotonin *turnover* in the brain. More specifically β-EP was found to increase the 5-HT metabolism in the brain stem,[247] hypothalamus,[247] and in other brain areas as well.[248,249] (D-Ala²)-Met-E displayed a similar action in the limbic forebrain[250,251] and the striatum.[250] An opposite effect, i.e., decrease in serotonin turnover has been observed only in the hippocampus upon β-EP treatment.[247] In a recent study, natural enkephalins and two synthetic analogues, DALA and FK 33-824 were found to increase the serotonin turnover in different brain areas. In these studies generally not the steady-state level of serotonin was found elevated upon i.c.v. treatment with opioid peptides but that of the 5-hydroxyindoleacetic acid (5-HIAA) pointing to an increased rate of serotonin catabolism (and anabolism as well). Only in one study [253] was the serotonin concentration also found elevated in the midbrain and in another[254] in the cerebral ganglia of an invertebrate species. Considering the increment of 5-HIAA levels (Table 1), these findings can be interpreted as manifestations of enhanced turnover rate. Actually, enkephalins not only increased the brain concentration of 5-HIAA but also facilitated the conversion of tryptophan to 5-hydroxytryptophan.[252] Thus, opioid peptides may accelerate the serotonin turnover via facilitation of tryptophan hydroxylase activity. As discussed in Chapter 2, Volume I, the similar action of morphine was attributed (among others) to increased availability of tryptophan, since the passage of the latter through the blood-brain barrier is one of the rate limiting steps in the serotonin metabolism. Interestingly, only β-EP was found to have such type of action,[249] but DALA did not increase its transport in one study,[250] while in another laboratory it did.[252] Consequently, not only the serotonin synthesis might be primarily affected but also an increased availability of tryptophan can explain the facilitated serotonin turnover just as in the case of opiates. However, in the experiments of Garcia-Sevilla et al.[252] the elevation by enkephalins of brain tryptophan concentration could only partially be reversed by naloxone. Thus, the effect cannot be mediated primarily by the opiate receptors.

Examining the doses applied, the structure-activity relationship is familiar. Of the natural enkephalins (giving them i.c.v.) 500 μg was needed,[252] β-EP facilitated the turnover in amounts of 5 to 25 μg,[1-3] DALA in somewhat higher doses,[250-252] and the extremely potent FK 33-824 was effective below 1 μg.[252] As shown in Table 1, these effects are naloxone sensitive.

Giving naloxone alone, the serotonin metabolism was decelerated, though the steady-state tryptophan level did not change.[249] Therefore, the opioid peptides might exert a tonic stimulatory action on the serotonin turnover but not by elevating the brain concentration of the corresponding precursor acid.

This enhancement of turnover may reflect either the facilitation of serotonin releasing pathways or conversely it might be a compensatory process due to the disuse of postsynaptic receptors.[259] So far we have no direct in vivo evidence for either an increased or decreased serotonin liberation and also the corresponding in vitro experiments (Chapter 1, Volume II) are rather inconclusive. There is a considerable body of evidence on the activation of serotoninergic mechanisms upon treatment with opioids. As described in Chapter 2, Volume I depletion of brain 5-HT or blockade of the serotoninergic receptors result in attenuation, while the latters' activation in potentiation of opiate analgesia. The same correlation seems to be valid in the case of opioid peptides. Namely 5-hydroxytryptophan given i.p.[281] or i.c.v. administered serotonin[282] equally potentiated the analgesic action of β-EP and (D-Ala²,Leu⁵)-E. On the other hand, the analgesic action of these peptides could be prevented by depleting the brain monoamines with reserpine, an effect reversible by serotonin.[282] Moreover, also the

Table 1
METABOLIC EFFECTS OF OPIOID PEPTIDES AND OPIATE ANTAGONISTS IN VIVO

Treatment	Species	Structure examined	Effect	Reversal by opiate antagonists	Ref.
Naloxone topically	Rat	Cerebral cortex	No effect on spontaneous Ach release	—	255
Naloxone i.v., i.p. 0.1-2.0 mg/kg	Rat	Cerebral cortex	No effect on spontaneous Ach release	—	255
Naloxone i.v. 0.1 mg/kg	Rat	Cerebral cortex	Facilitation of Ach release induced by thalamic or reticular stimulation	—	255
Naltrexone i.v. 0.1 mg/kg	Rat	Cerebral cortex	Facilitation of Ach release induced by thalamic or reticular stimulation	—	255
Met-E i.c.v. 5-20 /ug	Rat	Cerebral cortex	Decrease in Ach release	Yes	256
Leu-E i.c.v. 10 /ug	Rat	Cerebral cortex	Decrease in Ach release	Yes	256
Met-E i.v. 25 mg/kg	Rat	Cerebral cortex	No effect	—	256
β-EP i.c.v. 20 /ug	Rat	Striatum	Increase in DA turnover	Not examined	257
Met-E i.c.v. 100-200 μg	Rat	Striatum	No effect DA turnover	—	258
DALA i.c.v. 10-50 μg	Rat	Striatum and limbic structures	Increase in DA turnover	Not examined	258
Met-E i.c.v. 60 μg	Rat	Forebrain	No change in DA turnover	—	259
Met-E i.c.v. 60 μg	Rat	Median eminence	Decrease in DA turnover	Not examined	259, 260
β-EP i.c.v. 5-25 μg	Rat	Tuberculum olfactorium, n. accumbens, n. caudatus	Increase in DA turnover	Not examined	261
β-EP i.c.v. 5-25 μg	Rat	Median eminence	Decrease in DA turnover	Not examined	261
β-EP i.c.v. 13 nmoles	Rat	Midbrain	Decrease in DA turnover, increase in 5-HT concentration	Not examined	253
β-EP i.c.v. 13 nmoles	Rat	Striatum	No significant change in DA turnover	—	253
β-EP i.c.v. 0.029-2.9 nmoles	Rat	Cerebral corex, hippocampus, n. accumbens, globus pallidus	Decrease in Ach turnover	Yes	262, 263
β-EP i.c.v. 0.029-2.9 nmoles	Rat	n. caudatus	No change in Ach turnover	—	262, 263
β-EP into medial septum 0,7 nmoles	Rat	Hippocampus	Decrease in Ach turnover	Yes	263
β-EP into medial septum 0,7 nmoles	Rat	Cortex, n. caudatus	No change in Ach turnover	—	263

Table 1 (continued)
METABOLIC EFFECTS OF OPIOID PEPTIDES AND OPIATE ANTAGONISTS IN VIVO

Treatment	Species	Structure examined	Effect	Reversal by opiate antagonists	Ref.
β-EP i.c.v. 11.5 nmoles	Rat	As above	No change in either structure examined	—	262 263
β-EP i.c.v. 0.29-2.9 nmoles	Rat	n.caudatus	Decrease in GABA turnover	Yes	264
β-EP i.c.v. 0.29-2.9 nmoles	Rat	Globus pallidus, substantia nigra	Increase in GABA turnover	Yes	264
β-EP i.c.v. 5-20 μg	Rat	Limbic forebrain, striatum, diencephalon, brain stem, cerebral cortex	Increase in DA and 5-HT turnover, no change in NA turnover	Yes	248 249
β-EP i.c.v. 5-20 μg	Rat	Limbic forebrain, striatum, diencephalon, brain stem cerebral cortex	Increase in brain concentrations of tyrosine and tryptophan	Not examined	249
Naloxone i.p. 1-100 mg/kg	Rat	Limbic forebrain, striatum	Decrease in DA and 5-HT turnover, no change in brain concentrations of tyrosine and tryptophane	—	249
DALA i.c.v. 25 μg	Rat	Limbic forebrain	Increase in DA and 5-HT turnover	Yes	251 265
DALA i.c.v. 25 μg	Rat	Cerebral cortex	No change in NA turnover	—	265
(D-Ala²)-Met-E i.c.v. 25 μg	Rat	Limbic forebrain and striatum	Increase in 5-HT turnover	Yes	250
β-EP intracisternally 15 μg	Rat	Striatum	Increase in DA turnover	Yes	266
β-EP intracisternally 15 μg	Rat	Hypothalamus	Decrease in DA turnover	Not examined	266
Met-E i.c.v. 200 μg	Rat	Cerebral cortex	Slight decrease in catecholamine synthesis	Not examined	267
DALA i.c.v. 10 μg	Rat	Cerebral cortex	No change in catecholamine turnover	—	267
DALA i.c.v. 10 μg	Rat	Striatum	Increase in DA turnover but no effect on NA turnover	Not examined	267
Met-E i.c.v. 50-100 μg	Rat	n.caudatus	Increase in DA turnover	Yes	268
Met-E i.c.v. 100 μg	Rat	Whole brain	No change in 5-HT and 5-HIAA concentrations	—	268
(D-Ala²)-Met-E i.c.v. 12 μg	Rat	Whole brain	No effect on 5-HT and 5-HIAA concentrations	—	268
(D-Ala²)-Met-E i.c.v. 1.5-12 μg	Rat	n. caudatus	Increase in DA turnover	Yes	268

Table 1 (continued)
METABOLIC EFFECTS OF OPIOID PEPTIDES AND OPIATE ANTAGONISTS IN VIVO

Treatment	Species	Structure examined	Effect	Reversal by opiate antagonists	Ref.
(D-Ala²)-Met-E into "normal" n. caudatus 3-6 μg	Rat	n. caudatus	Increase in DA turnover	Not examined	268
(D-Ala²)-Met-E into kainic acid lesioned n. caudatus	Rat	n. caudatus	Increase in DA turnover	Not examined	268
β-EP intracisternally 15 μg	Rat	Brain stem, hypothalamus	Increase in 5-HT turnover	Yes	247
β-EP intracisternally 15 μg	Rat	Hippocampus	Decrease in 5-HT turnover	Yes	247
DALA i.c.v. 12.5-50 μg	Rat	Midbrain	Decrease in cGMP level	Not examined	269
DALA i.c.v. 12.5-50 μg	Rat	Striatum	Increase in cGMP level	Not examined	269
DALA into striatum 12-50 μg	Rat	Cerebellum	Decrease in cGMP level	Yes	270
DALA into cerebellum 20 μg	Rat	Cerebellum	No change in cGMP level	—	270
DALA into substantia nigra 20 μg	Rat	Cerebellum	No change in cGMP level	—	270
LPH (62-77) i.c.v. 100 ng	Rat	Paraventricular n., zona incerta, and n. tractus solitarus; other brain structures	Increase in DA turnover; No effect	Not examined	271
β-EP i.c.v. 10 μg	Rat	Hippocampus	Increase in Ach content	Yes	272
β-EP i.c.v. 10 μg	Rat	Striatum, cerebral cortex	No effect	—	272
DALA i.c.v. 10 μg	Rat	Hippocampus	Increase in Ach content	Yes	272
DALA i.c.v. 10 μg	Rat	Striatum, cerebral cortex	No effect	—	272
Naloxone s.c. 1.0 mg/kg	Rat	Striatum, cerebral cortex	No effect	—	272
β-EP i.c.v. 25-100 ng	Mouse	Brain synaptosomes	Decrease in Ca^{++} uptake	Yes	273
β-EP i.c.v. 9 μg	β-EP tolerant rat	Brain synaptosomes	Increase in CA^{++} uptake	Not examined	273
DALA into ventromedial thalamus 20 / ug	Rat	Cerebellum	No change in cGMP level	—	270
α-endorphin i.c.v. 100 ng	Rat	Medial septal n., dorsomedial n., subiculum, n. reticularis medullae oblongatae, region A_1, other brain structures	Decrease in NA turnover; No effect	Not examined; —	271
α-endorphin i.c.v. 100 ng	Rat	Striatum, septum, paraventricular n., n. interstitialis striae term., zona incerta	Decrease in DA turnover; No effect	Not examined; —	271

Table 1 (continued)
METABOLIC EFFECTS OF OPIOID PEPTIDES AND OPIATE ANTAGONISTS IN VIVO

Treatment	Species	Structure examined	Effect	Reversal by opiate antagonists	Ref.
		and amygdala other brain structures			
β-EP i.c.v. 100 ng	Rat	n. reticularis medullae oblongatae and n. tractus solitarius, other brain structures	Decrease in NA turnover No effect	Not examined —	271
β-EP i.c.v. 100 ng	Rat	Septum, zona incerta, n. tractus solitarius; other brain structures	Decrease in DA turnover No effect	Not examined —	271
LPH (62-77) i.c.v. 100 ng	Rat	Anterior hypothalamic n., periventricular thalamic n., and region A_7	Increase in NA turnover No effect	Not examined —	271
β-EP ventricle III 10 μg	Chick	Diencephalon	Decrease in GABA content, increase in GABA transaminase activity	Yes	274
β-EP ventricle III 10 μg	Chick	Brain stem	Decrease in GABA content, increase in GABA transaminase activity	Yes	274
β-EP ventricle III 10 μg	Chick	Brain hemispheres	No effect	—	274
β-EP i.c.v. 20 μg	Rat	Microsomal fraction of midbrain homogenate	Increased synthesis of phosphoinositides	Yes	275
Met-E intracardially 40-80 μg	Mytilus edulis	Cerebral ganglia	Increase in DA level, no change in 5-HT concentration	Yes	276
Leu-E intracardially 80 μg	Mytilus edulis	Cerebral ganglia	Increase in DA level, no change in 5-HT concentration	Yes	276
β-EP i.v. 0.4 mg/kg	Rat	Kidney	Increase in ornithine decarboxylase activity	Yes	277
β-EP i.v. 0.4 mg/kg	Rat	Gut, adrenal gland, heart, liver, thymus, pancreas, lung, spleen, striatum	No change in ornithine decarboxylase activity	—	277
Met-E topically	Anodonta cygnea	Cerebral ganglia	Increase in DA, 5-HT and cGMP levels, decrease in cAMP level	Yes	254

Table 1 (continued)
METABOLIC EFFECTS OF OPIOID PEPTIDES AND OPIATE ANTAGONISTS IN VIVO

Treatment	Species	Structure examined	Effect	Reversal by opiate antagonists	Ref.
β-EP intracisternally 15 μg	Rat	Hypothalamus	Decrease in DA turnover, no change in NA metabolism	Not examined	278
Met-E or Leu-E i.c.v. 500 μg	Rat	Different brain regions	Increase in DA turnover	Yes	255
Met-E or Leu-E i.c.v. 500 μg	Rat	Different brain regions	Increase in 5-HT turnover	No	255
Met-E or Leu-E i.c.v. 500 μg	Rat	Different brain regions	Increase in tryptophan content	Partially	255
Met-E or Leu-E i.c.v. 500 μg	Rat	Different brain regions	Increase in tyrosine level	No	255
DALA i.c.v. 4-256 μg	Rat	Different brain regions	Increase in DA and 5-HT turnover, elevation of the tryptophan content	Yes	252
DALA i.c.v. 125 μg	Rat	Different brain regions	No change in tyrosine level	—	252
FK 33-824 i.c.v. 0.003-1 μg	Rat	Different brain regions	Increase in DA and 5-HT turnover, elevation of the tryptophan content	Yes	252
FK 33-824 i.c.v. 1 μg	Rat	Different brain regions	No change in tyrosine level	—	252
β-EP i.c.v. 25 μg	Rat	Plasma	Increase in DA, NA, and adrenaline concentrations	Not examined	279
DALA i.c.v. 10 μg - 50 μg	Rat	Parietal cortex and hippocampus	Decrease in Ach turnover	Not examined	
(D-Ala[2])-Leu-E-amide i.c.v. 10 μg - 50 μg	Rat	Parietal cortex and hippocampus	Decrease in Ach turnover	Not examined	280
(D-Ala[2])-N-Me-Met[2]-E-amide i.c.v. 10 μg - 50 μg	Rat	Parietal cortex and hippocampus	Decrease in Ach turnover	Not examined	280
FK 33-824 i.c.v. 2 μg	Rat	Parietal cortex and hippocampus	Decrease in Ach turnover	Yes	280
(D-Ala[2],D-Leu[5])-E i.c.v. 50 μg	Rat	Parietal cortex and hippocampus	Decrease in Ach turnover	Not examined	280

Note: Ach: acetylcholine; DA: dopamine; NA: noradrenaline.

electrographic experiments (Chapter 2, Volume 2) pointed to the role of serotonin in opioid-induced analgesia. Consequently, it is reasonable to accept that the enhanced serotonin turnover reflects the activation of serotonin-releasing neurones. If so, serotonin is the only brain transmitter which is liberated at a higher rate upon treatment with opioids. As opioids generally inhibit the transmitter release via presynaptic action, the question arises whether this facilitation of serotonin production can be attributed to disinhibition, i.e., to inhibition of the liberation of another transmitter physiologically suppressing the serotonin neurones; question which remains to be answered in the future.

B. Effects on Catecholamine Turnover

As summarized in Table 1, numerous groups observed a facilitation of catecholamine turnover in the brain upon treatment with opioid peptides. [248,249,251,252,257,258,261,265-268,271,276] A close inspection of the data shows that this general statement must be restricted in several aspects. Actually only the dopamine turnover is enhanced in several parts of the brain but the noradrenaline metabolism is not. The two effects are naturally not easy to separate from each other, as discounting the last step the same enzymes are involved in both processes. To distinguish the two actions, generally either the relative concentrations of dopamine and noradrenaline metabolites were compared or the rate of 1-DOPA formation in brain areas rich in dopamine were compared to those containing predominantly noradrenaline.

An increase in the striatal concentrations of homovanillic acid (HVA) and dihydroxyphenylacetic acid (DOPAC) has been detected by several groups,[257,258,266,268] while nobody reported on a similar elevation of noradrenaline metabolite concentration either in the striatum or in other brain regions. Calderini et al.[267] compared the conversion of labeled tyrosine into dopamine and noradrenaline. DALA facilitated the conversion of labeled tyrosine to 1-DOPA in the striatum and in the limbic structures, but the noradrenaline synthesis was not found modified.[267] Accordingly, increased catecholamine turnover means only facilitated dopamine formation. Garcia-Sevilla et al.[249] came to similar conclusion in another way. Measuring the accumulation of 1-DOPA after inhibition of aromatic amino acid decarboxylase by NSD 1015 (3-hydroxybenzylhydrazine), the first step of catecholamine synthesis was found accelerated by β-EP (and morphine) in those areas of the brain (limbic forebrain, striatum) which are relatively rich in dopamine.[249] Aromatic amino acid decarboxylase, the enzyme responsible for the next step of dopamine synthesis is present in excess in the brain tissue. Consequently, though the possible effects of opioid peptides on the activity of dopamine-β-hydroxylase, synthetizing noradrenaline, has not been examined yet, the selective facilitation of dopamine turnover may probably be accepted since a more or less similar conclusion can be drawn from the experiments in which the catecholamine metabolism was assessed by measuring the rate of their depletion upon blocking the tyrosine hydroxylase activity.[261,265,266,271] In the latter study[265] the decline of striatal dopamine was enhanced upon i.c.v. administration of DALA, while the fall of noradrenaline concentration in the cerebral cortex was not modified.

Thus, enhancement of the tyrosine hydroxylase activity in selective brain regions might explain the effects of opioids. Of course, this facilitation might be due to the elevation of brain tyrosine level. In fact β-EP[249] and the natural enkephalins[252] were reported to elevate the brain concentration of tyrosine. However, in the latter study[252] two potent enkephalin analogues, DALA and FK 33-824 did not increase the brain level of this amino acid and the "positive" effect of natural enkephalins could not be antagonized by naloxone. Consequently, the enhancement of catecholamine metabolism cannot be attributed to opiate receptor induced increase in tyrosine supply.

A further important finding is that Met-E and DALA did not increase the striatal level of 3-methoxytyramine.[267] As 3-methoxytyramine is a dopamine metabolite formed outside of the presynaptic nerve terminals,[283] the lack of any effect on its level suggests that the increase in dopamine metabolism is an intracellular phenomenon. In other words, in spite of its increased turnover the dopamine release is not enhanced upon opioid treatment; very important conclusion, which must be taken into consideration upon interpreting the data on accelerated dopamine metabolism.

As shown by the data summarized in Table 1, this metabolic action of opioid peptides is naloxone sensitive, i.e., the action is mediated by opiate receptors similarly to that of the morphine congeners (Chapter 2, Volume I). Upon comparing the different opioid peptides, again the usual structure-activity relationship can be detected. The natural enkephalins acted only in high dosage[268] or they were inactive in the dose applied.[258,267] Examining, however, β-EP[249,257,266] or the potent enkephalin analogues, much smaller amounts are needed and also dose-effect relationships could be detected.[252,258,268] It must, however, be emphasized that above a certain dose level the increase of the amount (of β-EP) injected did not result in the expected further enhancement of effect, i.e., its dose-effect curves were bell-shaped.[249] Furthermore, examining their effects on 1-DOPA formation quantitatively, the maximal effects induced by opioids were relatively modest, i.e., they did not elicit greater action than doubling the original rate of dopamine turnover.[249,252] Thus, their action is probably not so strong than that of the classical neuroleptic agents, though so far nobody made direct comparison between the effects of neuroleptics and opioids in quantitative aspect.

Based on the rate of catecholamine depletion upon blocking the tyrosine hydroxylase activity, the dopamine turnover was found decelerated in the median eminence upon treatment with Met-E[259,260] or β-EP.[261,278] In other words, opioid peptides contrary to the striatal and limbic structures inhibit not only the dopaminergic synapses but also the dopamine turnover in the hypothalamus. For the time being it is extremely difficult to explain that the dopamine metabolism of hypothalamic neurones is influenced in the opposite direction. These unanimous findings[259-261,278] may account for some endocrine effects of opioid peptides. As discussed in Chapter 5, Volume II, opioid peptides increase the serum level of prolactin and growth hormone and the liberation of these substances is known to be tonically inhibited by dopamine. Consequently, the antidopamine action of opioid peptides gives satisfactory explanation for these endocrine actions.

As discussed in Chapter 5, Volume II, naloxone was found to decrease the serum prolactin level. As this effect was obviously mediated by displacement of the endogenous opioids, they might exert a tonic inhibitory action on the dopamine release from the tubero-infundibular neurones. On the other hand, endogenous opioids might induce a tonic acceleration of dopamine metabolism in other brain regions. According to Garcia-Sevilla et al.[249] naloxone and naltrexone treatments resulted in a significant deceleration of dopamine turnover in the striatum and limbic forebrain and also in slowing of the serotonin metabolism. Thus, endogenous opioids do exert a tonic action also in the more cranial parts of the brain but in opposite direction.

A general feature of the above-discussed studies is that comparing the different opioid peptides no qualitative differences have been detected (Table 1). Only the study of Versteeg et al.[271] represents an exception, who observed a decrement both in the dopamine and noradrenaline turnover in several well-defined areas of the brain upon treatment with β-EP and α-endorphin, while facilitation in some partly different regions after administering (des-Tyr[1])-γ-endorphin (LPH/62-77). Thus, β-EP and α-endorphin on the one hand and (des-Tyr[1])-γ-endorphin on the other hand influence the catecholamine turnover in opposite directions. This finding,[271] is difficult to reconcile with the previously mentioned data, but it confirms the behavioral studies of the same group on differential roles of these peptides in regulation of physiological functions (Chapter 3, Volume II).

The mechanism by which opioid peptides modify the dopamine turnover is not known. There are several lines of indirect evidence speaking against a direct action on the postsynaptic dopamine receptors.

Probably the most convincing ones are the observations of Biggio et al.,[268] who found an increase in the striatal DOPAC concentration upon local administration of DALA in animals with kainic acid lesioned caudate nucleus. As lesioning of the caudate nucleus with kainic acid resulted in selective destruction of cell bodies, leaving the dopamine releasing axons intact, DALA must have acted via presynaptic mechanisms.[268] This conclusion is in good agreement with the in vitro experiments, where opioids have no direct postsynaptic action (Chapter 1, Volume I). Moreover, the lack of effect of DALA on the methoxytyramine formation,[267] which is generated beyond the presynaptic nerve terminals[283] and the lack of specific interactions of opioids with postsynaptically acting drugs in behavioral experiments (Chapter 2, Volume I) speak for a presynaptic site of action.

Most authors of the above-discussed reports agree that the opioids act via presynaptic inhibition of dopamine release and its increased turnover is only a compensatory phenomenon. The much discussed decrease in dopamine release has been demonstrated only in vitro (Chapter 1, Volume II) and very little is known on the nature of the feedback mechanism(s) resulting in increased dopamine turnover. This feedback control cannot be mediated via the postsynaptic striato-nigral loop, since the kainic acid-induced lesion of caudate neurones did not abolish the process.[268] Probably the decrease in dopamine release results in decreased stimulation of the autoreceptors located on the presynaptic nerve terminals leading to liberation of the tyrosine hydroxylase of the intracellular feedback control. If accepting this speculation, a further hypothesis is needed to explain the lack of a similar feedback control of tubero-infundibular dopamine releasing neurones, where the decreased release is not accompanied by enhanced turnover.

Let us mention here that intracisternally administered β-EP increased the plasma concentration of dopamine, adrenaline, and noradrenaline, obviously as manifestations of increased sympathetic and/or adrenal medullary activity.[279]

C. Effects on Acetylcholine Turnover

The influence of opioids on cholinergic functions in the CNS seems to be a less controversial issue. Its turnover and release are not influenced in opposite directions, but the different brain areas are also in this relation differentially affected.

First, Jhamandas et al.[256] reported on a dose-dependent reduction in the acetylcholine release from the cortex of anesthetized rats upon i.c.v. application of Met-E, an effect also shown by Leu-E and fully reversible by naloxone just as in the case of classical opiates (Chapter 2, Volume I). In a simultaneous study Moroni et al.[262,263] observed that β-EP, given also i.c.v. in doses around analgesic ED_{50}, decreased the acetylcholine turnover in cerebral cortex, hippocampus, n. accumbens, and in globus pallidus but not in the caudate nucleus. The reduction in the cortical acetylcholine release upon enkephalin treatment[256] and the deceleration of its turnover in certain brain regions upon β-EP[262,263] presumably expresses the same process. As suggested by the authors,[262] the decelerated turnover was due to the decrease in the rate of firing of acetylcholine releasing cells. Interestingly α-endorphin had no effect,[262,263] but it must be added that the doses used (0.029-2.9/nmoles of β-EP and 11.5 nanomol of α-endorphin) are not equianalgesic. As discussed in Chapter 3, Volume II a few nmoles of the former are sufficient for analgesia, while much higher amounts of α-endorphin are needed for antinociception if it has such an action at all.

The data available to date are not sufficient to learn much about the structure-activity relationships. Nevertheless, even the observations published so far are enough to draw the conclusion that there is no correlation between the anticholinergic and analgesic potencies. Namely 5 to 20 μg of Met-E or Leu-E was sufficient to inhibit the cortical acetylcholine liberation for periods of 20 to 40 min,[256] while their alleged antinociceptive action appears at much higher dose levels and lasts only for several minutes if it can be detected at all (Chapter 3, Volume II). Disregarding that in one study the acetylcholine release[256] and in the others[262,263] its turnover has been measured, the effective doses were as follows; 5 to 20 μg Met-E,[256] 0.029 to 2.9 nmoles, i.e., about 1 to 10 μg β-EP,[263] 35 to 175 nmoles, i.e., 10 to 70 μg morphine,[263] and 11.5 nmoles, i.e., 20 μg α-endorphin.[263] Thus, the order of potency is β-EP > Met-E ≈ ≈ Leu-E > morphine > α-endorphin. As for the synthetic enkephalin analogues, in a recent study DALA, (D-Ala²,D-Leu⁵)-E, (D-Ala², Leu⁵)-E and (D-Ala²,N-Me-Met⁵)-E-amide were found to decrease the acetylcholine turnover in the parietal cortex and hippocampus in the i.c.v. doses of 10 to 50 ug, while FK 33-824 displayed similar action in the dose of 2 ug.[280]

However, there is hardly any type of opiate receptors which could account for this order of potency (Chapters 2, 5, 6, and 7, Volume III). Thus, probably it is not incorrect to suppose that more than one subclasses of opiate receptors are involved in mediation of this action of opioids, as the similar action of etorphine could be antagonized only with relatively high doses of naloxone.[280]

On the other hand, the acetylcholine turnover is not modified in all brain regions known to contain opiate receptors and endogenous opioids in considerable amounts. As seen above,[262,263] β-EP failed to modify the acetylcholine metabolism in the caudate nucleus, in the brain area the richest in endogenous opioids (Chapter 3, Volume III).

Also the above-mentioned enkephalin analogues failed to modify the striatal acetylcholine turnover.[280] Otherwise, in accordance with the previous studies[262,263] Botticelli and Wurtman[272] found a significant elevation of choline and acetylcholine contents in the hippocampus but not in the cerebral cortex and striatum upon i.c.v. treatment with β-EP or DALA. Significant alterations were observed in a structure containing only little amounts of opioids (Chapter 3, Volume III). Otherwise, this increase in the hippocampal acetylcholine content was ascribed to its decreased release.[272] The issue is further complicated by the well-established fact that opiate recepors of certain brain areas may modify the neurotransmitter release in other regions. The hippocampal acetylcholine turnover is decreased upon deafferentation of this structure[284] and also β-EP reduces its metabolism more effectively if it is administered into the medial septal nucleus.[263] Consequently, β-EP seems to act on cell bodies in the septum sending acetylcholine liberating nerve terminals into the hippocampus. Besides, also the cell bodies responsible for the opiate sensitive acetylcholine release from the cerebral cortex may be located somewhere else. Applying naloxone topically on the cerebral cortex, the spontaneous output of acetylcholine remained constant,[255] but it significantly increased the cortical acetylcholine liberation evoked by electrical stimulation of either the medial thalamus or the reticular formation.[255]

Thus, for the time being very little is known about effects of opioids on acetylcholine turnover besides that they are able to decrease the cholinergic functions in certain but surely not in all cholinergic pathways. Their contingent tonic action on acetylcholine release is still less clarified. As in one study[255] the spontaneous cortical liberation of acetylcholine has not been modified by naloxone (and naltrexone), administering them either topically or systemically, opioid peptides might have no significant role in regulation of the basal release.

D. Other Central Metabolic Effects

Also the GABA metabolism is modified by opioids. Nistico et al.[274] observed increased *GABA turnover* in chicks upon β-EP treatment in several brain areas (diencephalon, brain stem) while no changes in other regions. In rats Moroni et al.[264] found the GABA metabolism accelerated in the globus pallidus and substantia nigra and decelerated in the caudate nucleus. All these effects could be reversed by naloxone and naltrexone.[264,274]

In view of these in vivo data it is rather difficult to interpret the in vitro finding[285] according to Met-E inhibits the K^+-stimulated release of GABA from brain synaptosomes. If opioid peptides suppress by presynaptic mechanisms GABA release (just as the liberation of other transmitters), an autoreceptor-mediated decrement of turnover might be predicted. As in the latter study[285] whole brain homogenate was examined, it cannot be excluded that the GABA metabolism is slowed down specifically in those brain regions where its release is inhibited, but facilitated in other areas. Muscimol, the strong GABA agonist injected into the globus pallidus induced catalepsy, thus, increase in GABA turnover has been implicated in the cataleptogenic effect of opioid peptides.[264] As GABAergic neurones are known to participate in the nigro-striatal and striato-nigral circuitry,[264,274] these changes are probably connected with altered turnover of acetylcholine and dopamine in these structures. The exact nature of the interrelation of opioid peptides with GABA neurones requires further clarification.

Among others from this point of view are interesting the observations on the decrease in cGMP content in the cerebellum and the increase of cAMP level in the striatum upon i.c.v. treatment with DALA.[269] These effects are rather similar to those observed with the classical morphine congeners (Chapter 2, Volume I). The in vivo facilitation by opioids of adenylate cyclase in the striatum is difficult to explain, as their action on the basal enzyme activity in vitro is a matter of controversy, i.e., its activity has not been found facilitated (Chapter 1, Volume I). However, it seems to be sufficiently established that the postsynaptic dopamine receptors in the caudate nucleus play a major role in the regulation of cerebellar cGMP level, since their lesion by intrastriatally injected kainic acid results in a complete depletion of the cerebellar cGMP content.[270]

Consequently, it is easy to explain why induced DALA decrease in the cerebellar cGMP content if giving it into the striatum, but no such effect if administering it directly into the cerebellum or into the substantia nigra or ventromedial thalamus.[270] According to Biggio et al.[286] the striomesencephalic-cerebellar pathways are under tonic GABA mediated inhibitory control and the latter can be activated by cholinergic interneurones, which in turn are suppressed by dopamine. The above-discussed anti-dopamine and anticholinergic actions of opioids cannot be reconciled with this hypothesis, but if GABAergic mechanisms inhibit the efferent striatal neurones activating the mossy and climbing fibers via nigral relay units, the previously mentioned increase in GABA turnover in the globus pallidus might account for the decrease in cerebellar cGMP level.[270]

As discussed in Chapter 1, Volume I, opiates decrease the *Ca^{++}* uptake into the synaptosomes, which was regarded satisfactory explanation of their inhibitory action on liberation of certain neurotransmitters. Of the opioid peptides so far β-EP has been studied in this aspect by Guerrero-Munoz et al.[273] Injection of 25 to 100 ng β-EP i.c.v. to mice produced not only analgesia but also a dose-related inhibition of synaptosomal Ca^{++} uptake, and naloxone antagonized both effects.[273] Upon its repeated application tolerance developed not only to its analgesic but also to its Ca^{++} uptake inhibiting action.[273] Thus, β-EP mimics morphine also in its intricate mechanism of action. Though no in vivo data are available in relation of enkephalins, Ca^{++} ions may probably be supposed to have a similar role in mediation of their actions. As discussed in Chapter 1, Volume I, Ca^{++} ions inhibit the actions of opioids in vitro in several isolated organ preparations and also their effects in microelectrophysiological experiments (Chapter 2, Volume II).

A further interesting effect of β-EP is the facilitation of anorganic phosphoric acid incorporation into phosphatidylinositol and 1-phosphatidylinositol-3,4-biphosphate, which has also been attributed to the inhibition of Ca^{++} flux into the synaptosomes.[275]

The physiological significance of these findings[275] has not been clarified yet, but considering also the profound alteration of *phospholipid metabolism* upon opiate treatment (Chapter 2, Volume I), the importance of endogenous opioids in regulation of these processes should be accepted.

Finally let us mention here that β-EP has been reported to increase the *ornithine decarboxylase activity* in the kidney but not in other parenchymal organs[277] As this specific action could be demonstrated also in hypophysectomized animals, i.e., it is independent of ACTH release, it has been suggested to act as a specific trophic hormone of this organ.[277]

Summarizing the above-discussed data on the effects of opioids on metabolism of several transmitters and other substances, probably the following conclusions are justified.

1. The metabolic actions of opioid peptides are basically similar to those induced by opiates, at least in relation of neurotransmitters.
2. The modified turnover of acetylcholine, dopamine, serotonin, and GABA are presumably secondary phenomena due to the interactions of opioids with their receptors. Thus, increased turnover of serotonin probably reflects the stimulation of the corresponding receptors, while the increased dopamine metabolism seems to be presynaptically mediated compensatory process. The decrease in acetylcholine metabolism is almost surely connected with the anticholinergic action of opioids.
3. These metabolic effects are mediated by selective subgroups of opiate receptors which are at least in part independent of those responsible for the analgesic actions.
4. In different brain areas the turnover of the same neurotransmitters are influenced differentially by the opiate receptor mechanisms.

REFERENCES

1. **Simon, E. J. and Hiller, J. M.**, The opiate receptors, *Annu. Rev. Pharmacol. Toxicol.*, 18, 371, 1978.
2. **Duggan, A. W., Davis, J., and Hall, J. G.**, Effects of opiate agonists and antagonists on central neurons of the cat, *J. Pharmacol. Exp. Ther.*, 196, 107, 1976.
3. **Zieglgänsberger, W., French, E. D., Siggins, G. R., and Bloom, F. E.**, Opioid peptides may excite hippocampal pyramidal neurons by inhibiting adjacent inhibitory interneurons, *Science*, 205, 415, 1979.
4. **Vizi, E. S., Hársing, L. G., Jr., and Knoll, J.**, Presynaptic inhibition leading to disinhibition of acetylcholine release from interneurons of the caudate nucleus: effects of dopamine, β-endorphin and D-Ala^2,Pro^5-enkephalinamide, *Neuroscience*, 2, 953, 1977.
5. **McLennan, H.**, *Synaptic Transmission*, W. B. Saunders, Philadelphia, 1963.
6. **Florey, E.**, Neurotransmitters and modulators in the animal kingdom, *Fed. Proc. Fed. Am. Soc. Exp. Biol.*, 26, 1164, 1967.
7. **Stewart, J. M. and Channabasavaiah, K.**, Evolutionary aspects of some neuropeptides, *Fed. Proc. Fed. Am. Soc. Exp. Biol.*, 38, 2302, 1979.
8. **Kosterlitz, H. W. and Hughes, J.**, Some thoughts on the significance of enkephalin, the endogenous ligand, *Life Sci.*, 17, 91, 1975.
9. **Frederickson, R. C. A.**, Enkephalin pentapeptides — a review of current evidence for a physiological role in vertebrate neurotransmission, *Life Sci.*, 21, 23, 1977.
10. **Otsuka, M. and Takahashi, T.**, Putative peptide neurotransmitters, *Annu. Rev. Pharmacol. Toxicol.*, 17, 425, 1977.

11. **Snyder, S. H. and Goodman, R. R.,** Multiple neurotransmitter receptors, *J. Neurochem.*, 35, 5, 1980.
12. **Zieglgänsberger, W. and Fry, J. P.,** Actions of opioids on single neurons, in *Developments in Opiate Research,* Herz, A., Ed., Marcel Dekker, New York, 1978, 193.
13. **Herz, A., Schulz, R., and Wüster, M.,** Some aspects of opiate receptor function, in *First International Colloquium on Receptors, Neurotransmitters and Peptide Hormones,* Capri, Abstr. Vol. 54, 1979.
14. **Usdin, E. and Bunney, W. E., Jr., Eds.,** *Pre- and postsynaptic Receptors,* Marcel Dekker, New York, 1975.
15. **Atweh, S. F. and Kuhar, M. J.,** Autoradiographic localization of opiate receptors in rat brain. I. Spinal cord and lower medulla, *Brain Res.*, 124, 53, 1977.
16. **Kosterlitz, H. W. and Wallis, D. I.,** The action of morphine-like drugs on impulse transmission in mammalian nerve fibres, *Br. J. Pharmacol.*, 22, 499, 1964.
17. **Zieglgänsberger, W. and Bayerl, J.,** The mechanism of inhibition of neuronal activity by opiates in the spinal cord of cat, *Brain Res.*, 115, 111, 1976.
18. **Paton, W. D. M.,** The action of morphine and related substances on contraction and on acetylcholine output of coaxyally stimulated guinea pig ileum, *Br. J. Pharmacol.*, 11, 119, 1957.
19. **Trendelenburg, U.,** The action of morphine on superior cervical ganglion and on the nictitating membrane of the cat, *Br. J. Pharmacol.*, 12, 79, 1957.
20. **Kosterlitz, H. W., Waterfield, A. A., and Berthoud, V.,** Assessment of the agonist and antagonist properties of narcotic analgesic drugs by their actions on the morphine receptor in the guinea pig ileum, in *Narcotic antagonists,* (Adv. in Biochemical Psychopharmacology Vol. 8), Braude, M. C., Harris, L. S., May, E. L., Smith, J. J., and Villarreal, J. E., Eds., Raven Press, New York, 1974, 319.
21. **Thompson, J. W.,** The Cat's Nictitating Membrane as an Isolated Preparation, Ph.D. thesis, University of London, 1960.
22. **Cairnie, A. B., Kosterlitz, H. W., and Taylor, D. W.,** The effect of morphine on some sympathetically innervated effectors, *Br. J. Pharmacol.*, 17, 539, 1961.
23. **Henderson, G., Hughes, J., and Kosterlitz, H. W. W.,** The effects of morphine on the release of noradrenaline from the cat isolated nictitating membrane and guinea pig ileum myenteric plexus-longitudinal muscle preparation, *Br. J. Pharmacol.*, 53, 505, 1975.
24. **Hughes, J., Kosterlitz, H. W., and Leslie, F. M.,** Effect of morphine on adrenergic transmission in the mouse vas deferens. Assessment of agonist and antagonist potencies of narcotic analgesics, *Br. J. Pharmacol.*, 53, 371, 1975.
25. **LaMotte, C., Pert, C. B., and Snyder, S. H.,** Opiate receptor binding in primate spinal cord: distribution and changes after dorsal root section, *Brain Res.*, 112, 407, 1976.
26. **Schwartz, J. C., Pollard, H., Llorens, C., Malfroy, B., Gross, C., Pradelles, Ph., and Dray, F.,** Endorphins and endorphin receptors in striatum: relationship with dopaminergic neurons, in *The Endorphins,* Costa, E. and Trabucchi, M., Eds., Raven Press, New York, 1978, 245.
27. **Taube, H. D., Starke, K., and Borowski, E.,** Presynaptic receptor systems on the noradrenergic neurons of rat brain, *Naunyn-Schmiedeb. Arch. Pharmacol.*, 299, 123, 1977.
28. **North, R. A. and Tonini, M.,** Hyperpolarization by morphine of myenteric neurons, in *Opiates and Endogenous Opioid Peptides,* Kosterlitz, H. W., Ed., Elsevier/North-Holland Biomedical Press, Amsterdam, 1976, 205.
29. **North, R. A.,** Effects of morphine on myenteric plexus neurones, *Neuropharmacology,* 15, 719, 1976.
30. **Way, E. L. and Shen, F.-H.,** Catecholamines and 5-hydroxytryptamine, in *Narcotic Drugs,* Clouet, D. H., Ed., Plenum Press, New York, 1971, 229.
31. **Ho, I. K., Loh, H. H., and Way, E. L.,** Effect of cyclic AMP on morphine analgesia, tolerance and physical dependence, *Nature (London),* 238, 1972.
32. **Way, E. L.,** Brain neurohormones in morphine tolerance and dependence, in *Pharmacology and the Future of Man,* Vol. 1, Karger, Basel, 1973, 77.
33. **Hamprecht, B.,** Opioids and cyclic nucleotides, in, *Developments in Opiate Research,* Herz, A., Ed., Marcel Dekker, New York, 1978, 357.
34. **Henion, W. F., Sutherland, E. W., and Pasternak, T.,** Effects of derivatives of adenosine 3',5'-phosphate on liver slices and intact animals, *Biochim. Biophys. Acta,* 148, 106, 1967.
35. **Ho, I. K., Loh, H. H., and Way, E. L.,** Cyclic adenosine monophosphate antagonism of morphine analgesia, *Pharmacol. Exp. Ther.*, 185, 336, 1973.
36. **Way, E. L., Iwamoto, E. T., Khanna, S., Ho, K. J., Shen, F., and Loh, H. H.,** Precipitation of abstinence like syndrome in morphine dependent mice by pargyline, *J. Pharmacol. Exp. Ther.*, 199, 400, 1976.

37. **Costa, E., Carenzi, A., Guidotti, A., and Revuelta, A.,** Narcotic analgesics and the regulation of neuronal catecholamine stores, in *3rd Int. Catecholamine Symp. Biochem. Pharmacol.,* Usdin, E. and Snyder, S. H., Eds., Pergamon Press, London, 1973, 833.
38. **Carenzi, A., Cheney, D. L., Costa, E., Guidotti, A., and Racagni, G.,** Action of opiates, antipsychotics, amphetamine and apomorphine on dopamine receptors in rat striatum: in vivo changes of 3',5'-cyclic AMP content and acetylcholine turnover rate, *Neuropharmacology,* 14, 927, 1975.
39. **Clouet, D. H., Gold, D. J., and Iwatsubo, K.,** Effects of narcotic drugs on the cyclic adenosine 3',5'-monophosphate-adenylate cyclase system in rat brain, *Br. J. Pharmacol.,* 54, 541, 1975.
40. **Clouet, D. H. and Iwatsubo, K.,** Dopamine-sensitive adenylate cyclase of the caudate nucleus of rats treated with morphine, *Life Sci.,* 17, 35, 1975.
41. **Nakagawa, K. and Kuriyama, K.,** Morphine-induced changes of cyclic AMP metabolism and protein kinase activity in brain, *Jpn. J. Pharmacol.,* 26, 110, 1976.
42. **Puri, S. K., Cochin, J., and Volicer, L.,** Effect of morphine sulfate on adenylate cyclase and phosphodiesterase activities in rat corpus striatum, *Life Sci.,* 16, 759, 1975.
43. **Van Inwegen, R. G., Strada, S. J., and Robison, G. A.,** Effects of prostaglandins and morphine on brain adenyl cyclase, *Life Sci.,* 16, 1875, 1975.
44. **Collier, H. O. J.,** Morphine, prostaglandins and cyclic AMP, *Proc. 2nd International Cong. Hungarian Pharmacol. Soc.,* Budapest, 1974.
45. **Collier, H. O. J. and Roy, A. C.,** Morphine-like drugs inhibit the stimulation by prostaglandins of cyclic AMP formation by rat brain homogenate, *Nature (London),* 248, 24, 1974.
46. **Roy, A. C. and Collier, H. O. J.,** Prostaglandins, cyclic AMP and the biochemical mechanism of opiate agonist action, *Life Sci.,* 16, 1857, 1975.
47. **Laduron, P.,** Adenylcyclase: a possible target for morphine-like compounds and neuroleptics, *6th Int. Cong. Pharmacol.,* Helsinki, Abstr. 1477, 1975.
48. **Miller, R. J. and Iversen, L. L.,** Effect of psychoactive drugs on dopamine (3,4-dihydroxy-phenethylamine) -sensitive adenylate cyclase activity in corpus striatum of rat brain, *Biochem. Soc. Trans.,* 2, 256, 1974.
49. **Wilkening, D., Mishra, R. K., and Makman, M. H.,** Effects of morphine on dopamine-stimulated adenylate cyclase and on cyclic GMP formation in monkey brain amygdala, *Fed. Proc. Fed. Am. Soc. Exp. Biol.,* 35, 428, 1976.
50. **Klee, W. A. and Nirenberg, M.,** A neuroblastoma x glioma hybrid cell line with opiate receptors, *Proc. Natl. Acad. Sci. USA,* 71, 3474, 1974.
51. **Sharma, S. K., Klee, W. A., and Nirenberg, M.,** Dual regulation of adenylate cyclase accounts for narcotic dependence and tolerance, *Proc. Natl. Acad. Sci. USA,* 72, 3092, 197.
52. **Traber, J. and Hamprecht, B.,** Action of neurohormones and opiates on neuroblastoma x glioma hybrid cells in culture, in *Advances in Prostaglandin and Thromboxane Research,* Vol. 1., Samuelsson, B. and Padetti, R., Eds., Raven Press, New York, 1975, 337.
53. **Blume, A. J., Shorr, J., Finberg, J. P. M., and Spector, S.,** Binding of the endogenous nonpeptide morphine-like compound to opiate receptors, *Proc. Natl. Acad. Sci. USA,* 74, 4927, 1977.
54. **Klee, W. A.,** Opioid peptides as modulators of cyclic AMP levels, in *Modulators, mediators and specifiers in brain function, (Advances in Experimental Medicine and Biology,* Vol. 116,) Ehrlich, Y., Volavka, J., Davis, L. G. and Brunngraber, E. G., Eds., Plenum Press, New York, 1979, 225.
55. **Blume, A. J., Boone, G., and Lichtshtein, D.,** Regulation of the neuroblastoma x glioma hybrid opiate receptors by Na^+ and guanine nucleotides, in, *Modulators, mediators and specifiers in brain function,* (Advances in Experimental Medicine and Biology, Vol. 116), Ehrlich, Y., Volavka, J., Davis, L. G., and Brunngraber, E. G., Eds., Plenum Press, New York, 1979, 163.
56. **Klee, W. A. and Nirenberg, M.,** Mode of action of endogenous opiate peptides, *Nature (London),* 263, 609, 1976.
57. **Brandt, M., Gullis, R. J., Fischer, K., Buchen, C., Hamprecht, B., Moroder, L., and Wünsch, E.,** Enkephalin regulates the levels of cyclic nucleotides in neuroblastoma x glioma hybrid cells, *Nature (London),* 262, 311, 1976.
58. **Brandt, M., Buchen, C., and Hamprecht, B.,** Endorphins exert opiate-like action on neuroblastoma x glioma hybrid cells, *FEBS Lett.,* 80, 251, 1977.
59. **Wahlström, A., Brandt, M., Moroder, L., Wünsch, E., Lindeberg, G., Ragnarsson, U., Terenius, L., and Hamprecht, B.,** Peptides related to β-lipotropin with opioid activity: effects on levels of adenosine 3',5' cyclic monophosphate in neuroblastoma x glioma hybrid cells, *FEBS Lett.,* 77, 28, 1977.
60. **Goldstein, A., Cox, B. M., Klee, W. A., and Nirenberg, M.,** Endorphin from pituitary inhibits cyclic AMP formation in homogenates of neuroblastoma x glioma hybrid cells, *Nature (London),* 265, 362, 1977.
61. **Wilkening, D., Sabol, S. L., and Nirenberg, M.,** Control of opiate receptor - adenylate cyclase interactions by calcium ions and guanosine-5'-triphosphate, *Brain Res.,* 189, 459, 1980.

62. **Blume, A. J.,** Opiate binding to membrane preparations of neuroblastoma x glioma hybrid cells NG 108-15: effects of ions and nucleotides, *Life Sci.,* 22, 1843, 1978.
63. **Miller, R. J., Chang, K. -J., Leighton, J., and Cuatrecasas, P.,** Interaction of iodinated enkephalin analogues with opiate receptors, *Life Sci.,* 22, 379, 1978.
64. **Chang, K. -J., Miller, R. J., and Cuatrecasas, P.,** Interaction of enkephalin with opiate receptors in intact cultured cells, *Mol. Pharmacol.,* 14, 961, 1978.
65. **Simon, E. J. and Groth, J.,** Kinetics of opiate receptor inactivation by sulfhydryl reagents: evidence for conformational change in the presence of sodium ions, *Proc. Natl. Acad. Sci. USA,* 72, 2404, 1975.
66. **Pasternak, G. W., Snowman, A. M., and Snyder, S. H.,** Selective enhancement of ^{3}H-opiate agonist binding by divalent cations, *Mol. Pharmacol.,* 11, 735, 1975.
67. **Simantov, R., Snowman, A. M., and Snyder, S. H.,** Temperature and ionic influences on opiate receptor binding, *Mol. Pharmacol.,* 12, 977, 1976.
68. **Blume, A. J., Lichtshtein, D., and Boone, G.,** Coupling of opiate receptors to adenylate cyclase: requirement for Na^+ and GTP, *Proc. Natl. Acad. Sci. USA,* 76, 5626, 1979.
69. **Motomatsu, T., Lis, M., Seidah, H., and Chretien, M.,** Inhibition by beta-endorphin of dopamine-sensitive adenylate cyclase in rat striatum, *Biochem. Biophys. Res. Commun.,* 77, 442, 1977.
70. **Collier, H. O. J. and Francis, D. L.,** A pharmacological analysis of opiate tolerance/dependence, in *The Bases of Addiction,* Fishman, J., Dahlem Konferenzen, Berlin, 1978, 281.
71. **Walczak, S., Wilkening, D., and Makman, M. H.,** Interaction of morphine, etorphine and enkephalins with dopamine-stimulated adenylate cyclase of monkey amygdala, *Brain Res.,* 160, 105, 1979.
72. **Pert, C. B. and Taylor, D.,** Type 1 and type 2 opiate receptors: a subclassification scheme based upon GTP's differential effects on binding, in *Endogenous and Exogenous Opiate Agonists and Antagonists,* Way, E. L., Ed., Pergamon Press, New York, 1979, 87.
73. **Zukin, R. S., Walczak, S., and Makman, M. H.,** GTP modulation of opiate receptors in regions of rat brain and possible mechanism of GTP action, *Brain Res.,* 186, 238, 1980.
74. **Zukin, R. S. and Gintzler, A.R.,** Guanyl nucleotide interactions with opiate receptors in guinea pig brain and ileum, *Brain Res.,* 186, 486, 1980.
75. **Blume, A. J., Dalton, C., and Sheppard, H.,** Adenosine-mediated elevation of cyclic 3':5'-adenosine monophosphate concentrations in cultured neuroblastoma cells, *Proc. Natl. Acad. Sci. USA,* 70, 3099, 1973.
76. **Sabol, S. and Nirenberg, M.,** Regulation of adenylate cyclase of neuroblastoma x glioma hybrid cells by α receptors. I. Inhibition of adenylate cyclase activity mediated by α receptors, *J. Biol. Chem.,* 254, 1913, 1979.
77. **Traber, J., Fischer, K., Buchen, C., and Hamprecht, B.,** Muscarinic response to acetylcholine in neuroblastoma x glioma hybrid cells, *Nature (London),* 255, 558, 1975.
78. **Lefkovitz, R. J., De Lean, A., Hoffman, B. B., Stadel, J. M., Kent, R., Michel, T., and Limbrid, L.,** Molecular pharmacology of adenylate cyclase coupled alpha and beta-adrenergic receptors, in *Proc. 4th International Congress on Cyclic Nucleotides,* Brussels, 1980, in press.
79. **Levitzki, A. and Helmreich, E. J. M.,** Hormone-receptor-adenylate cyclase interaction, *FEBS Lett.,* 101, 213, 1979.
80. **Bradham, L. S., Holt, D.A., and Sims, M.,** The effect of Ca^{2+} on the adenyl cyclase of calf brain, *Biochim. Biophys. Acta,* 201, 250, 1970.
81. **Duffy, M. J. and Powell, D.,** Stimulation of brain adenylate cyclase activity by the undecapeptide substance P and its modulation by the calcium ion, *Biochim. Biophys. Acta,* 385, 275, 1975.
82. **Steer, M. L. and Lewitzki, A.,** The interaction of catecholamines, C^{2+} and adenylate cyclase in the intact erythrocyte, *Arch. Biochem. Biophys.,* 167, 371, 1975.
83. **Sanghoi, I. S. and Gershon, S.,** Commentary: brain calcium and morphine action, *Biochem. Pharmacol.,* 26, 1183, 1977.
84. **Hamprecht, B.,** Neuron models, *Angew. Chem.,* 88, 211, 1976.
85. **Brandt, M., Buchen, C., and Hamprecht, B.,** Calcium ions and opioid action in neuroblastoma x glioma hybrid cells, in *Endogenous and Exogenous Opiate Agonists and Antagonists,* Way, E. L., Ed., Pergamon Press, New York, 1979, 259.
86. **Hano, K., Kaneto, H., and Kakunaga, T.,** Significance of the calcium ion in morphine analgesia, *Jpn. J. Pharmacol.,* 14, 227, 1964.
87. **Harris, R. A., Loh, H. H., and Way, E. L.,** Effects of divalent cations, cation chelators and an ionophore on morphine analgesia and tolerance, *J. Pharmacol. Exp. Ther.,* 195, 488, 1976.
88. **Weger, P. and Amster, C.,** Weiteres zum Problem der Gewohnung an Morphin, *Arch. Exp. Pathol. Pharmakologie,* 181, 489, 1936.
89. **Shangvi, I. and Gershon, S.,** Morphine dependent rats. Blockade of precipitated abstinence by calcium, *Fed. Proc. Fed. Am. Soc. Exp. Biol.,* 35, 470, 1976.

90. **Shangvi, I. and Gershon, S.**, Morphine dependent rats. Blockade of precipitated abstinence by calcium, *Life Sci.*, 18, 649, 1976..
91. **Opmer, F. A. and van Ree, J. M.**, Competitive antagonism of morphine action in vitro by calcium, *Eur. J. Pharmacol.*, 53, 395, 1979.
92. **Hu, J., Huidoboro-Toro, J. P., and Way, E. L.**, Calcium antagonism of opiate action in the nontolerant and tolerant guinea pig ileum, in *Endogenous and Exogenous Opiate Agonists and Antagonists*, Way, E. L., Ed., Pergamon Press, New York, 1979, 263.
93. **Illés, P., Zieglgänsberger, W., and Herz, A.**, Calcium reverses the inhibitory action of morphine on neuro-effector transmission in the mouse vas-deferens, *Brain Res.*, 191, 511, 1980.
94. **Weiss, G. B.**, Cellular pharmacology of lanthanium, *Ann. Rev. Pharmacol.*, 14, 343, 1974.
95. **Harris, R. A., Iwamoto, E. T., Loh, H. H. and Way, E. L.**, Analgetic effect of lanthanium: cross-tolerance with morphine, *Brain Res.*, 100, 221, 1975.
96. **Harris, R. A., Loh, H. H., and Way, E. L.**, Antinociceptive effects of lanthanum and cerium in non-tolerant and morphine tolerant-dependent animals, *J. Pharmacol. Exp. Ther.*, 196, 288, 1976.
97. **Shikimi, T., Kaneto, H., and Hano, K.** Changes of brain calcium content in normal and morphinized mice after the administration of morphine, *Jpn. J. Pharmacol.*, 17, 135, 1967.
98. **Cardenas, H. L. and Ross, D. H.**, Morphine induced calcium depletion in discrete regions of the rat brain, *J. Neurochem.*, 24, 487, 1975.
99. **Yamamoto, H., Harris, R. A., Loh, H. H., and Way, E. L.**, Effects of acute and chronic morphine treatments on calcium localization and binding in brain, *J. Pharmacol. Exp. Ther.*, 205, 255, 1978.
100. **Ross, D. H., Medina, M. A., and Cardenas, H. L.**, Morphine and ethanol: selective depletion of regional brain calcium, *Science*, 186, 63, 1974.
101. **Harris, R. A., Yamamoto, H., Loh, H. H., and Way, E. L.**, Alterations in brain calcium localization during the development of morphine tolerance and dependence, in *Opiates and Endogenous Opioid Peptides*, Kosterlitz, H. W., Ed., Elsevier/North-Holland, Amsterdam, 1976, 361.
102. **Cardenas, H. L. and Ross, D. H.**, Calcium depletion of synaptosomes after morphine treatment, *Br. J. Pharmacol.*, 57, 521, 1976.
103. **Guerrero-Munoz, F., Guerrero, M. L., and Way, E. L.**, The effect of acute and chronic morphine administration in synaptosomal calcium uptake, in *Characteristics and Function of Opioids*, van Ree, J. M. and Terenius, L., Eds., Elsevier/North-Holland, Amsterdam, 1978, 311.
104. **Guerrero-Munoz, F., Cerreta, K. V., Guerrero, M. L., and Way, E. L.**, Effect of morphine on synaptosomal Ca^{++} uptake, *J. Pharmacol. Exp. Ther.*, 209, 132, 1979.
105. **Ross, D., Elizondo, M., and Cardenas, H. L.**, K^{+}-stimulated Ca^{++} uptake in synaptosomes after acute and chronic opiate treatment, *Fed. Proc. Fed. Am. Soc. Exp. Biol.*, 39, 844 Abstr., 3043, 1980.
106. **McCain, H. W., Yamamoto, H., and Way, E. L.**, Effect of morphine and β-endorphin on synaptosomal Ca^{++} fluxes in vitro, *Fed. Proc.* 39, 995, Abstr. 3828, 1980.
107. **Clark, A. G., Jovic, R., Ornellas, M. R., and Weller, M.**, Brain microsomal protein kinase in the chronically morphinized rat, *Biochem. Pharmacol.*, 21, 1989, 1972.
108. **Bonnet, K. A., Branchey, L. B., Friedhoff, A. J., and Ehrlich, Y. H.**, Long-term narcotic exposure reduces caudate cyclic nucleotide levels, protein phosphorylation and tyrosine hydroxylase activities, *Life Sci.*, 22, 2003, 1978.
109. **Ehrlich, Y. H., Bonnet, K. A., Davis, L. G., and Brunngraber, E. G.**, Decreased phosphorylation of specific proteins in neostriatal membranes from rats after long-term narcotic exposure, *Life Sci.*, 23, 137, 1978.
110. **Davis, L. G. and Ehrlich, Y. H.**, Opioid peptides and protein phosphorylation, in *Modulators, mediators and specifiers in brain function*, (Advances in Experimental Medicine and Biology, Vol. 116), Ehrlich, Y. H., Volavka, J., Davis, L. G., and Brunngraber, E. G., Eds., Plenum Press, New York, 1979, 233.
111. **O'Callaghan, J. P., Williams, N., and Clouet, D. H.**, The effect of morphine on the endogenous phosphorylation of synaptic plasma membrane proteins of rat striatum, *J. Pharmacol. Exp. Ther.*, 208, 96, 1979.
112. **Clouet, D. H. and Williams, N.**, The effect of opioids on the phosphorylation in vitro of synaptic membrane proteins, in *Endogenous and Exogenous Opiate Agonists and Antagonists*, Way, E. L., Ed., Pergamon Press, New York, 1979, 239.
113. **Bar, P. R., Schotman, P., and Gispen, W. H.**, Enkephalins affect hippocampal membrane phosphorylation, *Eur. J. Pharmacol.*, 65, 165, 1980.
114. **Dawson, G., McLawhon, R., and Miller, R. J.**, Opiates and enkephalins inhibit synthesis of gangliosides and membrane glycoproteins in mouse neuroblastoma cell line N4TG1, *Proc. Natl. Acad. Sci. USA*, 76, 605, 1979.
115. **Dawson, G., McLawhon, R., and Miller, R. J.**, Inhibition of sialogylcosphingolipid (ganglioside) biosynthesis in mouse clonal lines N4TG1 and NG 108-15 by β-endorphin, enkephalin, and opiates, *J. Biol. Chem.*, 255, 129, 1980.

116. **Miller, R. J., Dawson, G., and McLawhon, R.,** Opioid-induced changes in cell surface composition of neurotumor cell lines, in *Endogenous and Exogenous Opiate Agonists and Antagonists,* Way, E. L., Ed., Pergamon Press, New York, 1979, 267.
117. **Garvican, J. H. and Brown, G. L.,** A comparative analysis of the protein components of plasma membranes isolated from differentiated and undifferentiated mouse neuroblastoma cells in tissue culture, *Eur. J. Biochem.,* 76, 251, 1977.
118. **Truding, R., Shelanski, M. L., and Morell, P.,** Glycoproteins released into the culture medium of differentiating murine neuroblastoma cells, *J. Biol. Chem.,* 250, 9348, 1975.
119. **Greengard, P.,** Phosphorylated proteins as physiological effectors, *Science,* 199, 146, 1978.
120. **Paton, W. D. M.,** The action of morphine and related substances on contraction and on acetylcholine output of coaxially stimulated guinea pig ileum, *Br. J. Pharmacol.,* 11, 119, 1957.
121. **Cox, B. M. and Weinstock, M.,** The effect of analgesic drugs on the release of acetylcholine from electrically stimulated guinea pig ileum, *Br. J. Pharmacol.,* 27, 81, 1966.
122. **Cowie, A. L., Kosterlitz, H. W., and Watt, A. J.,** Mode of action of morphine-like drugs on autonomic neuro-effectors, *Nature (London),* 220, 1040, 1968.
123. **Kosterlitz, H. W. and Watt, A. J.,** Kinetic parameters of narcotic agonists and antagonists with particular reference to N-allylnoroxymorphone (naloxone), *Br. J. Pharmacol.,* 33, 266, 1968.
124. **Lees, G. M., Kosterlitz, H. W., and Waterfield, A. A.,** Characteristics of morphine sensitive release of neurotransmitter substances, in *Agonist and Antagonist Actions of Narcotic Analgesic Drugs,* Kosterlitz, H. W., Collier, H. O. J., and Villarreal, E. J., Eds., MacMillan, London, 1972, 142.
125. **Schaumann, W.,** Inhibition by morphine of release of acetylcholine from the intestine of the guinea pig, *Br. J. Pharmacol.,* 12, 115, 1957.
126. **Knoll, J. and Vizi, E. S.,** Effect of frequency of stimulation on the inhibition by noradrenaline of the acetylcholine output from parasympathetic nerve terminals, *Br. J. Pharmacol.,* 42, 263, 1971.
127. **Lord, J. A. H., Waterfield, A. A., Hughes, J., and Kosterlitz, H. W.,** Endogenous opioid peptides: multiple agonists and receptors, *Nature (London),* 267, 495, 1977.
128. **Waterfield, A. A., Smockum, R. W. J., Hughes, J., Kosterlitz, H. W., and Henderson, G.,** In vitro pharmacology of the opioid peptides, enkephalins and endorphins, *Eur. J. Pharmacol.,* 43, 107, 1977.
129. **Waterfield, A. A., Leslie, F. M., Lord, J. A. H., Lin, N., and Kosterlitz, H. W.,** Opioid activities of fragments of β-endorphin and of its leucine[65]-analogue. Comparison of the binding properties of methionine and leucine enkephalin, *Eur. J. Pharmacol.,* 58, 11, 1979.
130. **North, R. A. and Henderson, G.,** Action of morphine on guinea pig myenteric plexus and mouse vas deferens studied by intracellular recording, *Life Sci.,* 17, 63, 1975.
131. **Dingledine, R. and Goldstein, A.,** Effect of synaptic transmission blockade on morphine action in the guinea pig myenteric plexus, *J. Pharmacol. Exp. Ther.,* 196, 97, 1976.
132. **North, R. A.,** Effects of morphine on myenteric plexus neurones, *Neuropharmacology,* 15, 719, 1976.
133. **Sakai, K. K., Hymson, D. L., and Shapiro, R.** The mode of action of enkephalins in the guinea pig myenteric plexus, *Neurosci. Lett.,* 10, 317, 1978.
134. **Zieglgänsberger, W. and Fry, J. P.,** Actions of opioids on single neurons, in, *Developments in Opiate Research,* Herz, A., Ed., Marcel Dekker, New York, 1978, 193.
135. **Kosterlitz, H. W. and Taylor, D. W.,** The effect of morphine on vagal inhibition of the heart, *Br. J. Pharmacol.,* 14, 209, 1959.
136. **Lemaire, S., Magnan, I., Regoli, D.,** Rat vas deferens: a specific bioassay for endogenous opioid peptides, *Br. J. Pharmacol.,* 64, 327, 1978.
137. **Schulz, R., Faase, E., Wüster, M., and Herz, A.,** Selective receptors for β-endorphin on the rat vas deferens, *Life Sci.,* 24, 843, 1979.
138. **Jhamandas, K., Horn, V., and Sutak, M.,** Comparative effects of opiate agonists methadone, levorphanol and their isomers on the release of rat cortical Ach in vivo and in vitro, *Can. J. Physiol. Pharmacol.,* 53, 540, 1975.
139. **Jhamandas, K., Sawynok, J., and Sutak, M.,** Enkephalin effects on release of brain acetylcholine, *Nature (London),* 269, 433, 1977.
140. **Jhamandas, K. and Sutak, M.,** Effects of endogenous and exogenous enkephalins on brain acetylcholine: cross tolerance between morphine and enkephalins, in *Characteristics and Function of Opioids,* van Ree, J. M. and Terenius, L., Eds., Elsevier/North-Holland, Amsterdam, 1978, 359.
141. **Subramanian, N., Mitznegg, P., Sprügel, W., Domschke, W., Domschke, S., Wünsch, E., and Dembling, L.,** Influence of enkephalin on K^+-evoked efflux of putative neurotransmitters in rat brain, *Naunyn-Schmiedeb. Arch. Pharmacol.,* 299, 163, 1977.
142. **Harsing, L. G., Jr., Vizi, E. S., and Knoll, J.,** Increase by enkephalin of acetylcholine release from striatal slices of the rat, *Polish J. Pharmacol. Pharm.,* 30, 387, 1978.

143. **Vizi, E. S., Hársing, L. G., Jr., and Knoll, J.,** Presynaptic inhibition leading to disinhibition of acetylcholine release from interneurons of the caudate nucleus: effects of dopamine, β-endorphin and D-Ala²-Pro⁵-enkephalinamide, *Neurosci. Lett.*, 2, 953, 1977.
144. **Vizi, E. S., Rónai, A., Hársing, L. G., and Knoll, J.,** Inhibitory effect of dopamine on acetylcholine release from caudate nucleus, *Polish J. Pharmacol. Pharm.*, 29, 201, 1977.
145. **Vizi, E. S., Rónai, A., Hársing, L. G., Jr., and Knoll, J.,** Presynaptic modulation by norepinephrine and dopamine of acetylcholine release in the peripheral and central nervous system, in *Cholinergic Mechanisms and Psychopharmacology*, Jenden, D. J., Ed., Plenum Press, New York, 1977, 587.
146. **Clouet, D. H. and Ratner, M.,** Tyrosine hydroxylase activity in brains of morphine treated rats, *Fed. Proc. Fed. Am. Soc. Exp. Biol.*, 31, 2181, 1972.
147. **Cicero, T. J., Wilcox, C. E., Smithloff, B. R., Meyer, E. R., and Sharpe, L. G.,** Effects of morphine, in vitro and in vivo, on tyrosine hydroxylase activity in rat brain, *Biochem. Pharmacol.*, 22, 3237, 1973.
148. **Fukui, K., Shiomi, H., and Takagi, H.,** Effects of morphine on tyrosine hydroxylase activity in mouse brain, *Eur. J. Pharmacol.*, 19, 123, 1972.
149. **Carenzi, A., Cheney, D. L., Costa, E., Guidotti, A., Racagni, G.,** Action of opiates, antipsychotics, amphetamine and apomorphine on dopamine receptors in rat striatum: in vivo changes of 3',5'-cyclic AMP content and acetylcholine turnover in rats, *Neuropharmacology*, 14, 927, 1975.
150. **Clouet, D. H. and Williams, N.,** The effect of narcotic analgesic drugs on the uptake and release of neurotransmitters in isolated synaptosomes, *J. Pharmacol. Exp. Ther.*, 188, 419, 1974.
151. **Carmichael, F. J. and Israel, Y.,** In vitro inhibitory effects of narcotic analgesics and other psychotropic drugs on the active uptake of norepinephrine in mouse brain tissue, *J. Pharmacol. Exp. Ther.*, 186, 253, 1973.
152. **Hitzemann, R. J. and Loh, H. H.,** Effect of morphine on the transport of dopamine into mouse brain slices, *Eur. J. Pharmacol.*, 21, 121, 1973.
153. **Celsen, B. and Kuschinsky, K.,** Effects of morphine on kinetics of ^{14}C-dopamine in rat striatal slices, *Naunyn-Schmiedeb. Arch. Pharmacol.*, 284, 159, 1974.
154. **Trendelenburg, V.,** The action of morphine on the superior cervical ganglion and on the nictitating membrane of cat, *Br. J. Pharmacol.*, 12, 79, 1957.
155. **Thompson, J. W.,** The Cat's Nictitating Membrane as an Isolated Preparation, Ph.D. thesis, University of London, 1960.
156. **Cairnie, A. B., Kosterlitz, H. W., and Taylor, D. W.,** The effect of morphine on some sympathetically innervated effectors, *Br. J. Pharmacol.*, 17, 539, 1961.
157. **Henderson, G., Hughes, J., and Kosterlitz, H. W.,** The effects of morphine on the release of noradrenaline from the cat isolated nictitating membrane and the guinea pig ileum myenteric plexus-longitudinal muscle preparation, *Br. J. Pharmacol.*, 53, 505, 1975.
158. **Enero, M. A.,** Properties of the peripheral opiate receptors in the cat nictitating membrane, *Eur. J. Pharmacol.*, 45, 349, 1977.
159. **Knoll, J., Illés, P., and Medzihradszky, K.,** The action of enkephalins and enkephalin analogues on neurotransmission in the isolated nictitating membrane of the cat, *J. Pharm. Pharmacol.*, 30, 394, 1978.
160. **Illés, P. and Knoll, J.,** Opiate and dopamine receptors in the isolated nictitating membrane of the cat, *Pol. J. Pharmacol. Pharm.*, 30, 293, 1978.
161. **Rónai, A. Z., Berzétei, I., Székely, J. I., and Bajusz, S.,** The effects of synthetic and natural opioid peptides in isolated organs, in *Characteristics and Function of Opioids*, van Ree, J. M. and Terenius, L., Eds., Elsevier/North-Holland, Amsterdam, 1978, 493.
162. **Rónai, A. Z. and Berzétei, I.,** Similarities and differences of opioid receptors in different isolated organs, in *Endorphins' 78*, Gráf, L., Palkovits, M., and Rónai, A. Z., Eds., Publishing House of the Hungarian Academy of Science, Budapest, 1978, 237.
163. **Henderson, G., Hughes, J., and Kosterlitz, H. W.,** A new example of morphine-sensitive neuroeffector junction: adrenergic transmission in the mouse vas deferens, *Br. J. Pharmacol.*, 46, 764, 1972.
164. **Hughes, J., Kosterlitz, H. W., and Leslie, F. M.,** Effect of morphine on adrenergic transmission in the mouse vas deferens. Assessment of agonist and antagonist potencies of narcotic analgesics, *Br. J. Pharmacol.*, 53, 371, 1975.
165. **Montel, H. and Starke, K.,** Effects of narcotic analgesics and their antagonists on the rabbit isolated heart and its adrenergic nerves, *Br. J. Pharmacol.*, 49, 628, 1973.
166. **Taube, H. D., Endo, T., Bangerter, A., and Starke, K.,** Presynaptic receptor systems on the noradrenergic nerves of the rabbit pulmonary artery, *Naunyn-Schmiedeb. Arch. Pharmacol.*, 293, R2, 1976.
167. **Knoll, J.,** Neuronal peptide (enkephalin) receptors in the ear artery of the rabbit, *Eur. J. Pharmacol.*, 39, 403, 1976.

168. **Hughes, J., Smith, T. W., Kosterlitz, H. W., Fothergill, L. A., Morgan, B. A., and Morris, H. R.,** Identification of two related pentapeptides from the brain with potent opiate agonist activity, *Nature (London),* 258, 577, 1975.
169. **Rónai, A. Z., Gráf, L., Székely, J. I., Dunai-Kovács, Z., and Bajusz, S.,** Differential behaviour of LPH-(61-91)-peptide in different model systems: Comparison of the opioid activities of LPH-(61-91)-peptide and its fragments, *FEBS Lett.,* 74, 182, 1977.
170. **Lemaire, S., Berube, A., Derome, G., Lemaire, I., Magnan, J., Regoli, D., and StPierre, S.,** Synthesis and biological activity of β-endorphin and analogues. Additional evidence for multiple opiate receptors, *J. Med. Chem.,* 21, 1232, 1978.
171. **Wüster, M., Schulz, R., and Herz, A.,** Specificity towards the μ-, δ- and ε-opiate receptors, *Neurosci. Lett.,* 15, 193, 1979.
172. **Henderson, G. and North, R. A.,** Depression by morphine of excitatory junction potentials in the vas deferens of mouse, *Br. J. Pharmacol.,* 57, 341, 1976.
173. **Illés, P., Zieglgänsberger, W., and Herz, A.,** Calcium reverses the inhibitory action of morphine on neuroeffector transmission of the mouse vas deferens, *Brain Res.,* 191, 511, 1980.
174. **Montel, H., Starke, K., and Weber, F.,** Influence of morphine and naloxone on the release of noradrenaline from rat brain cortical slices, *Naunyn-Schmiedeb. Arch. Pharmacol.,* 283, 357, 1974.
175. **Taube, H. D., Starke, K., and Borowski, E.,** Presynaptic receptor systems on the noradrenergic neurones of rat brain, *Naunyn-Schmiedeb. Arch, Pharmacol.,* 299, 123, 1977.
176. **Taube, H. D., Borowski, E., Endo, T., and Starke, K.,** Enkephalin: a potential modulator of noradrenaline release in rat brain, *Eur. J. Pharmacol.,* 38, 377, 1976.
177. **Llorens, C., Martres, M. P., Baudry, M., and Schwartz, J. -C.,** Hypersensitivity to noradrenaline in cortex after chronic morphine relevance to tolerance and dependence, *Nature (London),* 274, 603, 1978.
178. **Korf, J., Burney, B. S., and Aghajanian, G. K.,** Noradrenergic neurones: morphine inhibition of spontaneous activity, *Eur. J. Pharmacol.,* 25, 165, 1974.
179. **Bird, S. J. and Kuhar, M. J.,** Iontophoretic application of opiates to the locus coeruleus, *Brain Res.,* 122, 523, 1976.
180. **Atweh, S. F. and Kuhar, M. J.,** Autoradiographic localization of opiate receptors in rat brain. II. The brain stem, *Brain Res.,* 129, 1, 1977.
181. **Pickel, W. M., Joh, T. H., Reis, D. J., Leeman, S. E., and Miller, R. J.,** Electron microscopic localization of Substance P and enkephalin in axon terminals related to dendrites of catecholaminergic neurons, *Brain Res.,* 160, 387, 1979.
182. **Brase, D. A. and Sampath-Khanna, S.,** Mouse strain differences in response to morphine, *Fed. Proc. Fed. Am. Soc. Exp. Biol.,* 35, 264, 1976.
183. **Sampath-Khanna, S., Brase, D. A., Loh, H. H., and Way, E. L.,** Opiate inhibition of ^{3}H-dopamine release from striatal tissue, *Pharmacologist,* 18, 545, 1976.
184. **Loh, H. H., Brase, D. A., Sampath-Khanna, S., Mar, J. B., Way, E. L., and Li, C. H.,** β-endorphin: in vitro inhibition of striatal dopamine release, *Nature (London),* 264, 567, 1976.
185. **Schwartz, J. C., Pollard, H., Llorens, C., Malfroy, B., Gros, C., Pradelles, Ph., and Dray, F.,** Endorphins and endorphin receptors in striatum: relationship with dopaminergic neurons, in *The Endorphins,* Costa, E. and Trabucchi, M., Eds., Raven Press, New York, 1978, 245.
186. **Seeman, Ph. and Lee, T.,** Neuroleptic drugs: direct correlation between clinical potency and presynaptic action on dopamine neurons, in *Antipsychotic Drugs: Pharmacodynamics and Pharmacokinetics,* Sedvall, G., Ed., Pergamon Press, Oxford, 1976.
187. **Vizi, E. S. and Volbekas, V.,** Inhibition by dopamine of oxytocin release from isolated posterior lobe of the hypophysis of the rat; disinhibitory role of β-endorphin/enkephalin, *Neuroendocrinology,* 31, 46, 1980.
188. **Kuschinsky, K. and Hornykiewicz, O.,** Effects of morphine on striatal dopamine metabolism: possible mechanism of its opposite effect on locomotor activity in rats and mice, *Eur. J. Pharmacol.,* 26, 41, 1974.
189. **Moleman, P. and Bruinvels, J.,** Morphine-induced striatal dopamine efflux depends on the activity of nigrostriatal dopamine neurons, *Nature (London),* 281, 686, 1979.
190. **Papeschi, R., Theiss, P., and Herz, A.,** Effects of morphine on the turnover of brain catecholamines and serotonin in rats — acute morphine administration, *Eur. J. Pharmacol.,* 34, 253, 1975.
191. **Samanin, R., Miranda, F., Mennini, T.,** Serotoninergic mechanism of narcotic action, in *Factors Affecting the Action of Narcotics,* Adler, M. L., Manara, L., and Samanin, R., Eds., Raven Press, New York, 1978, 523.
192. **Miranda, F., Invernizzi, R., and Samanin, R.,** Studies on the mechanism of the interaction of narcotic analgesics with brain serotonin, *Pharmacol. Res. Commun.,* 11, 455, 1979.
193. **Mennini, T., Pataccini, R., and Samanin, R.,** Effects of narcotic analgesics on the uptake and release of 5-hydroxytryptamine in rat synaptosomal preparations, *Br. J. Pharmacol.,* 64, 75, 1978.

194. **Messing, R. B., Flinchbaugh, C., and Waymire, J. C.** Tryptophan and 5-hydroxyindoles in different CNS regions following acute morphine, *Eur. J. Pharmacol.*, 48, 137, 1978.
195. **Perez-Crouet, J., Thoa, N. B., and Ng, L. K.**, Acute effects of heroin and morphine on newly synthesized serotonin in rat brain, *Life Sci.*, 17, 349, 1975.
196. **Larson, A. A. and Takemori, A. E.**, Effect of narcotics on the uptake of serotonin precursors by the rat brain, *J. Pharmacol. Exp. Ther.*, 200, 216, 1977.
197. **Larson, A. A. and Takemori, A. E.**, Effect of morphine on the rate of newly transported tryptophan and 5-hydroxytryptophan in brain, *J. Pharmacol. Exp. Ther.*, 205, 265, 1978.
198. **Garcia-Sevilla, J. A., Ahtee, L., Magnusson, T., and Carlsson, A.**, Opiate receptor mediated changes in monoamine synthesis in rat brain, *J. Pharm. Pharmacol.*, 30, 613, 1978.
199. **Ciofalo, F. R.**, Methadone inhibition of ^{3}H-5-hydroxytryptamine uptake by synaptosomes, *J. Pharm. Exp. Ther.*, 189, 83, 1974.
200. **Ho, A. K. S., Ng, L. K., Thoa, N. B., and Colburn, R. W.**, In vitro interactions on synaptosomal uptake of morphine and biogenic amines, in *Tissue Responses to Addictive Drugs*, Clouet, D. H. and Ford, D. H., Eds., Spectrum Publications, Inc., New York, 1976, 209.
201. **Warwick, R. O. and Schnell, R. C.**, Studies correlating morphine hypothermia with serotonin reuptake in the rat hypothalamus, *Eur. J. Pharmacol.*, 38, 329, 1976.
202. **Moffat, J. A. and Jhamandas, K.**, Effects of acute and chronic methadone treatment on the uptake of ^{3}H-5-hydroxytryptamine in rat hypothalamus slices, *Eur. J. Pharmacol.*, 36, 289, 1976.
203. **Slotkin, T. A., Seidler, F. J., and Whitmore, W. L.**, Methadone inhibits serotonin and norepinephrine uptake into rat brain synaptosomes and synaptic vesicles in vitro but not in vivo, *Eur. J. Pharmacol.*, 49, 357, 1978.
204. **Donzanti, B. A. and Warwick, R. O.**, Effect of methadone on serotonin uptake in rat periaqueductal grey slices, *Eur. J. Pharmacol.*, 59, 107, 1979.
205. **Green, J. P.**, Histamine, in *Handbook of Neurochemistry*, Vol. 4, Lajtha, A., Ed., Plenum Press, New York, 1970, 221.
206. **Subramanian, N. and Mulder, A. H.**, Potassium induced release of tritiated histamine from rat brain tissue slices, *Eur. J. Pharmacol.*, 35, 203, 1976.
207. **Hökfelt, T., Johansson, O., Kellerth, J. -O., Ljungdahl, A., Nilsson, G., Nygards, A., and Pernow, B.**, Immunohistochemical localization of substance P, in *Substance P*, van Euler, U. S. and Pernow, B., Eds., Raven Press, New York, 1977, 117.
208. **Cuello, A. C.**, Enkephalin and substance P containing neurons in the trigeminal and extrapyramidal systems, in *The Endorphins*, Costa, E. and Trabucchi, M., Eds., Raven Press, New York, 1978, 111.
209. **Otsuka, M. and Konishi, S.**, Electrophysiological and neurochemical evidence for substance P as a transmitter of primary sensory neurons, in *Substance P*, van Euler, U.S. and Pernow, B., Eds., Raven Press, New York, 1977, 207.
210. **LeBars, D., Menetrey, D., and Besson, J. M.**, Effects of morphine upon the lamina V type cell activities in the dorsal horn of decerebrate cat, *Brain Res.*, 113, 293, 1976.
211. **Henry, J. L.**, Substance P and pain: a possible relation in afferent transmission, in *Substance P*, van Euler, U.S. and Pernow, B., Eds., Raven Press, New York, 1977.
212. **Jancsó, G. and Jancsó-Gábor, A.**, Involvement of Hypothalamic Structures in Pain Responses and in Morphine Analgesia, XXVIII. Int. Cong. Physiol. Sci. Vol. XIV. 1890 A, 1980.
213. **Stewart, J. M., Getto, C. J., Neldner, K., Reeve, E. B., Krivoy, W. A., and Zimmermann, E.**, Substance P and analgesia, *Nature (London)*, 262, 784, 1976.
214. **Krivoy, W. A., Stewart, J. M., and Zimmermann, E.**, Analgesic actions of Substance P, in *Substance P*, von Euler, V. S. and Pernow, B., Eds., Raven Press, New york, 1977, 195.
215. **Malick, J. B. and Goldstein, J. M.**, Substance P: analgesia following intracerebral administration in rats, *Fed. Proc. Fed. Am. Soc. Exp. Biol.*, 36, 9941, 1977.
216. **Frederickson, R. C. A., Burgis, W., Harrell, C. E., and Edwards, J. D.**, Dual actions of substance P on nociception: possible role of endogenous opioids, *Science*, 199, 1359, 1978.
217. **Growcott, J. W. and Shaw, J. S.**, Failure of substance P to produce analgesia in the mouse, *Br. J. Pharmacol.*, 66, 129P, 1979.
218. **Stern, P. and Hadzovic, J.**, Pharmacological analysis of central actions of synthetic substance P, *Arch. Int. Pharmacodyn.*, 202, 259, 1973.
219. **Terenius, L.**, Effects of peptides and amino acids on dihydromorphine binding to opiate receptor, *J. Pharmac. Pharmacol.*, 27, 450, 1975.
220. **Chipkin, R. E. and Stewart, J. M.**, Effects of substance P and analogs on stimulated and non-stimulated guinea pig ileum, *Pharmacologist*, 19, 190, 1977.
221. **Galoyan, A. A., Akopyan, T. N., Karapetyan, R. O., Arutunyan, A. A., and Oganissyan, A. I.**, Enzymatic mechanisms of the formation of biologically active peptides in the hypothalamus, in *Endorphins'78*, Gráf, L., Palkovits, M., and Rónai, A. Z., Eds., Publishing House of the Hungarian Academy of Science, Budapest, 1978, 37.

222. **Schultzberg, M., Dreyfus, C. F., Gershon, M. D., Hökfelt, T., Elde, R., Nelsson, G., Said, S., and Goldstein, M.**, VIP-, enkephalin-, substance P-, and somatostatin-like immunoreactivity in neurons intrinsic to the intestine: immunohistochemical evidence from organotypic tissue culture, *Brain Res.*, 155, 239, 1978.
223. **Otsuka, M. and Konishi, S.**, Release of substance P-like immunoreactivity from isolated spinal cord of newborn rat, *Nature (London)*, 264, 83, 1976.
224. **Iversen, L. L., Jessell, T., and Kanazawa, I.**, Release and metabolism of substance P in rat hypothalamus, *Nature (London)*, 264, 81, 1976.
225. **Schenker, C., Mroz, E. A., and Leeman, S. E.**, Release of substance P from isolated nerve endings, *Nature (London)*, 264, 790, 1976.
226. **Jessell, T. M. and Iversen, L. L.**, Opiate analgesics inhibit substance P release from rat trigeminal nucleus, *Nature (London)*, 268, 549, 1977.
227. **Iwatsubo, K. and Kondo, Y.**, Inhibitory effect of morphine on the release of preloaded ^{3}H-GABA from rat substantia nigra in response to stimulation of caudate nucleus and globus pallidus, in *Characteristics and Function of Opioids*, van Ree, J. M. and Terenius, L., Eds., Elsevier/North-Holland, Amsterdam, 1978, 357.
228. **Mulder, A. H. and Snyder, S. H.**, Potassium-induced release of amino acids from cerebral cortex and spinal cord slices of the rat, *Brain Res.*, 76, 297, 1974.
229. **Clark, B. J. and Rocha E. Silva, M., Jr.**, An afferent pathway for the selective release of vasopressin in response to carotid occlusion and haemorrhage in the cat, *J. Physiol.*, 191, 529, 1967.
230. **Bisset, G. W., Chowdrey, H. W., and Feldberg, W.**, Release of vasopressin by enkephalin, *Br. J. Pharmacol.*, 62, 370, 1978.
231. **Huidoboro, F. and Huidoboro-Toro, J.**, Antidiuretic effect of morphine and vasopressin in morphine-tolerant and non-tolerant rats, differential effects on urine composition, *Eur. J. Pharmacol.*, 59, 55, 1979.
232. **Vizi, E. S. and Volbekas, V.**, Disinhibitory role of endorphins/enkephalins in the release of oxytocin and vasopressin controlled by dopamine: effects on axon terminals, in *Neuropeptides and Neural Transmission*, Marsan, C. A. and Traczyk, W. Z., Eds., Raven Press, New York, 1980, 257.
233. **Lincoln, D. W. and Clarke, G.**, Opiates: inhibitory factors in stimulus-secretion coupling, *Neuroscience Lett.*, Suppl. 5, S 23, 1980.
234. **Haldar, J. and Sawyer, W. H.**, Inhibition of oxytocin release by morphine and its analogs, *Proc. Soc. Exp. Biol. Med.*, 157, 476, 1978.
235. **Iversen, L. L., Iversen, S. D., and Bloom, F. E.**, Opiate receptors influence vasopressin release from nerve terminals in rat neurohypophysis, *Nature (London)*, 284, 350, 1980.
236. **Burnstock, G., Campbell, G., Satchell, D., and Smythe, A.**, Evidence that adenosine triphosphate or a related nucleotide is the transmitter substance released by non-adrenergic inhibitory nerves in the gut, *Br. J. Pharmacol.*, 40, 668, 1970.
237. **Burnstock, G.**, Purinergic nerves, *Pharmacol. Rev.*, 24, 509, 1972.
238. **McIlwain, H.**, Adenosine neurohumoral and regulatory roles in the brain, in *Central Nervous System*, Genazzani, E. and Hoiken, H., Eds., Springer-Verlag, Berlin, 1974, 3.
239. **Ginsborg, B. L. and Hirst, G. D. S.**, The effect of adenosine on the release of the transmitter from the phrenic nerve of the rat, *J. Physiol.*, 224, 629, 1972.
240. **Hedquist, P. and Fredholm, B. B.**, Effects of adenosine on adrenergic neurotransmission; prejunctional inhibition and postjunctional enhancement, *Naunyn-Schmiedeb. Arch. Pharmacol.*, 293, 217, 1976.
241. **Gintzler, A. R. and Mussachio, J. M.**, Interactions of morphine, adenosine triphosphate and phosphodiesterase inhibitors on the field-stimulated guinea pig ileum, *J. Pharmacol. Exp. Ther.*, 194, 575, 1975.
242. **Sawynok, J. and Jhamandas, K. H.**, Inhibition of acetylcholine release from cholinergic nerves by adenosine, adenine nucleotides and morphine; antagonism by theophylline, *J. Pharmacol. Exp. Ther.*, 197, 397, 1976.
243. **Hayashi, E., Mori, M., Yamada, S., and Kunitomo, M.**, Effects of purine compounds on cholinergic nerves: specificity of adenosine and related compounds on acetylcholine release in electrically stimulated guinea pig ileum, *Eur. J. Pharmacol.*, 48, 297, 1978.
244. **Fredholm, B. B. and Vernet, L.**, Morphine increases depolarization-induced purine release from rat cortical slices, *Acta Phys. Scand.*, 104, 502, 1978.
245. **Phillis, J. W., Jiang, Z. G., Chelack, B. J., and Wu, P. H.**, Morphine enhances adenosine release from the in vivo rat cerebral cortex, *Eur. J. Pharmacol.*, 65, 97, 1979.
246. **Shimo, Y. and Ishi, T.**, Effects of morphine on nonadrenergic inhibitory responses of the guinea pig taenia coli, *J. Pharm. Pharmacol.*, 30, 596, 1978.
247. **Van Loon, G. R. and de Souza, E.**, Effects of β-endorphin on brain serotonin metabolism, *Life Sci.*, 23, 971, 1978.

248. **Ahtee, L., Garcia-Sevilla, J. A., Magnusson, T., and Carlsson, A.,** Effects of morphine, naloxone, naltrexone and β-endorphin on monamine synthesis in rat brain, in *Proc. Int. Narcotic Res. Conf.*, van Ree, J. M. and Terenius, L., Eds., Elsevier/North-Holland, Amsterdam, 1978, 345.

249. **Garcia-Sevilla, J. A., Ahtee, L., Magnusson, T., and Carlsson, A.,** Opiate-receptor mediated changes in monoamine synthesis in rat brain, *J. Pharm. Pharmacol.*, 30, 613, 1978.

250. **Algeri, A., Consolazione, A., Calderini, G., Achilli, G., Puche Canas, E., and Garattini, S.,** Effect of the administration of (d-Ala)[2]-methionine-enkephalin on the serotonin metabolism in rat brain, *Experientia*, 34, 1488, 1978.

251. **Algeri, S., Calderini, G., Consolazione, A., and Lomuscio, G.,** Interaction of opioid peptides with monoaminergic neurones in the rat CNS, in *Proc. Int. Narcotic Res. Conf.*, van Ree J. M. and Terenius, L., Eds., Elsevier/North-Holland, Amsterdam, 1978, 347.

252. **Garcia-Sevilla, J. A., Magnusson, T., and Carlsson, A.,** Effects of enkephalins and two enzyme resistant analogues on monoamine synthesis and metabolism in rat brain, *Naunyn-Schmiedeb. Arch. Pharmacol.*, 310, 211, 1980.

253. **Izumi, K., Motomatsu, T., Chretien, M., Butterworth, R. G., Lis, M., Seidah, N., and Barbeau, A.,** β-endorphin induced akinesia in rats: effect of apomorphine and α-methyl-*p*-tyrosine and related modifications of dopamine turnover in the basal ganglia, *Life Sci.*, 20, 1149, 1977.

254. **Stefano, G. B. and Hiripi, L.,** Methionine enkephalin and morphine alter monoamine and cyclic nucleotide levels in the cerebral ganglia of the freshwater bivalve Anodonta cygnea, *Life Sci.*, 25, 291, 1979.

255. **Jhamandas, K. and Sutak, M.,** Morphine-naloxone interaction in the central cholinergic system: the influence of subcortical lesioning and electrical stimulation, *Br. J. Pharmacol.*, 58, 101, 1976.

256. **Jhamandas, K., Sawynok, J., and Sutak, M.,** Enkephalin effects on release of brain acetylcholine, *Nature (London)*, 269, 433, 1977.

257. **Van Loon, G. R. and Kim, C.,** Effect of β-endorphin on striatal dopamine metabolism, *Res. Commun. Chem. Pathol. Pharmacol.*, 18, 171, 1977.

258. **Algeri, S., Calderini, G., Consolazione, A., and Garattini, S.,** The effect of methionine-enkephalin and d-alanine-methionine-enkephalinamide on the concentration of dopamine metabolites in rat striatum, *Eur. J. Pharmacol.*, 45, 207, 1977.

259. **Ferland, L., Fuxe, K., Eneroth, P., Gustafsson, J. A., Hökfelt, T., and Skett, P.,** Interactions between neurons containing enkephalin-like immunoreactivity and catecholamine containing neurons. Effects of methionine-enkephalin on regional catecholamine levels and turnover, *Acta Endocrinol.*, 85, Suppl. 212, 73, 1977.

260. **Ferland, L., Fuxe, K., Eneroth, P., Gustafsson, J. A., and Skett, P.,** Effects of methionine-enkephalin on prolactin release and catecholamine levels and turnover in the median eminence, *Eur. J. Pharmacol.*, 43, 89, 1977.

261. **Fuxe, K., Ferland, L., Agnati, L. F., Eneroth, P., Gustaffson, J. A., Labrie, F., and Skett, P.,** Effects of intraventricular injection of β-endorphin on dopamine levels and turnover in the unanesthetized male rat. Evidence for an increase of dopamine turnover in the caudatus and the limbic forebrain and for a decrease of dopamine turnover in the median eminence, *Acta Pharmacol. Toxicol.*, 41, Suppl. 4, 48, 1977.

262. **Moroni, F., Cheney, D. L., and Costa, E.,** β-endorphin inhibits ACh turnover in nuclei of rat brain, *Nature (London)*, 267, 267, 1977.

263. **Moroni, F., Cheney, D. L., and Costa, E.,** The turnover rate of acetylcholine in brain nuclei of rats injected intraventricularly and intraseptally with alpha and beta-endorphin, *Neuropharmacology*, 17, 191, 1978.

264. **Moroni, F., Cheney, D. L., Peralta, E., and Costa, E.,** Opiate receptor agonists as modulators of γ-amino-butyric acid turnover in the nucleus caudatus, globus pallidus and substantia nigra of the rat, *J. Pharmacol. Exp. Ther.*, 207, 870, 1978.

265. **Algeri, S., Brunello, N., Calderini, G., and Consolazione, A.,** Effects of enkephalins on catecholamine metabolism in rat CNS, in *Advances in Biochemical Psychopharmacology*, Vol. 18, Costa, E. and Trabucchi, M., Eds., Raven Press, New York, 1978, 199.

266. **Van Loon, G. R. and Kim, C.,** β-endorphin-induced increase in striatal dopamine turnover, *Life Sci.*, 23, 961, 1978.

267. **Calderini, G., Consolazione, A., Garattini, S., and Algeri, S.,** Different effects of methionine-enkephalin and (D-ala[2]) methionine-enkephalinamide on the metabolism of dopamine and norepinephrine in rat brain: fact or artifact? *Brain Res.*, 146, 392, 1978.

268. **Biggio, G., Casu, M., Corda, M. G., di Bello, C., and Gessa, G. L.,** Stimulation of dopamine synthesis in caudate nucleus by intrastriatal enkephalins and antagonism by naloxone, *Science*, 200, 552, 1978.

269. **Harston, C. T.,** The effects of D-Ala[2]-Met[5]-enkephalin-amide on behavioral activity and cyclic nucleotides in the rat brain, *Dissertation Abstr. Intern.*, B 38, 6223, 1978.

270. **Biggio, G., Corda, M. G., Casu, M., and Gessa, G. L.,** Decrease of cyclic GMP in cerebellum by intrastriatal D-Ala²-met-enkephalinamide, *Life Sci.*, 23, 335, 1978.
271. **Versteeg, D. H. G., de Kloet, E. R., and de Wied, D.,** Effects of β-endorphin, α-endorphin and (des-tyr¹)-γ-endorphin on α-MPT-induced catecholamine disappearance in discrete regions of the rat brain, *Brain Res.*, 179, 85, 1979.
272. **Botticelli, L. J. and Wurtman, R. J.,** β-endorphin administration increases hippocampal acetylcholine levels, *Life Sci.*, 24, 1799, 1979.
273. **Guerrero-Munoz, F., Guerrero, M., Way, E. L., and Li, C. H.,** Effect of β-endorphin on calcium uptake in the brain, *Science*, 206, 89, 1979.
274. **Nisticò, G., di Giorgio, R. M., Rotiroti, D., Naccari, F., and Calatroni, A.,** Effects of intraventricular β-endorphin on GABA system in some areas of chick brain, *Res. Commun. Chem. Pathol. Pharm.*, 26, 469, 1979.
275. **Natsuki, R., Hitzemann, R. J., and Loh, H. H.,** Influence of morphine, β-endorphin and naloxone on the synthesis of phosphoinositides in the rat midbrain, *Res. Commun. Chem. Pathol. Pharm.*, 24, 233, 1979.
276. **Stefano, G. B. and Catapane, E. J.,** Enkephalins increase dopamine levels in the CNS of a marine mollusc, *Life Sci.*, 24, 1617, 1979.
277. **Haddox, M. K. and Russel, D. H.,** β-endorphin is a kidney trophic hormone, *Life Sci.*, 25, 615, 1979.
278. **Van Loon, G. R., Ho, D., and Kim, C.,** β-endorphin-induced decrease in hypothalamic dopamine turnover, *Endocrinology*, 106, 76, 1980.
279. **Van Loon, G. R. and Appel, N. M.,** β-endorphin-induced increases in plasma dopamine, norepinephrine and epinephrine, *Res. Commun. Chem. Pathol. Pharmacol.*, 27, 607, 1980.
280. **Wood, P. L. and Stotland, L. M.,** Actions of enkephalin μ and partial agonist analgesics on acetylcholine turnover in rat brain, *Neuropharmacology*, 19, 975, 1980.
281. **Hosobuchi, Y., Meglio, M., Adams, J. E., and Li, C. H.,** β-endorphin: development of tolerance and its reversal by 5-hydroxytryptophan in cats, *Proc. Natl. Acad. Sci. USA*, 74, 4017, 1977.
282. **Lee, R. L., Sewell, R. D. E., and Spencer, P. S. J.,** Importance of 5-hydroxytryptamine in the antinociceptive activity of the leucine-enkephalin derivative, D-Ala²-Leu⁵-enkephalin (BW 180C), in the rat, *Eur. J. Pharmacol.*, 47, 251, 1978.
283. **Gropetti, A., Algeri, S., Cattabeni, F., di Giulio, A. M., Galli, C. L., Ponzio, F., and Spano, P. F.,** Changes in specific activity of dopamine metabolites as evidence of a multiple compartmentation of dopamine in striatal neurones, *J. Neurochem.*, 28, 193, 1977.
284. **Wood, P. L., Cheney, D. L., and Costa, E.,** Modulation of the turnover rate of hippocampal acetylcholine by neuropeptides: possible site of action of α-melanocyte-stimulating hormone, adrenocorticotrophic hormone and somatostatin, *J. Pharmacol. Exp. Ther.*, 209, 97, 1979.
285. **Brennan, M. J. W., Cantrill, R. C., and Wylie, B. A.,** Modulation of synaptosomal GABA release by enkephalin, *Life Sci.*, 27, 1097, 1980.
286. **Biggio, G., Corda, M. G., Casu, M., and Gessa, G. L.,** Striato-cerebellar pathway controlling cyclic GMP content in the cerebellum: role of dopamine, GABA and enkephalins, in *Advances of Biochemical Psychopharmacology*, Vol. 18, Costa, E. and Trabucchi, M., Eds., Raven Press, New York, 1978, 227.

Chapter 2

ELECTROGRAPHIC ANALYSIS OF OPIOID MECHANISMS

J. I. Székely

TABLE OF CONTENTS

I. Introduction 34

II. Effects of Opiates on the Spontaneous Electroencephalographic (EEG) Activity 34

III. Effects of Opioid Peptides on the Spontaneous EEG Activity 39

IV. Effects of Opioid Peptides and Opiates on the Activity of Single Brain Neurones 44
- A. Cerebral Cortex 44
- B. Thalamus 53
- C. Amygdale 53
- D. Substantia Nigra and Striatum 53
- E. Nucleus Accumbens 56
- F. Hypothalamus and Hippocampus 56
- G. Periaqueductal Gray (PAG), Raphe Nuclei, Brain Stem, and Reticular Formation 58
- H. Locus Coeruleus (LC) 62
- I. Pons and Medulla 64
- J. Spinal Cord Ventral Horn Neurones 65
- K. Spinal Cord Dorsal Horn Neurones 69

V. Effects of Opioids on Spinal Mono- and Polysynaptic Reflexes 75

VI. The "Gate Control System" and the Opioids (Toward a Working Hypothesis on Spinal Modulation by Enkephalins of Pain Reflexes) 81
- A. The Role of Primary Afferent Hyperpolarization 84
- B. Substance P and Enkephalins 86

VII. Effects of Opioids on Neuronal Activity In Vitro 88

VIII. Conclusions 95

References 99

I. INTRODUCTION

Upon examining the electrographic effects of opioids mostly microiontophoretic application and unit activity recording have been given preference to systemic drug administration and registration by macroelectrodes; a reasonable choice if the pharmacologists are interested in the intricate mechanism of action and in the relative contributions of individual brain areas to the pharmacological effects of the substances under study. However, there are several aspects of opiate actions which can be studied only by macroelectrophysiological techniques even in the age of microelectrophysiology. First of all, the eventual changes in the activation level and the appearance of pathological (epileptiform) electrical activity are easier to record by macroelectrodes. Especially if their relationship to the general locomotion or motor manifestation of epilepsy is under study.

II. EFFECTS OF OPIATES ON THE SPONTANEOUS ELECTROENCEPHALOGRAPHIC (EEG) ACTIVITY

The first reports on the effects of opiates on spontaneous EEG were published in the late forties and the early fifties. Thus, morphine was one of the first psychopharmaca subjected to detailed EEG-analysis. In these studies[1] mostly synchronizing effect was found, i.e., decrease in the frequency and increase in the amplitude of spontaneous EEG activity was recorded by macroelectrodes in rats, rabbits, dogs, and humans but not in cats upon treatment with morphine-like drugs. Qualified as CNS depressant pharmaca just this synchronized EEG activity was expected. For this reason the sometimes oversimplified statements on the synchronizing effect of opiates had not been defined for a long time. Meanwhile numerous data accumulated according to which both synchronization and desynchronization can be elicited by opiates and at higher dose level the EEG activity becomes pathological. (In these and the subsequently reviewed studies higher amounts of opiates were applied than needed for analgesia.)

The rather voluminous literature on the EEG effects of opiates is unanimous in relation to the parallelism between the motor and electrographic actions. If such a parallelism really existed desynchronization should appear under the experimental conditions known to result in hyperactivity, i.e., if giving them in very low doses, when only their behavioral disinhibitory action prevails (A) or during the phase of late locomotor excitation after dissipation of initial motor depression (B). Moreover, EEG arousal may be anticipated upon repeated treatment as tolerance develops to the motor depressant (and analgesic) effects but not to their stimulant actions (C). Finally EEG arousal might be expected in those species known to respond with general behavioral excitation to morphine treatment (D).

A. As for the supposed electrographic arousal upon administering small doses of morphine-like agents, no data are available. This may be simply due to the lack of the proposed correlation. Nevertheless, the lack of such data also might be attributed to certain methodological difficulties. If the animals are awake, a further slight electrographic arousal is difficult to detect, unless the EEG records are quantitatively analyzed, a procedure not frequently used at the time of the first studies on the EEG actions of opiates. In doing so, i.e., continuously recording the sleep-waking cycle in rats by fast Fourier transformation of the EEG, insomnia was observed by Havlicek et al.[2] upon injecting 0.1 μg β-EP i.c.v. This dose is much smaller than needed for analgesia and behavioral depression, therefore, this finding may indicate the existence of EEG activation at subanalgesic dose level. Unfortunately the contingent similar action of opiates has not been studied by these authors.[2] In this aspect probably the observations of Albus and Herz[3] are relevant, who reported on an initial EEG activation upon i.c.v. injection of morphine in rabbits. Considering the slow penetration of i.c.v. administered hydrophil morphine molecules, this mode of application may probably imitate the systemic administration of the same substance in very low dosage. It is a very interesting finding considering that morphine has been unanimously reported to cause synchronization in rabbits upon peripheral administration applying whether low or high doses.[1,4-6] In cats, first Leimdorfer[7] then Navarro and Eliott[5] observed EEG activation, i.e., increase in frequency and decrease in amplitude in the cortical leads and disappearance of the slow waves in different subcortical structures, moreover, onset of regular theta rhythm in the hippocampus upon i.p. administration of a small dose (1 mg/kg) of morphine. Giving, however, higher doses also in this species slow waves of high amplitude characterized the picture.[5]

B. First Colasanti and Khazan[8] stressed that in rats the initial period of high voltage slow EEG activity and behavioral stupor occurring after administration of narcotics was uniformly succeeded by a secondary phase of EEG and behavioral arousal. This secondary arousal is obviously the electrographic manifestation of the late locomotor stimulation. During this period the desynchronization was interrupted by several brief episodes of slow wave activity associated with behavioral stupor.[8] Thus, the electrographic and motor effects are well correlated even if the subsequent time sequences are compared.

Moreover, just as in relation of spontaneous locomotion, on increasing the dose the duration of both the initial depression and the subsequent stimulation was lengthened.[8] As for the mixed agonists-antagonists, also these substances (nalorphine, pentazocine, cyclazocine) elicited this biphasic pattern in the EEG-activity, but in contrast with all narcotic agonists, the secondary phase of behavioral stimulation elicited by them was of longer duration than the primary phase of locomotor depression and EEG synchrony.[8,9]

Depletion of the brain 5-HT content with *p*-chlorophenylalanine resulted in significant lengthening of the arousal reaction[10] just as in the case of spontaneous locomotion (Chapter 2, Volume I). However, depletion of the brain catecholamines by α-methyl-*p*-tyrosine did not modify significantly the above-discussed pattern of EEG activity.[8]

C. Upon repeated application tolerance develops to the initial phase of EEG synchronization but not to the subsequent desynchronization examining it whether in heroin and methadone addicts[11] or in animal experiments.[12] Thus, dissipation of the analgesic and behavioral depressant actions of the opiates is accompanied by gradual disappearance of the synchronizing effect. Taking into consideration that this correlation seems to be valid not only in animal experiments but also in humans,[11] a simple conclusion arises: the synchronized EEG activity is a reliable indicator of both the analgesic and euphoric actions of opiates.

D. Just for the above-mentioned reason it would be very interesting to know the EEG effects of morphine in cats and mice, i.e., in species, which respond with behavioral activation to opiates. As for mice, no data are available. In cats according to one study,[7] no synchronizing effect can be elicited by either dose of morphine, while in another work synchronization was recorded in higher dosage.[5] Therefore, the electrographic response of cats to opiates hardly fits into the general scheme, which probably explains that this species, otherwise so frequently used in electrophysiological studies, has been neglected if opiates were examined. Let us mention here that in our experiments both cats and mice[13] showed a certain behavioral stupor immediately upon morphine treatment, while the well-known symptoms of feline mania (in cats) and those of the running fit (in mice) appeared only with a latency of 15 to 30 min (Chapter 2, Volume I). Consequently, it would be very interesting to analyze separately the EEG changes during this brief period of initial behavioral depression.

Returning to the typical electrographic actions of opiates, first of all the depressant actions were studied in detail.[4-6,14,15] Giving morphine to rats or rabbits up to the doses of 30 to 50 mg/kg, the ensuing EEG picture is similar to that seen during sleep. This deceleration of EEG activity can be detected both by simple inspection of the records and also by automatic frequency analysis.[4-6,8,14,15] The onset of the large, slow waves is, however, not the first manifestation of electrographic deactivation. The typical sleep-like EEG pattern is introduced by spindle activity just as in the natural drowsy state. The spindles are of high frequency and appear periodically mostly in the cortex. The subsequent slow waves of high amplitude arise generally simultaneously in the cerebral cortex and in the different thalamic[4] and hypothalamic[14] leads, in the mid-brain,[4] and in the amygdala.[16] Probably hippocampus is the only exception, where according to the tracings shown in the report of Albus and Herz[3] a decrement of amplitude may be seen. (A finding, clearly visible but not discussed specifically by the authors.[3]) This finding only further confirms the basic concept of Green and Arduini[17] on the reciprocal pattern of electrocortical and hippocampal activities, i.e., that the cortical desynchronization is generally accompanied by hippocampal synchronization and vice versa.

Giving morphine in higher dosage (generally more than 50 mg/kg) the sleep-like EEG activity is gradually replaced by pathological waves.[4-6,8,14,15] First isolated spikes or intermittent bursts of spikes are superimposed upon or interspersed with the slow waves. These spikes appear both at cortical and subcortical levels. Nevertheless, spikes can be observed generally first in the cortical leads, sometimes for a short while over one hemisphere[15] then they spread to the other side and subsequently to the subcortical regions. The next step is the onset of spike and wave complexes (''petit-mal''-like spike and dome sequences), which finally give way to typical ''grand-mal'' seizure-activity. In rabbits the seizure activity appears first in the associative cortex then it spreads to the other cortical regions and to the different subcortical areas.[4] During the ''grand-mal''-like seizures convulsive disorders may be observed in all brain structures examined with the exception of the cerebellar and lumbar spinal leads.[4] These epileptic discharges are more or less synchronous in the different cortical and subcortical leads. After subsidence of the fit the tracings are almost isoelectric or mostly an activity of low voltage and slow frequency prevails till the onset of the next epileptic outburst.[15] The actions of other narcotics such as of codeine,[4] methadone,[15] and levorphanol[6] are similar.

Seizure activity may also be elicited by topical application of morphine (0.2 to 1.0% solution) to the cortex.[4]

The electrical manifestation of convulsive activity is accompanied by motor symptoms such as twitching of the ears, muzzle, or the head, generally synchronously with the cortical spikes. At the time of the generalized seizure activity clonic or clonic-tonic movements may be seen all over the body.

Accordingly the appearance of sleep-like synchronization and the convulsive EEG pattern are two subsequent steps of opiate action, which are difficult to separate from each other.[3,6,8,14-16] Consequently, the decreased level of arousal and increased susceptibility to seizures seem to be interrelated.

Of the subcortical structures hippocampus[5] and medial thalamus[4,6] are the first in joining the convulsive activity. In rabbits the epileptiform discharges appear almost simultaneously in the motor cortex and in the dorsal hippocampus.[5] According to Corrado and Longo[4] and Gangloff and Monnier[6] the electrical seizures induced by morphine are predominantly cortico-diencephalic. For this reason they[4,6] suppose that first the rhinencephalic structures and the intralaminar nuclei of the thalamus are activated resulting in spikes or spike and wave discharges. Later during the "grand-mal"-like clonic-tonic seizures obviously the other subcortical structures would be step by step involved.

As for the effects of opiates on EEG activity in humans, the data available refer mostly to addicts and not to "experimentally naive" healthy subjects obviously due to ethical considerations. In the experiments of Gibbs and Maltby[18] morphine caused at first an increase in voltage of the 20 c/s activity and a shift toward the slower frequencies only later. Consequently, also in humans the very first effect of morphine is a transient desynchronization if giving it in sufficiently low dose. This action is, however, followed by lowering of the α-frequency and appearance of δ-waves, i.e., a progressive slowing of the record.[1,19] As in these[1,18,19] studies the subjects did not fall asleep but no δ-waves can be seen in the EEG record of normal adults, these waves should be considered pathological. They are probably introductory manifestations of the epileptic action of opiates. Epileptiform EEG activity has been found in human addicts also upon methadone treatment.[20]

However, it has repeatedly been reported that in postaddicts the usual doses of morphine or methadone generally does not elicit striking changes in the EEG records, at least no one that can be detected by simple inspection, though in some cases a certain slowing of the activity could be recorded.[1] Thus, also in humans tolerance develops to the EEG effects of opiates.

Probably more informative are the experiments connected with the effects of opiates upon the electrographic arousal reactions to different kinds of stimulation. By measuring the threshold of the EEG activation (desynchronization) to different classes of sensory stimulation, the eventual differential actions of morphine-like drugs on specific sensory pathways may be clarified. As reported by Silvestrini and Longo,[21] administering small amounts of morphine (5 to 10 mg/kg i.v.) to unanesthetized rabbits, a selective depression of arousal reaction to painful stimuli could be noticed, while the desynchronization to other kinds of sensory stimulation remained unaltered. At the same time the threshold of arousal reaction to electrical stimulation of the anteromedial thalamic nuclei was enhanced but the similar reaction to mesencephalic stimulation did not change.[21] At higher dose levels, however, the arousal reaction to all kinds of sensory or electrical stimulation was found inhibited.[21] These data indicate a selective action of morphine on the integration of pain reactions at the thalamic level.

In other simultaneously published works[3,6] inhibition of the "attention reaction" to any kind of external stimuli was observed upon opiate treatment also in unanesthetized rabbits. Moreover, also the electrographic arousal elicited by stimulation of the mesencephalic reticular formation was suppressed by morphine and levorphanol.[3,6] The apparent contradiction in that relation that which kinds of stimuli lose their arousing capacities may probably be explained by differences in the doses applied.[3,6,21] Nevertheless, these studies show[3,6,21] that opiates suppress all kinds of arousal reactions at fairly high dose levels, while that induced by noxious stimuli in much lower amounts. Of course, the question arises whether opiates inhibit the primary pain pathways or only the arousal reaction elicited by them. In a careful study Chin and Domino[22] found no consistent changes in the primary evoked potentials upon morphine treatment if electrical stimulation of the tooth pulp or the sciatic nerve served for painful stimuli. Also these observations show that morphine does not inhibit the pain transmission in the primary sensory pathways.

On the other hand, opiates were reported to enhance the generation of convulsive potentials in many different experimental models. Morphine produced a definite enhancement of cortical "recruiting" responses upon stimulation of various thalamic areas in dogs[22] and in rabbits[6] as well. Levorphanol had a similar effect.[6] In rabbits morphine drastically increased the spikes elicited by low frequency electrical stimulation of the hippocampus in all cortical and subcortical areas, while stimulating the midbrain reticular formation in similar ways an inhibition of evoked spikes was observed at low dose level and potentiation in higher doses.[6]

Consequently, the antinociceptive action might be explained by a selective inhibition of nonspecific mesencephalic-thalamic activating system, while convulsive phenomena can be initiated both from the hippocampus and from the thalamo-cortical recruiting system. A third suspected site of seizure generation is the amygdala. Teitelbaum et al.,[16] injecting morphine or levorphanol unilaterally into the anterior amygdala of rats, observed spike and wave generation at the injection site. After a time lag of 15 to 30 min a similar pattern of epileptoid activity appeared in the contralateral untreated amygdala, i.e., a mirror focus developed.[16]

The suspected role of above-mentioned forebrain structures in generation of epileptiform activity upon administering high amounts of opiates does not necessarily mean that also the previously discussed synchronizing and analgesic actions are generated in these areas. Thus Albus and Herz[3] found no striking effects on the spontaneous EEG in rabbits if morphine was injected into the separated anterior parts of the ventricular system or into the third ventricle. Administering it, however, into the fourth ventricle distinct synchronization, analgesia, and relative unresponsiveness to electrical stimulation of brain stem reticular formation was detected. Essentially the same observation has been made in dogs upon perfusing fentanyl through the fourth ventricle.[23]

Actually in both laboratories[3,23] analgesia and sleep-like EEG activity could be elicited by separate activation of the opiate receptors located in the caudal parts of the brain. Of course, if the drugs are administered into the ventricular space, it is difficult to specify exactly the brain area involved in the action, but the above-mentioned studies clearly show that the epileptogenic activity of opiates is mediated by forebrain structures, while the analgesic action and the corresponding decrement in the activation level is elicited by brain stem areas. From this point of view the experiments of Yeung et al.[24] are of special interest, who observed synchronized cortical activity but no analgesia in rats if morphine was given into the medial thalamus; while injecting it into the PAG or intrathecally they observed inhibition of the tail-flick response without alteration of the desynchronized cortical EEG activity. These data obtained in dogs[23] and in rabbits[3] seem to contradict those recorded in rats.[24] Unless we suppose that the opiates via different brain structures simultaneously modify the spontaneous EEG activity, the excitability of certain pathways, and the polysynaptic nociceptive reflexes. Accordingly, opiates should inhibit the polysynaptic nociceptive reflexes already at spinal and medullary levels with simultaneous blockade of the sensory input into the reticular formation responsible for the maintenance of cortical activity. A direct action on the mesencephalic PAG and adjoining reticular formation would explain both the analgesic effect via the descending reticulospinal pathways (probably with the raphe nuclei as relay centers) and the synchronizing action due to the inhibition of ascending reticulo-thalamo-cortical activating pathways. Accordingly, stimulation by opiates of certain forebrain structures and neuronal circuitries resulting in epileptiform EEG activity would be an action entirely independent of those elicited via more caudal structures.

Finally let us mention here that the electrographic effects of morphine and those induced by Win 35,197-2, a selective agonist of κ-receptors, were found to be partially different.[25] In another study[9] the morphine and ethylketocyclazocine induced EEG alterations showed differential sensitivities towards naloxone. Thus also the macroelectrographic studies confirm the multiplicity of opiate receptors. However, in the latter study[9] the ethylketocyclazocine-induced effects were reversed more easily with naloxone than the morphine elicited synchronization contrary to the data obtained by the Martin group (Chapter 2, Volume I), who found opposite differences in sensitivity to opiate antagonists in spinal dogs.

III. EFFECTS OF OPIOID PEPTIDES ON THE SPONTANEOUS EEG ACTIVITY

Almost all data on the effects of natural enkephalins and their synthetic analogues on the spontaneous EEG activity were obtained upon their i.c.v. application, which is probably not surprising in view of the relatively weak biological potency and susceptibility to enzymatic degradation of the former and the relative inability to pass the blood-brain barrier of the latter (Chapters 3, 5, 6, Volume III).

The multiphasic and complex effects of i.c.v. administered natural enkephalins were described in detail by Urca et al.[26] and Frenk et al.[27] Injecting 100 μg of them to conscious rats epileptiform activity was found: first intermittent spike bursts and sharp waves were seen then a short period of desynchronized EEG followed by a second buildup of pathological activity.[26,27] This second phase of epileptiform activity consisted of sharp waves, spike and wave complexes, isolated spikes, and polyspikes accompanied with characteristic motor manifestations such as wet dog shakes, myoclonic twitches of the trunk, head, or limbs.[26, 27] Giving them in "subconvulsive" dosage, spindles of mixed frequency, or hypersynchronous slow waves with occasional spike and wave complexes were seen.[26,27] All these effects could be reversed by naloxone, and surprisingly the convulsive effects of Leu-E were stronger than that of the Met-E.[26,27] A more or less similar picture was seen upon i.c.v. application of a more potent synthetic analogue DALA.[28]

Inhibition of the arousal reaction also can be observed upon i.c.v. treatment with natural enkephalins or with their more potent synthetic analogues.[29]

Actually rather similar findings were reported on the EEG actions of β-endorphin. As mentioned previously giving it in a very small dose (0.1 μg), a distinct alerting effect has been observed, the animals remained awake for a period lasting at least 2 hr.[2] In higher doses a drowsy state was seen with spindle activity, and by further enhancement of the dose the typical sleep pattern of low frequency high amplitude synchronous waves were seen then epileptiform activity.[2]

Henriksen et al.[30] and others[2] analyzing the electrographic action of β-EP observed effects very similar to those recorded upon enkephalin treatment. First high voltage paroxysmal waves appeared in the cortical and in the subcortical leads (predominantly in the hippocampus and the amygdala) lasting about a minute then a period of desynchronized activity ensued with almost isoelectric record from the hippocampus and with isolated spikes in other subcortical leads first of all in the amygdala.[2,30] The second buildup consisting of spikes and polyspikes mostly synchronous in the cortical and subcortical leads lasted for several minutes and returned repeatedly depending on the dose applied.[2,30] As for the motor effects, first an increase in the motor activity was observed then a period of immobility, finally myoclonic jerks of the trunk and extremities and characteristic body shake movements (wet dog shake) were seen accompanying the main phase of convulsive activity.[2,30] Of course, these motor manifestations of the convulsions were obscured by the β-endorphin induced immobility and catalepsy.[2,30]

Thus, whether the natural enkephalins or β-EP were examined the electrographic alterations consisted of the same three to four subsequent phases: (1) initial behavioral stupor accompanied by high voltage activity and occasional spikes, (2) behavioral arousal with desynchronization, (3) grand mal-like convulsive EEG activity with myoclonic jerks, (4) and finally the phase of exhaustion characterized by low frequency and low amplitude EEG activity.[2,26-28,30]

Aloisi et al.[29] distinguish only two phases, an initial period of excitation, followed by synchronized and epileptic activity (observable both in rats and rabbits upon i.c.v. treatment). Otherwise, also these authors found the EEG-effects of morphine, natural enkephalins, and those of the latter synthetic derivatives qualitatively similar. The extremely potent analogue FK 33-824 proved to be the most potent also in induction of EEG effects.[29]

Unfortunately no systematic study has been made concerning the EEG changes accompanying the subsequent phase of excitation, if it exists at all. As discussed previously, the electrographic depression induced by opiates is always followed by a late period of hyperactivity[8] corresponding to the locomotor facilitation (Chapter 2, Volume I). Unfortunately in the cases of opioid peptides only the initial depressant and epileptic EEG effects were analyzed. It requires further examination whether really no late hyperexcitation exists at all, or this phenomenon, regarded as nonspecific rebound-like effect was simply neglected in these otherwise very informative studies.[2,6-28,30] β-EP given i.v. in doses as high as 20 mg/kg failed to elicit the epileptiform activity.[30] However, as discussed in Chapter 3, Volume II, it is more than questionable if peripherally administered β-EP is capable of eliciting any central effect at all due to its relative inability to pass the blood-brain barrier.

The depressant and epileptic activities of opioid peptides can be reversed by naloxone just as those of the opiates.[2,26-28,30] This antagonizability by naloxone, i.e., the participation of opiate receptors does not necessarily mean that the same structures and the same subclass of opiate receptors mediate the epileptogenic and the much debated analgesic effects of the natural enkephalins. Opiates induce epileptiform discharges in much higher doses than needed for analgesia (Chapter 2, Volume II). β-EP[30] and natural enkephalins are much stronger epilepsy inducing agents than analgesics. The natural enkephalins induce seizure activity well below the i.c.v. dose of 100 μg,[26,27] whereas their antinociceptive action is doubtful even in much higher amounts (Chapter 3, Volume II). As for analgesia Met-E seems to be the stronger analgesic (if it has such activity at all, Chapter 3, Volume II), but Leu-E is the more potent in eliciting the above-discussed EEG-alterations.[26,27] And similarly, the minimal effective dose of β-EP to induce convulsive EEG activity was smaller (1 nanomole) than needed for analgesia (at least 5 nanomole) in the same animals.[30] In this experiment[30] the minimal doses of β-EP necessary for induction of analgesia and epileptoid EEG activity, respectively, were close to each other.

Upon repeated application of morphine or Met-E an increased susceptibility to their seizure inducing effects was observed.[27] Though the gradual decrease in threshold to epilepsy inducing stimuli is a well-known phenomenon, establishing even certain models of experimental epilepsy (kindling), the same process in relation of opioids is at least unusual. No other case is known where the sensitivity to any effect of an opioid would be increased upon repeated administration. Nevertheless, even this observation[27] proves that different mechanisms mediate the analgesic and epileptogenic actions of opioids. Both effects must be mediated by opiate receptors, since the β-EP-induced epileptiform activity could be only moderately inhibited by diazepam and phentoin, i.e., by typical antiepileptic agents, or by haloperidol, amphetamine, apomorphine, and scopolamine.[30]

Urca and his co-workers reported[26,27] that injecting Met- or Leu-E into structures dorsal to the lateral ventricle produced seizure activity, while aiming them to loci ventral to the aqueduct analgesia developed. Moreover, Leu-E-induced epileptiform EEG activity in a dose as low as 1 μg if administering it into the medial thalamus.[27] Let us mention here that opiate-induced analgesia is always accompanied by increased unit activity in the PAG, but injecting enkephalins into the lateral ventricle, rats showing analgesia, also responded with an increment in the PAG multiple unit firing while those failing to display analgesia did not.[26] Nevertheless, all these animals had typical epileptiform manifestations. Consequently, different brain areas and different neurone populations should mediate the analgesic and epileptiform effects. Considering the data mentioned in the preceding chapter, obviously certain forebrain structures may be suspected to participate in these events. It has been recently reported that injection of morphine or enkephalins into different sites of the limbic system such as amygdala[16,31] or medial thalamus[27,32] resulted in hypersynchronized and epileptiform activity but not in analgesia. Frenk et al.[32] injected the same amounts of Met-E (120 μg) into the PAG or into the dorsomedial nucleus of the thalamus and simultaneously recorded the EEG, general motor activity, and the eventual analgesic effect in the tail flick test. Administering Met-E into the ventromedial portion of the PAG or into the adjacent regions of the ventral tegmentum, analgesia was observed without motor seizures or any alteration in the cortical EEG activity.[32] Injecting, however, the same amount of Met-E into the dorsomedial thalamic nucleus, convulsive electrographic activity was found accompanied with myoclonic twitches and wet dog shakes but no antinociception.[32] These seizures were indistinguishable from those induced by i.c.v. administered Met-E, Leu-E, or morphine.[26,27]

In this context it is worth mentioning that electrical stimulation of the dorsomedial thalamus gives rise to seizures but not to analgesia, while stimulation of the PAG results in analgesia without concomitant convulsive phenomena.[33,34]

The medial thalamus and amygdala are, however, not necessarily the primary sites in induction of epileptiform activity by opioid peptides. The β-EP-induced epileptoid activity persisted even after bilateral amygdalectomy or medial dorsal lesions.[30] Thus, attention has been focused on the hippocampus, a structure which is otherwise probably the most frequently implicated one in generation of the human epilepsy.

In an elegant study Taylor and her co-workers[35] examined the effect of opioid peptides on the electrical activity of ocular hippocampal transplants. The obvious advantage of this technique is the exclusion of all possible influence exerted by other brain structures. Microiontophoretic application of Met-E, β-EP, or DALA elicited a marked increase in the amplitude followed in most cases by clear-cut epileptiform activity.[35] Consequently, hippocampus is capable to generate epileptoid seizures without the participation of any other structure, and this mechanisms may be activated by opioid peptides. This finding[35] and the failure of medial thalamic lesions or amygdalectomy to prevent the β-EP-induced epilepsy[30] points to the possible primary role of the hippocampus in this process. (The role of opioid peptides in hippocampal epilepsy will be discussed in another chapter.)

The alleged primary role of hippocampus or generally that of the limbic structures has a further important implication. All these regions contain more or less opiate receptors and some of them are rather rich in enkephalins, while there is only a hardly detectable amount of β-EP (Chapters 1 and 3, Volume III). Consequently, if opioids are really involved in the genesis of epilepsy, only enkephalins can be taken into consideration.

The well-established role of the hippocampus in the pathology of human epilepsy only further substantiates the possible importance of this structure in the opiate-induced seizure activity. Some data, however, suggest that the epileptiform activity induced by opioids can only be erroneously mixed up with the similarly termed human disease. Giving Leu-E directly into the medial thalamus, seizure activity can be elicited already with the minute amount of 1 μg.[27] Injecting it, however into the hippocampus[36] or into the lateral ventricles,[26] much higher amounts are needed for elicitation of the convulsive phenomena.

Nevertheless, also certain species differences must be counted with. Thus, as reported by Elazar et al.,[36] in cats using the "encéphale isolé" preparation 0.2 to 1.0 μg of Leu-E was sufficient to elicit well-developed epileptiform alterations in the EEG if giving it into the hippocampus.[36] This seizure activity consisted of high amplitude spikes and clonic-tonic generalized convulsions.[36] Injecting Leu-E into the hippocampus of rats anesthetized by urethane, much higher doses were needed to elicit similar effects.[36]

All the authors quoted above agree that spindling and hypersynchronization then the appearance of different pathological waveforms and finally the onset of generalized seizure activity represent subsequent grades of the same process. Consequently, the primary role should rather be attributed to those brain structures, responsible for the cortical EEG synchronization. As discussed in the standard neurophysiological textbooks, it is the nonspecific nuclei of the thalamus which generate the cortical synchronized activity. Electrical stimulation of these nuclei is known to produce the so-called "recruiting" and "augmenting" potentials. Accordingly, the cortical synchronization is considered a process generated not in the hippocampus but in the thalamo-cortical circuitry. As mentioned above,[24] injecting morphine into the medial thalamus induces high voltage slow-wave activity just as the opioid peptides do.[27,32] Furthermore, morphine facilitates the cortical recruiting potentials elicited by electrical stimulation of the corresponding thalamic nuclei.[22] Hence the possibility emerges that the above-discussed epileptiform activity arises by facilitation of physiological mechanisms resulting in waves similar only in their outlook and not in their genesis to the real epileptic discharges.

The very fact that electrical stimulation or local administration of opioids elicit different responses from the medial thalamus and the PAG (seizures and analgesia, respectively) may indicate differences in the opioid mechanisms. Actually, this divergence of the opiate receptor mediated actions prompted several authors to suppose that different subgroups were responsible for the differential effects.[26,27,32]

However, it must be taken into consideration that identical opiate receptors might mediate different effects in the different brain areas depending on the physiological function of cells influenced by them and on the neurotransmitter the release of which is modified upon their application. (Just as local application of a cholinergic agonist may result in very different effects depending on the brain site chosen.) On the other hand, the differences in relative potencies of morphine, Leu-E, and Met-E in induction of analgesia and epileptoid activity cannot be reconciled with any hypothesis based on the uniformity of opiate receptors. Just as the differences in the relative potencies of adrenaline, noradrenaline, and isopropylnoradrenaline in different pharmacological assays and their differential antagonizability give the basic evidence for existence of different adrenergic receptors.

Nevertheless, if the epileptiform activity is attributed to a separate subclass of opiate receptors, probably its interaction with other transmitters also is different from those of the classical opiate receptors. In this aspect only one work[37] can be quoted, where the DALA-induced myoclonic twitches were found potentiated by haloperidol and inhibited by ergometrine and not modified by methysergide. The inability of methysergide to antagonize this convulsive action suggests that contrary to analgesia, serotoninergic facilitation is not involved in this effect.

As discussed in Chapter 1, Volume III morphine is regarded a rather selective agonist of the μ-receptors, while natural enkephalins, especially Leu-E are those of the δ-receptors and β-EP is roughly equiactive agonist of both populations. Considering the very weak analgesic (Chapter 3, Volume II) and rather strong epileptogenic[26,27] potencies of the enkephalins, especially of Leu-E, it is logical to suppose that the analgesic action is mediated by μ-and the epileptogenic effect by δ-receptors. Moreover, in view of the scarcity of EEG and behavioral effects of natural enkephalins if giving them into the PAG, the low level of β-EP in the limbic structures, which are otherwise so rich in opiate receptors and enkephalins (Chapters 1 and 3, Volume III), and finally the differences in antagonizability of enkephalins, β-EP, and morphine derivatives in vitro (Chapter 1, Volume I and Chapter 5, Volume III) and in vivo (Chapter 3, Volume III), probably the following tentative conclusions may be drawn.

The analgesic and epileptogenic effects of opiates and opioid peptides are mediated by two different receptor populations, the first located in the PAG and/or related structures, the other in limbic areas such as hippocampus, amygdala, and in the medial thalamus.

The first subgroup of opiate receptors shows the pharmacological characteristics of the μ-type, while in the limbic structures δ-receptors dominate.

A pathological hyperfunction of the limbic δ-receptors may play role in the genesis of certain kinds of human epilepsy.

A potent and specific δ-receptor antagonist if ever developed might be a novel drug in the treatment of epilepsy.

Concerning the last supposition even the usefulness of naloxone as an antiepileptic agent arises, since this drug is an antagonist of both types of opiate receptors (Chapter 2, Volume I). However, no such clinical data are known. Moreover, naloxone, contrary to these considerations, was found to have convulsive action of its own. It potentiated the audiogenic seizures in rats[38] and induced tonic-clonic convulsions alone, if given in high doses to rats, rabbits, and cats.[39]

The rather scanty data concerning the tolerance development to the epileptogenic effects of opioid peptides are conflicting. In the experiments of Frenk et al.[27] an increment in the epileptogenic potency of morphine and natural enkephalins was detected in animals previously treated with opioids. 30 μg of morphine injected i.c.v. did not elicit epileptiform activity in drug naive animals, but it did cause seizures in rats treated some days earlier with a higher epileptogenic dose of Met-E (200 μg).[27] In another experiment of these authors[26] the same dose of naloxone blocked the seizures in seven of eight naive rats but it did so only in three of non-naive animals that had seizures on the previous day upon morphine treatment. These findings suggest that previous exposure to seizures leads to sensitization to the epileptogenic action of opiates.

Elazar et al.,[36] however, obtained opposite data. In acute experiments repeated administration of Leu-E to rats and to cats resulted in a loss of its effectiveness to elicit epileptiform EEG alterations. The loss of epileptogenic potency could not be attributed to a general exhaustion, decrement of excitability, as penicillin a frequently used local epileptogenic agent retained its ability to induce seizures if applying it through the same cannula.[36] It is another question whether the phenomenon, which is better to call tachyphylaxis has anything to do with the tolerance development. Nevertheless, it must be stressed that this tachyphylaxis was found unilateral, just as in the previous studies of Teitelbaum et al.[16,31] with morphine. After repeated unilateral morphine treatment administering it into the contralateral hippocampus or into other brain areas, its efficacy was found unaltered.[36] Thus, the process in its early stage (whether called tachyphylaxis or acute tolerance) is strictly local.

Giving it in low doses, the β-EP-induced cortical synchronization was accompanied with a shift of the hippocampal θ-rhythm to higher frequency.[30,40] This reciprocal pattern of cortical and hippocampal EEG activities is a well-established physiological phenomenon, i.e., β-EP changed the electrical activity just as under natural conditions.[17] This shift of hippocampal rhythm to higher frequencies has been observed also by de Wied et al.[40] giving β-EP i.c.v. in the minute amount of 20 μg. Thus the cortical synchronization was accompanied by hippocampal desynchronization. Consequently, it should have acted via the midbrain reticular formation, which regulates the arousal level in the forebrain. The effect of β-EP was observed applying rather small doses, smaller amounts than necessary to induce analgesia. Hence the possibility arises that the opioid peptides present in the brain stem take part in the physiological regulation of wakefulness, since the midbrain reticular formation contains opiate receptors and opioid peptides in fairly high concentrations (Chapters 1 and 3, Volume III). However, if endogenous opioids are really involved in physiological regulation of the arousal level, first of all β-EP and μ-receptors must be counted with. The mesencephalic reticular formation contains relatively higher amounts of β-EP than enkephalin. More exactly as for β-EP containing fibers, it is one of the richest areas of the brain (Chapter 3, Volume II), while of enkephalins the highest amount can be detected in the limbic-diencephalic-telencephalic structures (Chapter 3, Volume III).

Of course, if a pharmacologically active naturally occurring substance is supposed to maintain a certain biological function, the opposite effect should be obtained by its depletion, or upon administering its antagonist. Consequently, if β-EP or more generally endogenous opioid peptides were involved in physiological regulation of normal activation level, naloxone-like agents should disturb the normal sleep-wakefulness cycle. Surveying the rather voluminous literature on the effects of naloxone and related agents, hardly any data can be found in this relation.

If endogenous opioids exert a tonic inhibitory influence on the cortical arousal mechanisms, naloxone-like substances should elicit desynchronization in otherwise untreated subjects. Volavka et al.[41] observed a slight EEG deactivation upon per os naltrexone treatment in healthy male volunteers. For this reason, either the tonic sedative action of endogenous opioids should be contested, or the "hidden" agonist potency of this relatively pure opiate antagonist explains the unexpected finding.[41] Thus, endogenous opioids are probably not involved in the physiological regulation of activation level or in its cyclic alteration. The other possibility is, however, that such changes cannot be detected by simple observation of the experimental subjects. Therefore, further studies are warranted where the effect of naloxone-like agents on the sleep-wakefulness cycle were clarified by the aid of continuous monitoring of EEG, EMG, and other parameters. It cannot be excluded that using reliable quantitative electrographic methods, certain subtle changes could be detected of little pharmacological but great physiological significance.

IV. EFFECTS OF OPIOID PEPTIDES AND OPIATES ON THE ACTIVITY OF SINGLE BRAIN NEURONES

Needless to emphasize that more reliable information can be expected upon recording the activity of single brain neurones of known anatomical localization and physiological function. Actually the microelectrophysiological experiments contributed to a significant degree to our present knowledge on the mode of action of opioid peptides. Nevertheless, the obvious limitation of this type of experimentation cannot be neglected.

A. Cerebral Cortex

The predominant effect of opioids (opiates and opioid peptides) on cortical unit activity is a clear-cut depression[42-55] but excitatory effect has also been observed. (All these data refer to the rat.) Opiates and opioid peptides mostly inhibit the spontaneously discharging cortical neurones and block the excitation elicited by electrophoretic application of glutamate or acetylcholine.[42-55] In several cases this depression of firing was followed by a rebound excitation upon switching off the phoretic current.[51,53] The rebound excitation, however, was of short duration (10 to 40 sec) and generally followed by a second phase of reduced activity.[51] The depressant action was observed mostly in the sensorymotor cortex,[43,52] and in the frontal cortex,[42,44,49,51] but also over the parietal area[44,52] and in the gyrus cinguli.[50] No units were found inhibited upon Met-E or morphine iontophoresis in the posterior cortex[42,49] and in the postcruciate cortex of cats.[56] Otherwise, no data are available on the relative density of opiate sensitive neurones in the cortex, only Palmer et al.[44] reported that the vast majority of responding cells were found more than 0.5 mm below the surface of the cortex. According to these authors the responsive cells could also be activated antidromically, i.e., by stimulation of the pyramidal tract.[44]

The inhibitory action has been uniformly found sensitive to naloxone thus, the effect is mediated by opiate receptors (Table 1).

Many authors reported (Table 1) that a part of the cortical units was actually stimulated by opioids, though the majority of cells reacted with depression. This opioid-induced excitatory action was found naloxone insensitive,[43,46] therefore this effect was considered as a nonspecific action. However, Nicoll et al.[52] found both the depressant and excitatory responses elicited either by (nor)morphine or Met-E antagonizable by naloxone.

With the exception of one study[98] spontaneously firing or chemically stimulated units were examined and not specifically those ones reacting with excitation to noxious stimuli. Obviously such units are difficult to find in the cerebral cortex. Nevertheless, if not those units are specifically examined which react selectively to nociceptive stimulation, also the conclusion should be more cautious; this depression of the cortical unit activity might refer rather to the general depressant character of opioids than to their analgesic action. Only one group observed the inhibition of stimulus-induced firing of cortical units upon systemic application of morphine or etorphine. In experiments of Biscoe et al.[98] the unit discharges to electrical stimulation of sciatic nerve were inhibited in the deep pyramidal cell layer of the cerebral cortex in rats if these opiates were administered intravenously. Applying them, however, locally by iontophoresis, only weak and inconsistent effects were observed. Consequently, these effects probably reflect only the inhibition of the arousal reaction elicited by nociceptive stimuli but not the selective analgesic action.

Giving naloxone alone, in one study inhibition was found, attributed to its local anesthetic effect by the authors themselves.[52] In other laboratories, however, naloxone generally failed to modify the spontaneous activity.[46,99] Clear-cut facilitation has been observed only in one laboratory.[53] Obviously facilitation by naloxone is expected if the cells are tonically inhibited by the endogenous opioid peptides. However, the data on the effect of naloxone on the spontaneous firing of cortical cells[46,52,53,99] are not less contradictory than the observations on its behavioral actions in otherwise drug naive animals (Chapter 3, Volume II).

As for the tolerance development, this phenomenon was observed first by Satoh et al.[43] According to these authors[43] tolerance develops to the depressant actions of morphine but not to its excitatory effect. Not only tolerance has been observed but also cross-tolerance between opiates and opioid peptides, i.e., the inhibitory effects of enkephalins were abolished in animals rendered tolerant by s.c. implantation of morphine pellets.[51] Moreover, in these tolerant animals more neurones reacted with excitation to opiates and opioid peptides than in drug naive ones.[43]

In one laboratory[43] tachyphylaxis developed in acute experiments upon repeated application of morphine, while others[52] observed no such gradual diminution of either the inhibitory or excitatory actions of normorphine, β-endorphin, and Met-E, if these substances were repeatedly applied to the same cells for periods lasting more than 1 hr. Of course the acute tolerance, i.e., tachyphylaxis and chronic tolerance are different biological phenomena. Unfortunately, the presently used microelectrode techniques are not suitable to record the activity of the same units longer than 1 to 2 hr.

Table 1
EFFECTS[a] OF OPIOID PEPTIDES AND OPIATE ANTAGONISTS ON UNIT[b] ACTIVITY IN VIVO

Treatment	Species	Structure	Effect	Reversal by narcotic antagonists	Ref.
Naloxone i.v.	Rat	Hypothalamus (area preoptica and supraoptica)	No effect in "naive" animals, facilitation in morphine dependent animals	—	57
Naloxone iont.	Rat	Cerebral cortex	Facilitation in both "naive" and morphine dependent animals	—	58
Naloxone iont.	Rat	Medial thalamus	Facilitation in "naive" and particularly in morphine dependent animals	—	58
Met-E iont.	Rat	Frontal cortex	Inhibition	Yes	42
Met-E iont.	Rat	Candate nucleus	Inhibition	Yes	42
Met-E iont.	Rat	PAG[c]	Inhibition	Yes	42
Met-E iont.	Rat	Thalamus	Inhibition of nociceptive stimulation induced firing	Yes	59
Met-E iont.	Rat	Brain stem	Inhibition	Yes	60
Met-E iont.	Cat (decerebrated)	Medulla (n. reticularis gigantocellularis)	Inhibition	No	61
Leu-E iont.	Cat (decerebrated)	Medulla (n. reticularis gigantocellularis)	Inhibition	No	61
Met-E iont.	Cat	Spinal cord Renshaw cells	Excitation	Yes	62
Met-E iont.	Rat	Thalamus	Inhibition of spontaneous and noxious stimulation induced firing	Partially	54
Met-E iont.	Cat	Spinal cord Renshaw cells	Excitation	Yes	63
Leu-E iont.	Cat	Spinal cord Renshaw cells	Excitation	Yes	63
Met-E iont.	Rat	Brain stem	Excitation	Not mentioned	63
Leu-E iont.	Rat	Brain stem	Excitation	Not mentioned	63
Met-E iont. (into substantia gelatinosa)	Cat (spinal)	Spinal cord dorsal horn (laminae IV, V)	Selective inhibition of firing elicited by noxious stimulation	Yes	64 65
Met-E-amide-iont. (into substantia gelatinosa)	Cat (spinal)	Spinal cord dorsal horn (laminae IV, V)	Selective inhibition of firing elicited by noxious stimulation	Yes	64 65
Met-E iont.	Rat	Cortex	Inhibition	Partially	47

Table 1 (continued)
EFFECTS[a] OF OPIOID PEPTIDES AND OPIATE ANTAGONISTS ON UNIT[b] ACTIVITY IN VIVO

Treatment	Species	Structure	Effect	Reversal by narcotic antago-nists	Ref.
Met-E iont.	Rat	Thalamus	Inhibition of both spontaneous and noxious stimulation induced firing	Partially	47
Met-E iont.	Rat	Medulla	Inhibition	Partially	47
Met-E iont.	Rat	Cerebral cortex	Inhibition	Yes	48
Met-E iont.	Rat	Striatum	Inhibition (or excitation)	Yes (not mentioned)	48
Met-E iont.	Rat	Frontal cortex	Inhibition	Yes	49
Met-E iont.	Rat	Posterior cortex	No effect	—	49
Met-E iont.	Rat	Candate nucleus	Inhibition	Yes	49
Met-E iont.	Rat	PAG	Inhibition	Yes	49
Met-E iont.	Rat	Midbrain reticular formation	Inhibition	Yes	49
(des-OH)-Met-E	Rat	Cerebral cortex	No effect	—	49
Met-E iont.	Cat (decerebrated)	Brain stem reticular formation	Inhibition or no effect	Not mentioned	66
Leu-E iont.	Cat (decerebrated)	Brain stem reticular formation	Inhibition or no effect	Not mentioned	66
Met-E iont.	Cat (decerebrated)	n. raphe magnus	Inhibition, no effect, or facilitation	Not mentioned	66
Met-E iont.	Cat (decerebrated)	PAG	Mostly inhibition	Not mentioned	66
α-endorphin iont.	Cat (decerebrated)	PAG	Inhibition or excitation	Not mentioned	66
Met-E iont.	"Naive" rat	Frontal cortex	Inhibition	Yes	51
Met-E iont.	Morphine dependent rat	Frontal cortex	No effect or excitation	Not mentioned	51
Leu-E iont.	Morphine dependent rat	Frontal cortex	Inhibition	Not mentioned	51
Naltrexone (intraarterially)	Rat	Superior cervical ganglion	Inhibition of ganglionic transmission	—	67
Met-E iont.	Rat	Hippocampus	No effect	—	68
Leu-E iont.	Rat	Hippocampus	No effect	—	68
β-EP iont.	Rat	Thalamus	Inhibition	Yes	52
Met-E iont.	Rat	Thalamus	Inhibition	Yes	52
β-EP iont.	Rat	Hippocampus	Excitation	Yes	52
Met-E iont.	Rat	Hippocampus	Excitation	Yes	52
β-EP iont.	Rat	Brain stem	Inhibition or excitation	Yes	52
Met-E iont.	Rat	Brain stem	Inhibition	Yes	52
β-EP iont.	Rat	Cerebral cortex	Mostly inhibition	Yes	52
Met-E iont.	Rat	Cerebral cortex	Inhibition	Yes	52
β-EP iont.	Rat	Candate nucleus	Inhibition	Yes	52
Met-E iont.	Rat	Candate nucleus	Inhibition	Yes	52
Met-E iont.	Rat	n. accumbens	Inhibition	Yes	69
Leu-E iont.	Rat	n. accumbens	Inhibition	Yes	69

Table 1 (continued)
EFFECTS[a] OF OPIOID PEPTIDES AND OPIATE ANTAGONISTS ON UNIT[b] ACTIVITY IN VIVO

Treatment	Species	Structure	Effect	Reversal by narcotic antagonists	Ref.
Met-E iont.	Rat	Locus coeruleus	Inhibition	Yes	70
Met-E iont.	Rat	Medulla	Inhibition	Yes	71
Leu-E iont.	Rat	Medulla	Inhibition	Yes	71
(D-Ala[2],D-Leu[5])-E	Rat	Medulla	Inhibition	Yes	71
DALA iont.	Rat	Medulla	Inhibition	Yes	71
(D-Ala[2]D-Leu[5])-E	Rat	Brain stem (reticular formation)	Inhibition	Yes	72
Met-E i.c.v.	Conscious rat	PAG	Increase in multiple unit activity	Yes	26
Met-E iont. (near to the units recorded)	Cat (spinal)	Spinal cord dorsal horn (lamina V)	Inhibition of spontaneous and noxious or non-noxious stimulation induced firing	Partially[d]	73
Met-E-amide iont. (near to the units recorded)	Cat (spinal)	Spinal cord dorsal horn (lamina V)	Inhibition of spontaneous and noxious or non-noxious stimulation induced firing	Partially[d]	73
Met-E iont. (into substantia gelatinosa)	Cat (spinal)	Spinal cord dorsal horn (lamina V)	Inhibition of spontaneous and noxious stimulation induced firing	Yes	73
Met-E-amide iont. (into substantia gelatinosa)	Cat (spinal)	Spinal cord dorsal horn (lamina V)	Inhibition of spontaneous and noxious stimulation induced firing	Yes	73
Met-E iont.	Rat	Locus coeruleus	Inhibition	Not mentioned	74
Naloxone i.v.	Cat (spinal)	Spinal cord ventral root	Potentiation of C-fiber (nociceptive) reflex response	—	75
Naltrexone i.v.	Cat (spinal)	Spinal cord ventral root	Potentiation of C-fiber (nociceptive) reflex response	—	75
Met-E iont.	Rat	Cerebral cortex	Excitation or inhibition	Partially[e]	46
Leu-E iont.	Rat	Cerebral cortex	Excitation	No	46
Met-E iont.	Rat	Brain stem	Mostly excitation	Yes	46
Leu-E iont.	Rat	Brain stem	Excitation	Yes	46
Met-E iont.	Cat	Spinal cord dorsal horn (laminae IV, V)	No effect	—	46
Leu-E iont.	Cat	Spinal cord dorsal horn (laminae IV, V)	No effect	—	46
Met-E iont.	Cat	Spinal cord Renshaw cells	Excitation	Yes	46
Leu-E iont.	Cat	Spinal cord Renshaw cells	Excitation	Yes	46

Table 1 (continued)
EFFECTS[a] OF OPIOID PEPTIDES AND OPIATE ANTAGONISTS ON UNIT[b] ACTIVITY IN VIVO

Treatment	Species	Structure	Effect	Reversal by narcotic antagonists	Ref.
Met-E iont. (into substantia gelatinosa)	Cat	Spinal cord dorsal horn (laminae IV, V)	No effect	—	46
Met-E iont.	Rat	Medullary reticular formation	Inhibition	Yes	76
Leu-E iont.	Rat	Medullary reticular formation	Inhibition	Yes	76
(D-Ala^2)-Met-E	Rat	Medullary reticular formation	Inhibition	Yes	76
Met-E iont.	Rat	Hippocampus (pyramidal cells)	Excitation	Yes	77
(D-Ala^2)-Met-E iont.	Rat	Hippocampus (pyramidal cells)	Excitation	Yes	77
β-EP	Rat	Hippocampus (pyramidal cells)	Excitation	Yes	77
Met-E iont.	Rat	Hippocampus (interneurones)	Inhibition	Yes	77
(D-Ala^2)-Met-E	Rat	Hippocampus (interneurones)	Inhibition	Yes	77
β-EP	Rat	Hippocampus (interneurones)	Inhibition	Yes	77
Met-E iont.	Rat	Cerebral cortex	Inhibition	Yes	44
DALA iont.	Rat	Cerebral cortex	Inhibition	Yes	44
N-acetyl-(D-Ala^2, Lys^3)-Met-E-amide iont.	Rat	Cerebral cortex	No effect	—	44
Met-E iont.	Rat	n. accumbens	Inhibition	Not mentioned	78
Leu-E iont.	Rat	n. accumbens	Inhibition	Not mentioned	78
(D-Ser^2,Met^5)-E-methyl-esther iont.	Rat	n. accumbens	Inhibition	Not mentioned	78
Naloxone iont.	"Naive" rat	Locus coeruleus	No effect	—	79
Naloxone iont.	Morphine dependent rat	Locus coeruleus	Facilitation	—	79
(D-Ala^2,D-Leu^5)-E iont.	Rabbit	Thalamus	Inhibition of nociceptive stimulation induced firing	Yes	80
Met-E iont.	Cat (spinal)	Spinal cord dorsal horn (laminae II, III, V)	Inhibition of both spontaneous and noxious stimulation induced firing	Yes	81
Met-E-amide iont.	Cat (spinal)	Spinal cord dorsal horn (laminae II, III, V)	Inhibition of both spontaneous and noxious stimulation induced firing	Yes	81

Table 1 (continued)
EFFECTS[a] OF OPIOID PEPTIDES AND OPIATE ANTAGONISTS ON UNIT[b] ACTIVITY IN VIVO

Treatment	Species	Structure	Effect	Reversal by narcotic antagonists	Ref.
Met-E iont. (into substantia gelatinosa)	Cat (spinal)	A δ afferent sural fibers	Primary afferent depolarization. Potentiation of the decrease in threshold elicited by stimulation of the superficial peroneal nerve	Yes	83
Met-E iont. (into substantia gelatinosa)	Cat (spinal)	A δ afferent sural fibers	Increase in the threshold of antidromic activation	Yes	83
Naloxone i.v.	Cat	Trigeminal ganglion or tooth pulp (trigeminal units)	No effect on primary afferent depolarization	—	84
FK 33-824 i.v.	Cat (spinal)	Spinal cord	Increase in excitability of primary afferents, depression of γ-motoneurone activity and segmental polysynaptic reflexes	Yes	85
Met-E iont.	Rat	Locus coeruleus	Inhibition	Yes	86
Met-E iont.	Rat	Hippocampus	No effect, sometimes inhibition or excitation	No	87 88
Met-E iont.	Rat	Medulla oblongata (n. reticularis paragigantocellularis)	Facilitation of spontaneous and noxious stimulation induced firing	Yes	89
Leu-E iont.	Rat	Medulla oblongata (n. reticularis paragigantocellularis)	Facilitation of spontaneous and noxious stimulation induced firing	Yes	89
Met-E iont.	Rat	Medulla oblongata (n. reticularis gigantocellularis)	Inhibition	Yes	89
Met-E iont.	Rat	Spinal cord dorsal horn (laminae IV, V, VI)	Inhibition of the spontaneous and peripheral stimulation induced discharges, but no change in the resting membrane potential	Yes	90

Table 1 (continued)
EFFECTS[a] OF OPIOID PEPTIDES AND OPIATE ANTAGONISTS ON UNIT[b] ACTIVITY IN VIVO

Treatment	Species	Structure	Effect	Reversal by narcotic antagonists	Ref.
Leu-E iont.	Rat	Spinal cord dorsal horn (laminae IV, V, VI)	Inhibition of the spontaneous and peripheral stimulation induced discharges, but no change in the resting membrane potential	Yes	90
Met-E iont.	Rat	Hippocampal transplant in the eye	Facilitation	Yes	35
β-EP	Rat	Hippocampal transplant in the eye	Facilitation	Yes	35
Naloxone i.v.	Rat	Spinal cord dorsal horn	Facilitation of firing induced by C-fiber stimulation	—	91
Try-Arg iont. (kyotorphin)	Rabbit (spinal)	Spinal cord dorsal horn (lamina V)	Inhibition of firing induced by noxious stimulation	Partially	92
FK 33-824 i.v.	Cat (spinal or intact)	Spinal cord dorsal root	Depression of the slow negative component of dorsal root potential	Yes	93
(D-Ala^2,Pro^5)-E-amide i.v.	Cat (spinal or intact)	Spinal cord dorsal root	Depression of the slow negative component of dorsal root potential	Yes	93
Naloxone	Cat (spinal or intact)	Spinal cord dorsal root	No effect	—	93
(D-Ala^2)-Met-E iont.	Rat	Hippocampus (CA 1 pyramidal neurones)	Excitation	Yes	94
N-acetyl-Met-E iont.	Rat	Hippocampus (CA 1 pyramidal neurones)	No effect	—	94
Met-E iont.	Rat	Hypothalamus (n. ventromedialis)	Inhibition	Yes	95
Met-E iont.	Rat	Mesencephalic reticular formation	Inhibition of nociceptive stimulation induced firing	Yes	96
Met-E iont.	Rat	Mesencephalic reticular formation	No consistent effect on the spontaneous firing	Partially[e]	96
Met-E (microinjected into the dorsal horn)	Spinal cat	Spinal cord ventral root	Inhibition of C fiber (nociceptive) reflex response	Yes	97
Met-E (microinjected into the ventral horn)	Spinal cat	Spinal cord ventral root	Facilitation of C fiber (nociceptive) reflex response	No	97

Table 1 (continued)
EFFECTS[a] OF OPIOID PEPTIDES AND OPIATE ANTAGONISTS ON UNIT[b] ACTIVITY IN VIVO

Treatment	Species	Structure	Effect	Reversal by narcotic antagonists	Ref.
Naltrexone i.v.	Spinal rat	Spinal cord ventral root	Facilitation of C fiber (nociceptive) reflex response	—	97

Note: iont. = iontophoretically.

[a] Examined by intra- or extracellular microelectrodes.
[b] Units firing spontaneously or activated by glutamate, homocysteic acid and or acetylcholine.
[c] Periaqueductal gray matter.
[d] Antagonism upon local microinjection but not by i.v. application of naloxone.
[e] Antagonism of the depressant action but no reversal of the excitatory action.

Interestingly, nobody found qualitative differences between opiates and opioid peptides, i.e., the direction of response to morphine and to Met-E were found mostly similar.[42,44,46,47,49,52] Consequently, also these electrophysiological findings confirm that the opiates and the opioid peptides exert their actions via the same receptor population.

Nevertheless, certain quantitative differences were reported from several laboratories. In the experiments of Nicoll et al.[52] the responses elicited by endorphins and normorphine were of longer duration than the effects of Met-E. On the other hand, Met-E and Leu-E proved to be nearly equipotent,[51,52] but α-endorphin was found considerably less potent than β-endorphin.[52]

In another study[44] DALA was found much more active in inhibition of the frontal cortical neurones than its natural congener, Met-E. However, another enkephalin analogue, *N*-acetyl-(D-Ala^2, Lys^3, Met^5)-E-amide displayed a rather weak action. Nevertheless, this analogue was found very weak in other assays as well.[44]

Upon interpreting the data on the quantitative differences in the effects of opiates and opioid peptides, it should be borne in mind that there are serious limitations in establishing the relative potencies of compounds applied microiontophoretically, especially if electroosmosis is used. Obviously the pH, the molar concentrations of the drug, the current intensity may considerably modify the strength of effects under study.

These electrophysiological data reflect the relative stronger biological potencies of β-EP and those of the metabolically stabilized enkephalin analogues if comparing them to α-endorphin or natural enkephalins, respectively. It is, however, difficult to decide whether the increased potencies of stabilized enkephalin analogues are due to their increased resistance to enzymatic breakdown or to increased affinity to the opiate receptors. Of course this question arises in relation of their other biological effects too and its solution requires other tests like measuring their action upon iontophoretic application.

Very little is known on the intricate mode of action of opioids, i.e., it is not clarified whether they depress the cortical (and other) neurones directly or by interfering with the action(s) of certain endogenous neurotransmitters. From this point of view are very interesting the recent data of Palmer and Hoffer,[55] who reported that antipsychotic agents known to block catecholamine receptors, such as spiroperidol, α-flupenthixol and (+)-butaclamol reversibly antagonized the enkephalin-induced depression of spontaneous cortical unit activity. As the DALA-induced depression was not antagonized by iontophoretically applied Mg,$^{++}$ i.e., by an agent inhibiting the synaptic transmission, they concluded that intact catecholaminergic transmission was necessary for full expression of enkephalin-elicited neuronal depression and the site of catecholamine-enkephalin interaction was postsynaptic.[55]

B. Thalamus

Contrary to the above-discussed "mixed" effects on cortical units both opiates and opioid peptides seem to modify the thalamic neurones unidirectionally. Depressant action has been detected in rats,[52,59,47,54,100] in rabbits[80] and in cats,[101] as well. In one study, however, morphine-induced naloxone-sensitive weak excitation.[56] As for the spontaneous activity, it was found diminished,[47,52,80] or unaltered.[100,101] Selecting those neurones which responded with firing to nociceptive stimulation, a significant inhibition of painful stimulus elicited activation was reported in all these species.[47,54,59,80,100] Exciting, however, the thalamic neurones with non-nociceptive stimuli, no inhibition was observed.[100] In all these experiments the inhibitory action could be reversed by naloxone. Since the opioids were administered by iontophoresis, these effects can be attributed to locally present opiate sensitive neurones.

Moreover, naloxone was found to activate several units in the medial thalamus just as in the cerebral cortex.[58] Even more interestingly, in morphine-dependent animals a higher proportion of thalamic units were found facilitated by naloxone than in naive animals.[58] These data confirm the conclusion drawn from other, nonelectrophysiological experiments according to medial thalamus is involved in generation of opiate withdrawal symptoms.

If opioids exert a direct inhibitory action on the medial thalamic neurones, the generation of synchronized activity upon their application[24] can be easily explained. A decreased rate of firing of thalamic neurones might increase the probability of phase locking of these neurones into the thalamo-cortical reverberating circuitry, the main source of synchronized cortical macroelectrographic activity. It must, however, be noted that local application of opioids into the medial thalamic nuclei did not result in analgesia (Chapter 2, Volume I).

C. Amygdala

As for the amygdala, a partially naloxone reversible facilitation was observed in one laboratory[102] upon morphine administration. Another group[103] found mixed effects on spontaneous firing and a selective inhibition of discharges elicited by noxious stimuli. It is worth mentioning that the morphine-induced facilitation of neurones of the amygdala was only partially reversed by naloxone.[102] The morphine-induced facilitation in the amygdala could be reversed by diazepam and chlordiazepoxide,[102] a finding pointing to a possible antagonism between the opiate and benzodiazephine receptor mediated mechanisms.

D. Substantia Nigra and Striatum

The modulation by opioids of the nigro-striatal system is a rather controversial issue due to the squarely contradictory data reported in this aspect. Upon morphine administration the nigral unit activity has been found facilitated in some laboratories[104-106] but others observed inhibition.[50] Both the inhibition and facilitation proved to be naloxone-sensitive.[50,104,105] The caudate neurones are inhibited upon treatment with morphine,[42,50,52,105,107] enkephalins,[42,48,49,52] and β-EP[52] as well. In one study[48] enkephalins excited some striatal units.

Morphine increased the nigral activity administered peripherally or locally into the caudate nucleus just as haloperidol did. Naloxone (given i.v.) blocked the effect of morphine whether administered peripherally or intracaudally, but it did not alter the similar action of haloperidol.[104] However, 1-DOPA and apomorphine (i.v.) reversed the facilitation of nigral units upon treatment with either morphine or haloperidol as reported by Iwatsubo and Clouet.[104] Lee et al.[105] observed a simultaneous activation of nigral neurones and inhibition of the striatal units.[105] As reported by this group,[105] the inhibitory action of morphine on caudate neurones could be reversed not only by naloxone but also by haloperidol and pimozide.

Considering that dopamine acts as an inhibitory transmitter in the nigro-striatal tract[105] and opioids stimulate the nigral units but inhibit the striatal neurones and both effects can be reversed not only by naloxone but also by dopamine receptor blockers, the conclusion seems to be very simple. Enhancement of nigral activity would result in increased dopamine release in the striatum, which in turn should inhibit the caudate neurones. The previous findings on the high density of opiate receptors in the substantia nigra (Chapter 1, Volume III) and the contralateral turning upon intranigral administration of opiates and opioid peptides[108,109] further confirm this hypothesis. Moreover, the previously discussed increase in striatal dopamine turnover (Chapter 2, Volume I) may also be a manifestation of facilitated dopamine release in the caudate nucleus.

However, other considerations rather refute such a proposition. If opioids acted via enhancement of dopamine release in the striatum, why just an inhibition of dopamine liberation has been detected in vitro (Chapter 1, Volume II). Moreover, opiates and opioid peptides generally inhibit the transmitter release, at least in all in vitro preparations selectively sensitive to enkephalins, endorphins, and opiates (Chapter 1, Volume I and Chapter 1, Volume II). The increase in the striatal dopamine turnover might be not only a manifestation of increased dopamine release but also a secondary consequence of dopamine receptor blockade. Furthermore, injection of morphine or haloperidol directly into the striatum also increased the nigral unit activity.[104] This latter effect may be a feedback phenomenon, since facilitation was elicited by a dopamine receptor blocking agent (haloperidol), i.e., by blockade of the receptor population where dopamine is the natural agonist. Of course, this feedback mechanism may be mediated by pre- and postsynaptic mechanisms as well. The facilitation of the nigral units by morphine was more easily reversed by peripherally administered apomorphine or 1-DOPA than that induced by haloperidol.[104] Therefore, the findings of Iwatsubo and Clouet[104] suggest that haloperidol and opioids act via different mechanisms. Haloperidol might elicit the facilitation of nigral elements by postsynaptic mechanism, since it is known to inhibit the stimulation by dopamine of adenylate cyclase, an enzyme generally considered to be an integral part of the postsynaptic dopamine receptor complex. Opioids, however, do not inhibit the basal activity of adenylate cyclase and their effect on its activation by dopamine is also a controversial issue, at least in vitro (Chapter 1, Volume II). Thus, several further theoretical possibilities emerge.

Opioids block the striatal presynaptic dopamine receptors, i.e., they interrupt the autoreceptor mediated negative feedback process. Such a mechanism of action could explain the enhanced activity of nigral cells which send dopamine releasing terminals into the striatum and also the increased dopamine turnover and the lack of any effect on the postsynaptically located adenylate cyclase as well. Hence, the netto effect would be a facilitation of dopaminergic transmission, which explains the inhibition of caudate neurones, since dopamine is an inhibitory transmitter in this structure.

However, numerous data indicate[110,111] that dopamine receptors become hypersensitive upon acute and/or chronic opiate treatment. Among others, the apomorphine-induced gnawing is potentiated after morphine administration[110] and the binding of labeled domperidone to striatal dopamine receptors as well.[111] Such changes may be observed generally after blockade of dopamine receptors which become rapidly hypersensitive to their agonists. Consequently, these effects cannot be explained by presynaptically mediated increase in the dopaminergic transmission.

A further possibility is that both haloperidol and morphine facilitate nigral elements by striatal postsynaptic mechanisms, but by different ones, as opioids contrary to the neuroleptics do not inhibit unambiguously the dopamine receptors coupled to adenylate cyclase (Chapter 1, Volume II). As the caudate units are depressed upon treatment with opioids (see above), the facilitation of the nigral units might be due to disinhibition, i.e., to the inhibition of the descending striato-nigral inhibitory pathway.

Nevertheless, the existence of postsynaptically located dopamine receptors which were not connected with adenylate cyclase has not been experimentally verified yet. Moreover, the majority of different kinds of experimental data indicate that opioids act presynaptically. However, according to a recent study both haloperidol and morphine lost their ability to stimulate the nigral neurones upon systemic (i.v.) application if the striatal neurones had previously been destroyed by kainic acid. This finding[106] indicates a direct postsynaptic action. Thus, probably a further alternative mechanism can be proposed.

Let us suppose that opioids exert presynaptic inhibitory action both in the substantia nigra and the striatum, as they are generally localized presynaptically, and the inhibition by enkephalins of the acetylcholine and dopamine release, i.e., liberation of the principal transmitters of these structures have already been demonstrated (Chapter 1, Volume II). A presynaptic inhibition in the nigral may decrease the release of the transmitters mediating the feedback striato-nigral inhibitory control of unit activity in the nigra. If the feedback control is brought about by acetylcholine (as suggested in the pharmacological textbooks), the well-known anticholinergic effects of opiates might explain this loss of inhibitory control of nigral units. Thus, the presynaptic inhibition can result in increased activity of nigral neurones. A similar opioid mediated presynaptic inhibition of dopamine liberation in the striatum might be responsible for the reduction in dopamine release in the striatum and for the decreased firing of the striatal neurones as well. In a broader aspect such a reduction in the dopaminergic transmission might explain the behavioral symptoms reminiscent of dopamine receptor blockade and the secondary receptor supersensitivity as well.

Moreover, if opiates exert inhibitory action on striatal dopaminergic transmission presynaptically, it is easy to understand why they lose their ability to suppress the striatal neurones in animals pretreated with pimozide or haloperidol.[105] Reduction of dopamine release will have obviously no significant consequences if the postsynaptic receptors are already blocked.

As the dopaminergic transmission is attenuated in the striatum, behavioral symptoms develop which are similar to those induced by direct postsynaptic receptor blockade. Of course, they may be reversed by dopamine agonists and opiate agonists as well. After a while the unused dopaminergic receptors become supersensitive. Accordingly, after dissipation of the opiate action, when the dopamine release is resumed, it elicits exaggerated excitation of its receptors, i.e., behavioral stimulation. Furthermore, the higher amount of opioid has been administered the longer will last the inhibition of synaptic transmission, consequently also the degree of receptor supersensitivity will be enhanced. A higher degree of receptor supersensitivity would, however, induce more intense and more prolonged excitatory syndrome.

From this point of view it would be very interesting to learn about the effect of opioids if applying them in subanalgesic dose, which are known to induce slight but distinct behavioral excitation (Chapter 2, Volume I). If the aforesaid argumentation were correct and the extrastriatally mediated behavioral effects might be disregarded, opioids should increase the nigral activity at a lower dose level than inhibit the caudate units. No such study is known yet. Moreover, it is surely an oversimplification of the issue if only the dopaminergic and cholinergic transmissions are counted with in relation of the nigro-striatal feedback system. In this context it is probably worth mentioning that the effects of morphine on the acoustic evoked potentials in the substantia nigra could be reversed by the serotonin receptor blocking agent cinanserine while in the caudate nucleus by a serotonin depleting substance *p*-chlorophenylalanine.[112] Thus, either a certain role of intrastriatal serotoninergic cells must be taken into consideration or the above-discussed striatal electrophysiological events are influenced by extrastriatal serotonin liberating neurones.

Since no report is known on the activation of nigral or striatal units upon painful stimuli, even these electrographic data show that this system has not much to do with the narcotic analgesia and conversely the physiological functions of nigral and striatal enkephalinergic neurones are not related to regulation of the pain sensation.

E. Nucleus Accumbens

According to several studies[69,78] morphine, the natural, and synthetic enkephalins alike inhibit the firing of neurones located in this mesolimbic structure. The effect is consistent with the data on the behavioral depression elicitable by local application of opioids (Chapter 2, Volume I) and with the supposed role of opiate receptors in this extrastriatal dopaminergic area in mediation of the behavioral effects of opioids.

F. Hypothalamus and Hippocampus

The effect of opiates on hypothalamic neurones was first examined by Eidelberg and Bond.[57] These authors observed an inhibitory action in naive animals but excitation in morphine-dependent rats.[57] Thus, the effects observed reminded of those recorded in the cerebral cortex.[43] For the reasons mentioned above, these findings are probably not related to the specific actions of opiates, namely in this experiment[57] morphine was applied systemically and the units recorded were in the area preoptica and supraoptica.

Probably more informative are the data presented by Kerr et al.,[113] who found that systemic application of morphine increased the neuronal activity in the ventromedial nucleus of the hypothalamus ("satiety center", VMH) and decreased the firing in the lateral hypothalamic area ("hunger center", LHA). The latter inhibitory effect in the LHA preceded the facilitation observed in the VMH by 1 to 2 sec.[113] Consequently, the primary action might be the inhibition of the units located in LHA, while the facilitation in VMH was probably a secondary disinhibitory effect due to the well-known reciprocal innervation of the two neighboring centers.

These findings have recently been confirmed and extended by Ono et al.[95] These authors applied the substances iontophoretically and selected such neurones which generally reacted also to glucose. Due to its role in regulation of the satiety, glucose mostly inhibited the LHA units and stimulated those located in the VMH.[95] The majority of units examined were influenced in the same direction by glucose, morphine, and Met-E, the latters' effect being naloxone sensitive.[95]

These data might explain the well-known inhibitory action of opioids on spontaneous food-intake. Their selective effect on the glucosensitive neurones shows that the suppression of the consumptive behavior is not only a secondary consequence of the general "drive reduction" due to their primary euphoric action. Considering the presence of opiate receptors and endogenous opioid peptides in the hypothalamus (Chapters 1 and 3, Volume III), their physiological role in regulation (inhibition) of food intake should be seriously considered.

The changes in the activity of neurones located in VMH and LHA during opiate dependence also has been studied.[113] In addicted animals during morphine withdrawal the firing rate in the LHA was greater than that recorded in the same area of experimentally naive animals, while the spontaneous activity of VMH units did not change upon dependence development.[113] Examining the influence of methadone in morphine-dependent animals, i.e., simulating the methadone substitution therapy in this model, both VMH and LHA units were found facilitated and this effect of methadone could be reversed by morphine.[113] These findings are not easy to interpret. Nevertheless, the increased firing of neurones in both areas probably may be regarded an excitatory manifestation of opiate withdrawal syndrome due to the substitution of the previously administered strong agonist (morphine) with a weaker one (methadone).

Of the forebrain structures the hippocampus has been studied the most intensively in regard of the influence of opioids. This interest may probably be attributed to the anomalous effects observed and to its alleged role in the genesis of the epileptiform activity.

In the first reports no or mostly inhibitory actions were described.[68,87] Segal,[68] recording the unit activity in the pyramidal cell layer of area CA_1 of dorsal hippocampus, observed only insignificant alterations in the spontaneous activity. Only a small fraction of cells was found inhibited or facilitated upon iontophoretic application of morphine, or excited by Met-E and Leu-E.[68] Moreover, both morphine and Leu-E antagonized the glutamate and aspartate-induced firing in a part of the cells.[68] Fry et al.[87,88] found the majority of the hippocampal units unaffected by morphine and Met-E. Only some of them were inhibited, but this inhibition was rarely reversed rather potentiated by naloxone. Comparing the two stereoisomers levorphanol and dextrorphan, most units were influenced in the same direction by both and if stereospecific effects were observed, they were inhibitory.[87,88] Hence the effects seemed to be nonspecific, and their prevalence was generally attributed to the relatively low concentration of opiate receptors and endogenous opioid peptides in this limbic region (Chapters 1 and 3, Volume III). Moreover, contrary to the observations made on the cortical and hypothalamic neurones, both the "specific" inhibitory effects exerted on the striatal units and the nonspecific facilitation of hippocampal units were found attenuated in morphine-tolerant rats, i.e., in these animals enkephalin induced mostly depression and not excitation in the hippocampus.[88]

During the subsequent more intensive studies it turned out that not only these unspecific effects were characteristic of this structure. First Nicoll et al.[52] reported on their "surprising" finding according to the hippocampal pyramidal cells were excited by iontophoretic application of normorphine, Met-E, and β-EP. Furthermore, no significant differences were found between the stimulant effect of morphine and that of the opioid peptides, their excitatory actions being sensitive to naloxone whether applied systemically or locally.[52]

These apparent contradictions were resolved by simultaneous recording of the activities of two different types of hippocampal neurones, i.e., the discharges of the mostly quiescent pyramidal cells displaying only bursts of firing and that of the continuously firing interneurones. As demonstrated elegantly by Zieglgänsberger et al.,[77] Met-E, (D-Ala²)-Met-E, and β-EP elicited a burst-like discharges of the pyramidal cells, but inhibited the spontaneously discharging interneurones. A similar excitation of the pyramidal neurones could be elicited by the GABA receptor antagonist bicuculline, while naloxone iontophoresis inhibited the facilitation induced by opioid peptides.[77] Thus, these authors[77] concluded on the abolishment by opioids of a tonic inhibitory control of presumably GABA releasing interneurones.[77] Consequently, the hippocampal interneurones should tonically inhibit the activity of pyramidal units via GABA-ergic synapses and these inhibitory interneurones were inhibited by opioids through an opiate receptor mediated input.

These data were confirmed and extended by Dunwiddie et al.[94] In this study active enkephalin analogues increased the amplitude of the so-called "population spike response", a potential reflecting the number of synchronously discharging pyramidal cells, without any change in the simultaneously recorded EPSP. Accordingly, there was an increased probability of pyramidal cell discharge in response to synaptic input, but this was not the result of a direct increase in excitatory synaptic drive. Consequently, the pyramidal neurones are not directly activated by the opioids.

The facilitation by Met-E and β-EP of single pyramidal cells has recently been demonstrated in hippocampal transplants "in oculo". In the experiments of Taylor et al.[35] hippocampal tissue cut out of rat embryo was transplanted into the eye of adult animals and examined at least a month later in acute experiments, where not only the characteristic stimulation of pyramidal unit activity but also the generation of epileptiform activity could be demonstrated.[35] Thus, both the characteristic microelectrographic effect (enhanced firing) and the actions detectable by macroelectrodes (epileptiform waveforms) may be generated by opioids under experimental conditions which exclude any interference of structures lying outside of the hippocampus.[35]

The question arises how to reconcile the first negative findings[68,87,88] with the latter ones which convincingly prove the indirect facilitatory action of opioids.[35,52,77,94,114] Probably two explanations may be given. First, the continuously firing interneurones may easily be confounded with the pyramidal cells. Secondly, as shown by the in vitro studies (Chapter 2, Volume II), the opioids do exert also an unspecific depressant effect on these cells, which obviously may confound the selective but indirect stimulant action.

Probably it is reasonable to mention in this chapter that the recurrent motor seizure activity, induced by hippocampal administration of kainic acid was followed by an initial decrease then increase in the local Met-E content.[115] This observation of Hong et al.[115] points to an eventual causal relationship between epileptic activity and E turnover, nevertheless, the biphasic change in hippocampal Met-E content may equally be the cause or just only the consequence of seizure activity.

G. Periaqueductal Gray (PAG), Raphe Nuclei, Brain Stem, and Reticular Formation

Of the mesencephalic structures PAG has been the most intensively studied. The reason of this interest can be attributed to the previous findings according to local application of opiates in minute amounts into this regions results in the same analgesic and behavioral effects which can be elicited by systemic application (Chapter 2, Volume I). Though all the authors concerned agree that the neurones in the PAG are specifically sensitive to opiates and opioid peptides, data are conflicting as for the direction of responses elicited by them (see below and Table 1). This discrepancy may have different explanations. First, examining different species the observed effects elicited from the same area also may be different. As shown in Table 1, opioids induce more frequently inhibitory responses in the rat than in the cat. Secondly, frequently the area studied was not exactly defined. PAG means sometimes strictly the cranial part of the substantia grisea centralis, i.e., the cells examined were presumably located somewhere at the level of substantia nigra.[42] In other studies[26,116] the same definition includes also the central area at the level of raphe nuclei. This means that the expression refers not only to the central-most mesencephalic area but also to the corresponding regions in the pons. A further disturbing factor is that the frontal and horizontal planes of the rat and cat brains are differently defined in the currently used various stereotaxic maps. In several cases speaking of PAG the neurones located a bit more laterally, i.e., in the adjacent reticular formation are also included[117,118] and very frequently only "brain stem neurones" are mentioned without closer identification. A further source of the discrepancies is that the neuronal responses might be different even opposite in anesthetized and in conscious animals. Finally it must be taken into consideration that some authors expected *â priori* inhibitory effect, while others considered the stimulant action "specific".

First Frederickson et al.,[42,49] and Gent and Wolstencroft[66] reported that iontophoretically administered morphine and Met-E inhibited the spontaneous activity of PAG neurones in rats[42,49] and in decerebrated cats[66] as well. Shima et al.[117] reported that in slightly anesthetized cats the firing of PAG neurones elicited by noxious stimulation could be inhibited by systemic administration of morphine. Probably in many other studies, unless more exactly specified, the recorded "brain stem neurones" or "pontin" cells also belonged to the PAG, an area which is functionally surely not restricted to the mesencephalon. In these studies (Table 1) mostly inhibition was detected.

In those experiments where the *raphe nucleus* has been distinctly specified, no effect[45] or occasional inhibition[66,118] has been mentioned upon morphine or Met-E iontophoresis. However, in one study[119] an activation of the raphe neurones has been detected upon i.v. application of morphine. This work of Anderson et al.[119] was performed in cats just as the experiments of Gent and Wolstencroft,[66] who also observed that in some cases Met-E or morphine iontophoresis induced facilitation. The other studies reporting on inhibition or no effect were made in rats.[45,118]

A further interesting finding of Anderson et al.[119] is that facilitation by morphine of the raphe cells, otherwise responsive to nociceptive stimuli, could be reversed by naloxone. Thus, it is the first site (progressing in caudal direction) where opioids elicit naloxone reversible excitation. Since electrical stimulation of the raphe nuclei elicits analgesia[121] and electrical lesion of this structure attenuates the analgesic effect of opiates (Chapter 2, Volume I), actually these neurones are expected to be activated by opioids. And in all other cases where the morphine-induced excitation was found naloxone resistant, the observations might reflect either disinhibition or aspecific effects. The raphe nuclei might contain neurones activating descending pain reflex inhibitory pathways and also cells projecting cranially supposedly depressed by the opioids. From this point of view it is important to know that the n. raphe magnus has direct anatomical connections with the spinal cord (laminae I, II, and VI).[119,120] Moreover, as the above-mentioned activation was demonstrated only upon i.v. application of morphine[119] but not by its iontophoretic application,[66] at least in cats this activation might be brought about via other brain structures as well. Such a brain area might be the mesencephalic PAG, from where analgesia can be elicited by local microinjection (Chapter 2, Volume I). Probably the naloxone sensitive inhibitory action reflects the general depressant character of opioids, while those neurones activated by them are difficult to detect. Considering that the pain reactions (pain reflexes) are present also in spinal animals and spinalization leads to the attenuation of morphine analgesia (Chapter 2, Volume I), it follows that opiates should act at least partially via activation of certain descending pathways since local application of opiates into the raphe complex has been found to induce analgesia.[121-123] Consequently, certain units are expected to be activated upon treatment with opioids. Probably opioids primarily inhibit those neurones which tonically inhibit others sending ascending and descending axons responsible for modulation of the sensory transmission as proposed by Yaksh.[124] The question arises where are these neurones? Beside some lower medullary centers (see below) they are expected to be found in the PAG and in the raphe nuclei for the above-discussed reason. The obvious contradictions seem to be resolved by Urca et al.,[26,116] who compared the effects of Met-E and morphine given to anesthetized and unanesthetized animals. Administering the opioids to awake rats, they observed a facilitation of the multiple unit activity, which correlated well with the analgesic effect. More exactly, administering Met-E i.c.v., activation of the units in the PGA was observed in those animals which also displayed an analgesic action.[26] In another study morphine given systemically also facilitated the activity in the PAG.[125] Finally in a recent experiment[116] the effects of opioids on the activity of dorsal raphe area of the PAG was compared in awake and in anesthetized rats. The same dose of morphine increased the activity in waking state but decreased it in anesthetized animals.[116]

The implications of the last findings are far-reaching. The term "analgesic action" has no real meaning in anesthetized animals, otherwise unresponsive to noxious stimuli. Thus, all the data obtained in anesthetized rats and cats might be irrelevant just in relation of the mode of action. However, the observed facilitation in conscious animals corresponds to the expectations based on the alleged activation by opioids of descending (and ascending) pain-inhibitory pathways. It is difficult either to confirm directly or to refute this hypothesis. Iontophoretic application is the best way to record the activity of isolated units and this can be carried out most reliably in anesthetized animals. On the other hand, Urca and Liebeskind[116] recorded the so-called multiple unit activity and using this technique the activities of single neurones are pooled. Therefore, the net increase observed by them does not exclude certain elements that were inhibited. As it will be discussed below, there are some more caudally located groups of neurones which were found facilitated with opioids even in anesthetized animals.

Otherwise, not only in the PAG but also in the more laterally located midbrain reticular formation neurones can be detected which are selectively responsive to noxious stimuli and this activation can be suppressed by opioids.

In several publications no further information was given on the anatomical localization that they were situated in the *brain stem*. Thus, they must be reviewed separately. In rats these cells were mostly excited upon iontophoretically given morphine.[46,63,99,126-131] In other studies or at other units mostly inhibition has been found.[52,60,72,129-131] It must be emphasized, however, that in all these studies both excitation and inhibition have been observed, only the proportion of neurones reacting with increase or decrease in the firing rate was different. As for the opioid peptides (Met-E and β-EP), their effects were found mostly depressant in some studies,[52,60,72,96] while others observed mostly facilitation.[63] According to Bradley et al.[60] Met-E inhibits all brain stem cells which also may be depressed by morphine, but there are units which are facilitated by morphine but inhibited by Met-E. In some studies[60,96] naloxone reversed the depressant effects of both compounds but did not antagonize the facilitatory action. However, in the study of Nicoll et al.[52] both the stimulant and the inhibitory actions of Met-E and β-EP could be reversed by naloxone. Nobody found the morphine-induced excitation naloxone-reversible, while some authors[99,126] could reverse the morphine- or levorphanol-induced depression by opiate antagonists.[131]

Met-E and morphine frequently inhibited the facilitation of unit activity caused by acetylcholine, serotonin, and noradrenaline,[63,126,130,127] though some neurones were found stimulated by morphine, Leu-E, and Met-E as well.[63] The glycine induced depression could be antagonized by morphine but not by Met-E.[63]

Summarizing these seemingly heterogeneous data two conclusions seem to be justified.

- Opioid peptides induce inhibition more frequently than morphine does.
- Both the stimulant and depressant actions of the enkephalins and β-EP are naloxone sensitive, but the morphine-induced facilitation is not.

Hence the question arises, how to explain the naloxone-insensitive facilitation by morphine of certain neurones in the brain stem and in some other brain regions as mentioned above. Again several alternative explanations emerge.

Probably morphine acts on two different receptor populations and only one of them, the naloxone sensitive is depressed by it, just as put forward by Jacquet and Lajtha based on behavioral experiments (Chapter 2, Volume I). This supposition is further confirmed by other reports according to some neurones were excited by morphine but inhibited both by etorphine and Met-E.[60]

According to the preliminary communication of Goldinger et al.,[132] some opiate agonists inhibit the binding of labeled strychnine to the synaptosomal membranes of rat spinal cord. This observation suggests that morphine congeners behave as competitive antagonists of the glycine receptors. This supposition might give satisfactory explanation for the naloxone-insensitive excitatory action of morphine and some of its derivatives. According to a further explanation morphine would excite some neurones disinhibiting them by occupying the enkephalin receptor sites. In this case, naloxone is expected to only further potentiate the action of morphine, which has actually been reported.[99,126] In the experiments of Bramwell and Bradley[99] nalorphine and naloxone not only failed to antagonize the morphine-induced depression of certain units, but they potentiated the stimulant effect of each other. Furthermore, naloxone and nalorphine excited mostly those neurones which were also stimulated by morphine.[99] Such a cooperation between opiate agonists and antagonists is easiest to explain by supposing the displacement of endogenous ligand(s).

What makes the problem even more complicated is that these two different mechanisms do not exclude each other, they may actually coexist. Moreover, if accepting the second possible explanation, i.e., that opiates excite some units by occupying the enkephalin (or endorphin) binding sites, the possibility of a specific but naloxone insensitive effect emerges. Of course, in this case the neurones must be inhibited by enkephalins or by other natural opioid peptides. Accordingly, Bradley et al.[60] did find neurones which were inhibited by Met-E and facilitated upon morphine iontophoresis and only the action of the former could be reversed by naloxone. However, in the same experiment etorphine inhibited all the units. Later also in the brain stem neurones were found which could be depressed both by etorphine and (D-Ala^2, D-Leu^5)-E but facilitated by morphine.[71] Thus, the difficulty reappears. Could it be that etorphine and morphine act on partly different neurones, while etorphine and Met-E on the same ones? This finding can be explained, however, more simply: etorphine being a stronger agonist than morphine, not only displaces the endogenous opioids from their receptors but also exerts the agonist action of its own. From this point of view the data of Nicoll et al.[52] are disturbing. They[52] found almost all brain stem neurones inhibited both by normorphine and Met-E, but as for β-EP, it stimulated one third of the units examined and the others were inhibited by it.

Probably the tolerance studies may be helpful in this question, as generally tolerance develops to the specific effects of opiates, but not to their unspecific actions independent of opiate receptors.

In two studies (from the same laboratory) a significant decrease in the number of brain stem neurones excited by morphine has been observed upon repeated application or previous tolerance development to opiates.[128,129] However, no significant change in the depressant effect of morphine has been detected.[128] Consequently, just the facilitation seems to be the opiate specific effect, at least in relation of the brain stem neurones. Otherwise, tachyphylaxis developed to the depressant effect of morphine if the cortical units were examined,[43,51] while no tachyphylaxis developed to the depressant effects of either morphine[129] or etorphine[72] on the brain stem neurones. Tachyphylaxis developed, however, to the depressant effect of (D-Ala^2, D-Leu^5)-enkephalin-induced inhibition in the brain and to the similar effect of Met-E in the cortex.[26,72]

The development of tachyphylaxis to (D-Ala^2, D-Leu^5)-enkephalin but not to the effect of etorphine[72] confirms the possibility that there are different kinds of opiate receptors in the brain stem. As the action of opioid peptides differed from those of the morphine and etorphine as well and neither the latters' actions were identical, the existence of at least three different subtypes of opiate sensitive cells might be proposed in the brain stem; one of them mediating the stimulatory action of morphine, an effect to which tolerance develops upon repeated administration. A further subtype would be responsible for the inhibitory action of opiates (morphine, levorphanol, etorphine), i.e., for an effect found resistant to tolerance development. As tachyphylaxis develops to the similar inhibitory action of opioid peptides also the existence of a third population might be proposed. Let us repeat here, that the differences in the electrographic effects of opioid peptides and morphine congeners could be attributed to the preferential interaction of the former and latter with the so-called δ- and μ-receptors, respectively, (Chapter 1, Volume III). As for the differential actions of morphine and some of its derivatives, to date it is very difficult to find any comprehensive explanation. In this aspect again the data demonstrated by Jacquet and Lajtha (Chapter 2, Volume I) might be relevant, who proposed the existence of two distinct morphine-sensitive receptor populations in the midbrain, one of them mediating the depressant effects, the other the stimulant actions, the former being naloxone sensitive, the latter not. Consequently, these electrographic data could be regarded as evidence of the much debated hypothesis of the authors. However, no publication is known, where Jacquet and Lajtha themselves would have referred to these electrographic data indirectly confirming their hypothesis.

H. Locus Coeruleus (LC)

Of the more precisely defined brain stem structures probably the LC deserves special attention. The majority of the experiments on the participation of this area in the electrophysiological effects of opioids has been made in Aghajanian's laboratory. As reported first by Korf et al.,[45] the LC units were inhibited by morphine iontrophoresis and this inhibition could be reversed by naloxone. Naloxone in itself had no effect.[70] The naloxone-specific inhibition of LC units has been confirmed and extended to other opiates and opioid peptides.[133-136] These cells also can be inhibited by Met-E.[70,137] Consequently, contrary to the other brain stem neurones this population seems to be influenced by opiates and opioid peptides in the same direction. A further important property of these cells is that they can be activated by noxious stimuli and this facilitation is also attenuated by morphine.[45] Therefore, LC may be one of the important sites mediating the analgesic action of opioids. However, this area consists mainly of noradrenaline containing cells. These noradrenaline containing cells can be activated by acetylcholine (via muscarinic receptors) and by substance P.[86] Interestingly, however, scopolamine pretreatment did not modify either the activation by substance P or the inhibition by Met-E.[86] Consequently, acetylcholine, Met-E and substance P represent separate and independent inputs acting directly on the noradrenaline containing neurones. Recently Strahlendorf et al.[136] observed inhibition of the LC unit firing upon electrical stimulation of the arcuate nucleus. Since the cell bodies located in the arcuate nucleus of the hypothalamus contain β-EP (Chapter 3, Volume III) probably not Met-E but β-EP is the physiological opioid ligand on the opiate receptors of LC.

In the light of these experiments it is easy to understand that centrally administered noradrenaline attenuates the opiate analgesia, while α-adrenergic receptor blockade potentiates the effect of morphine (Chapter 2, Volume I). However, the data of Sasa et al.[138] may contradict such a generalization. These authors showed that morphine abolished the inhibition of relay neurones in the spinal trigeminal nucleus produced by electrical stimulation of the LC.[138] I.c.v. administered noradrenaline also inhibited these relay neurones.[138] Thus, inhibition by morphine of the noradrenaline containing units of the LC may result in disinhibition of certain sensory relay cells. Such a finding seems to be inconsistent with the analgesic action of morphine. Of course, it is not sure that these cells, disinhibited by morphine via the LC, are activated by noxious stimuli under physiological conditions.

Otherwise, the LC units represent the only population the activity of which reflects both the tolerance development and the stimulation upon withdrawal. As reported by Aghajanian,[79] the inhibitory effect of morphine upon LC units is significantly attenuated in animals repeatedly implanted s.c. with morphine pellets. Administering naloxone locally or systemically to these animals, the LC units were facilitated which could be regarded the electrographic manifestation of withdrawal.[79] In this study the morphine-sensitive units also could be suppressed by clonidine (i.e., by an α_2-adrenergic receptor agonist) and the latter's effect was reversed by piperoxane (i.e., by an α_2-adrenergic receptor antagonist).[79] Interestingly the tolerance development to the inhibitory effect of morphine did not modify the similar action of clonidine.[79] Consequently, there is an α_2-receptor mediated inhibitory input to the LC cells. This input is independent of opiate receptors as seen above.[79] These findings are in good agreement with the clinical studies (Chapter 9, Volume III) according to the opiate withdrawal symptoms can be relieved by clonidine and they also explain its antinociceptive action (Chapter 2, Volume I).

Consequently, the inhibition of brain stem noradrenergic system might have a decisive role, but what about the serotonin producing cells of the brain? The behavioral data and the interaction experiments indicated that of the brain monoamines serotonin was primarily involved in the action of opiates (Chapter 2, Volume I). Hence, rather the brain serotonin neurones are expected to be excited by opiates. However, as mentioned previously the raphe cells were found either uninfluenced[45,139] or just inhibited[45,66,118] upon treatment with opioids, and excitation was detected only in one study.[119]

From this point of view the observations made by Segal[133] seem to be of crucial importance. He found the spontaneous activity of LC units suppressed by serotonin iontophoresis. Electrical stimulation of low intensity of the dorsal raphe nucleus had similar effect, and selective destruction by 5,7-dihydroxytryptamine of serotonin cells of the raphe prevented the morphine-induced inhibition of the firing in the LC elicited by noxious stimuli.[133] Consequently, a serotoninergic connection must be supposed between the dorsal raphe nucleus and the LC. Of course, the role of this pathway in mediation of morphine analgesia requires further clarification.

The observed excitatory response to nociceptive stimulation can represent only rostrally projecting neurones. Moreover, the widespread occurrence of such cells, mostly in the PAG and the neighboring reticular formation may explain the strong arousing potency of noxious stimuli. Furthermore, as certain cortical and/or other forebrain structures are responsible for the pain sensation, the accompanying aversive subjective state and the autonomic reactions as well, inhibition of the ascending pathways by the corresponding brain stem relay units may contribute to the attenuation of many subjective and objective components of the pain.

However, opiates and opioid peptides also inhibit certain spinal reflexes, first of all those elicited by nociceptive stimuli (Chapter 2, Volume II). As these reflexes are preserved also in spinal animals, supraspinal mechanisms do not seem to play essential role in their normal mediation. Nevertheless, also these spinal reflexes are inhibited by opioids (Chapter 2, Volume II) and their antinociceptive effect is significantly attenuated even if not totally blocked in spinal animals (Chapter 2, Volume I). Consequently, the importance of descending inhibitory pathways cannot be overlooked. Yet it is difficult to decide that where are the neurones projecting to the spinal cord which are activated by opioids. As discussed in Chapter 8, Volume III, stimulation of the PAG and the raphe nuclei results in powerful analgesia. Specifically, the electrical stimulation of the n. raphe magnus in the rat[140] or the n. centralis inferior raphae in the cat[141] inhibits the activation of dorsal horn interneurones in the spinal cord upon nociceptive stimulation. Therefore, the raphe units are expected to be stimulated by opioids as discussed above, but contrary to the expectation rather inhibition has been observed.

I. Pons and Medulla

As for the more or less exactly defined parts of *pons* and *medulla,* i.e., in medullary neurones,[71,142] in the medullary reticular formation[76] and in caudal trigeminal neurones,[143] mostly depressant effects have been observed upon iontophoretic application of opioid peptides and opiates. Both the spontaneous activity and the noxious stimulation induced discharges were found inhibited in these experiments, the effects being naloxone sensitive. Furthermore, nobody found qualitative differences between the actions of opiates and those of the opioid peptides. The duration of action is shorter in the cases of natural enkephalins than with the stabilized enkephalin analogues and etorphine.[71] The relatively short duration of action of Met-E is consistent with its supposed role as a neurotransmitter, since all such natural substances are rapidly inactivated. Practically identical data have been obtained in rats[71,76,142-144] and in cats.[61]

Opiates and Leu-E inhibit the spontaneous activity or the facilitated firing upon noxious stimulation in the *n. reticularis gigantocellularis* (NRG) i.e., in an integral relay center of spino-reticulo-thalamic pathways.[61,89,144] Thus, opioids may exert their inhibitory action on pain transmission via this neurone population too. Interestingly this effect of morphine, Met-E, and Leu-E could not be antagonized by local iontophoretic application of naloxone,[61] while in another study the morphine's effect was reversed by i.v. administered naloxone.[144] For the time being it is difficult to explain this controversy. Of course it is not certain that exactly the same neurone population was examined in these studies.[61,144] Such an explanation may be trivial but probably not so in the case of the medullary reticular formation. There is another area, called *n. reticularis paragigantocellularis* (NRPG), where the majority of the units showed increased firing upon iontophoretic application of morphine and Met-E.[89] Again the natural enkephalins displayed relatively short-lived action. The excitatory effects of both substances could be reversed by naloxone.[89] In experiments of Azami et al.[145] such NRGP neurones were selected, which reacted with excitation to peripheral noxious stimuli. About one third of these neurones gave decreased response to the painful stimuli if previously morphine had been applied iontophoretically, the majority of these units, however, displayed enhanced reaction upon morphine treatment.[145] Naloxone also inhibited the stimulation by painful stimuli.[145]

Considering that microinjecting morphine, Met-E, or Leu-E into the NRPG elicits analgesia, and this is the brain site from where analgesia may be elicited by the smallest amounts of opioids,[139,146] the above-mentioned findings are not surprising. The very fact that the excitation induced by noxious stimulation of NRPG units is further enhanced by opioids suggests that a feedback control of spinal transmission of nociceptive reflexes is mediated by this area. Probably these are the very neurones which meet the main criteria of opioid-stimulated descending pathways, and actively inhibit the spinal pain reflexes. And let us recall here that microinjecting morphine into this region results in analgesia, i.e., in an inhibition of tail pinch reaction without attenuation of the affective responses accompanying the painful stimuli.[139,146]

Besides NRPG, the raphe complex and the PAG are the best known areas from which analgesia may be elicited by local application of opioids (Chapter 2, Volume I) or by electrical stimulation (Chapter 8, Volume III). Their contingent participation in the descending inhibition of pain reflexes is, however, a bit questionable. In these regions opioids rather depressed the spontaneous activity and that elicited by noxious stimuli, which may explain the suppression of pain sensation but not the active inhibition of spinal reflexes. As mentioned there, Urca et al.[26,116] suggested that only those experiments should be considered which were made in awake animals, since anesthesia might suppress just the facilitation of pain suppressing activity. However, these NRPG units were found activated in anesthetized animals[89,145] indicating that the working mechanisms of medullary and higher centers were different.

Moreover, the distinct reticulo-spinal connections of NRPG are well known[89,145] while those between the raphe and the PAG neurones are not unequivocally clarified. Thus, the theoretical possibility arises that the PAG and the raphe nuclei exert their activity via the NRPG. Moreover, if assuming a tonic inhibitory control of NRPG units by the raphe nuclei and/or the PAG, an inhibition of these cranially located neurones might result in disinhibition of the formers.

In this aspect the experiments of Mohrland and Gebhart[147] may be informative. These authors[147] observed the facilitation of spontaneous activity and the increase in firing induced by noxious stimuli in the medullary reticular formation if morphine was applied locally into the PAG. Consequently, opioids seem to suppress an inhibitory control of medullary neurones via midbrain structures.

J. Spinal Cord Ventral Horn Neurones

As for the spinal neurones, both excitatory and inhibitory effects have been reported upon administration of opiates and opioids. These effects were differently modified by naloxone. Moreover, also qualitative differences have been observed upon comparing the actions of opiates and opioid peptides. Data are particularly confusing if not only the modification of spontaneous activity is recorded but also the alterations in unit activity induced by nociceptive and non-nociceptive stimuli, or by electric shocks applied on the ventral and dorsal roots, or by iontophoretic application of biogenic amines are considered. To make the survey easier, the data on the Renshaw cells and on those located in the dorsal horn will be discussed separately.

First Kruglov[148] reported that narcotic analgesics reduced the motoneurone inhibition elicited through polysynaptic spinal segmentary pathways. As also the antidromic inhibition of these cells were attenuated, he[148] concluded on a blockade of the synapses between interneurones upon morphine treatment. Felpel et al.[149] reported that the recurrent inhibition of *Renshaw cell* activity induced by ventral root stimulation was attenuated upon i.v. morphine treatment. However, no change has been found in their spontaneous activity and in their excitability by nicotine either. (The motor-axon collaterals have nicotinic cholinergic synapses on the Renshaw cells.)

Thus, narcotic agonists seemed to suppress the tonic recurrent inhibition of spinal motoneurones at the Renshaw cell-motoneurone inhibitory synapses, a conclusion drawn by Duggan and Curtis.[150] These authors,[150] applying morphine locally by iontophoresis observed a facilitation of the spontaneous activity of Renshaw cells, moreover, morphine similarly to acetylcholine increased the number of spikes elicited by electrical stimulation of the ventral root. They also supposed a presynaptic inhibition of acetylcholine release from the motor-axon collaterals, resulting in an increase in the excitability of the postsynaptic membrane.[150] These experiments give no information on how to explain the increased spontaneous activity of these units. Though, even today no final answer can be given to this question, the basic finding, i.e., facilitation by morphine of these nicotinic cholinergic neurones, has been repeatedly confirmed.

The facilitation by homocysteate and acetylcholine of Renshaw cells was found augmented by simultaneous morphine iontophoresis and also etorphine increased the spontaneous firing of these neurones.[56,62,151,152] The acetylcholine action was especially enhanced by morphine.[152] These effects might be a common feature of opiates as also levorphanol increased the spontaneous firing of these cells, while dextrorphan did not.[56,151] Hence, an opiate specific increase in activity and excitability of Renshaw cells might be proposed. However, things are far from being so simple. Etorphine, contrary to morphine, did not activate the Renshaw cells, did not abolish the glycine-induced depression, moreover, it reduced the facilitation induced by acetylcholine and homocysteate.[152] In other studies made in cats and rats, all the three morphinans tested (levorphanol, dextrorphan, and levallorphan) reduced the sensitivity both to acetylcholine and homocysteate though the first is pure narcotic agonist, which increases the spontaneous activity of Renshaw cells,[56,151] the second an inactive enantiomer and the third a pure antagonist. Therefore, their effect was attributed rather to their local anesthetic action.[151] The spontaneous activity was not affected by dextrorphan and levallorphan in some experiments,[151] while in others dextrorphan had a depressant action.[56]

The data are even more confusing in relation of the antagonism by naloxone. Excitation by morphine and also by acetylcholine were suppressed by naloxone, while the stimulant action of homocysteate was not affected.[56,151,152] Naloxone given alone moderately facilitated the Renshaw cells in a concentration adequate to reduce the action of morphine[56] and it also reduced the excitatory action of nicotine.[151] Naloxone had no effect on the depressant action of glycine and did not inhibit the reversal by morphine of glycine-induced depression.[151] Also, the levorphanol-induced excitation of spontaneous activity could be antagonized by naloxone.[56] The effects of naloxone on spontaneous and acetylcholine induced activities of Renshaw neurones are very similar to those of morphine and levorphanol.[56,151] Naloxone acts like a mixed-agonist-antagonist, which shows its agonist potency if giving it alone; a very strange finding if considering the otherwise pure antagonist character of naloxone. As the reversal of glycine-induced depression of Renshaw cells by morphine is not modified upon naloxone treatment,[151] it cannot be regarded as an opiate specific effect.

Further strange findings were reported by Davies,[153] who examined the effects of morphine in dependent rats. It excited a significantly larger proportion of Renshaw cells (and spinal interneurones) in dependent animals than in naive ones. Furthermore, contrary to the findings of Satoh et al.[43,53] in relation of cortical neurones, the sensitivity of Renshaw cells to acetylcholine was not enhanced in dependent animals.[153] In agreement with these findings[43,53] a larger proportion of Renshaw neurones could be facilitated by acetylcholine in dependent rats than in control animals.[153] A further astonishing finding was that naloxone proved to be more effective in antagonizing the acetylcholine-induced activation of Renshaw cells in dependent animals than in naive ones.[153] This may be related to the enhanced in vivo potency of naloxone to antagonize the analgesic action of opiates and to precipitate abstinence syndrome as tolerance and dependence develop.

However, it must be emphasized that all these data were obtained in cats. In the only study made in another species, namely in rats, only a small proportion of Renshaw cells were excited by morphine and this excitation could not always be reversed by naloxone.[153] However, the Renshaw cells were excited by acetylcholine and this effect reversed by naloxone in this species too.[153]

Now the question arises, what is the action of enkephalins like, if the behavior of morphine congeners is so irregular in relation to the Renshaw neurones?

As reported by Davies and Dray[46,62,63] electrophoretic application of Met-E resulted in excitation of Renshaw cells, the duration of the action being similar to that of morphine and antagonizable by naloxone.

Nevertheless, contrary to morphine Met-E did not increase the sensitivity of Renshaw cells to acetylcholine.[62] If it was ejected for longer periods, the sensitivity to acetylcholine decreased.[62] At the same time the excitatory effect of glutamate was not affected.[62] Moreover, Met-E did not modify the depressant effect of glycine.[62] No significant differences were found between Met-E and Leu-E.[46] In acute experiments upon repeated application no desensitization was observed.[62]

Accordingly, enkephalins did not mimic morphine in relation to the previously-mentioned various interactions. The only common in the effects of opiates and opioid peptides is the naloxone sensitive facilitation of Renshaw cells.

Enkephalins are not the only naturally occurring peptides which excite the Renshaw neurones. As reported by Davis and Dray[155] and by others,[156] also substance P excited these cells and also its action might be reversed by naloxone. Moreover, substance P, similarly to the enkephalins, reduced the acetylcholine elicited excitation but did not modify the action of morphine,[155,156] and substance P similarly to the enkephalins did not antagonize the excitatory actions of d,l-homocysteic and glutamic acids.[156] (The inhibition of the effects of this amino acid is characteristic for those agents interacting with the glycine and GABA receptors.) The very fact that substance P did not modify the effect of morphine on Renshaw cells[155] proves that these substances act on different receptors. Thus, the Renshaw cells are activated by opiates, enkephalins, substance P, and acetylcholine as well. All these substances might exert their action via opiate mechanisms since their effects can be reversed by naloxone; a very strange phenomenon, which has been attributed to the enkephalin releasing property of acetylcholine and substance P by North.[157] However, there is no evidence that opiate receptors were located on the Renshaw cells (Chapter 1, Volume III) and generally the enkephalins are supposed to modulate the release of classical neurotransmitters and not inversely. Just North[157] suggested, that rather an "inhibition of local inhibitory interneurones must be seriously considered". The data are even more difficult to interpret in relation to the morphine congeners.

Then how to explain the activation of cholinergic neurones, the facilitated transmission in cholinergic synapses by agents known to act just in opposite direction in most in vitro and in vivo models? Why is the action of a pure opiate antagonist (naloxone) similar to that of morphine? How to explain the observed differences in the effects of morphine and other opiate agonists, such as etorphine and levorphanol? How to interpret the increased agonist potency of morphine or the enhanced antagonist efficacy of naloxone in morphine dependent animals?

There are many alternative explanations to these peculiar findings, but none of them is completely consistent with the present concepts on the mechanisms of action of opioids.

Theoretically an enhancement of the action of tonically released acetylcholine may be brought about by inhibition of acetylcholinesterase. Such an action has been demonstrated in very high in vitro concentration (Chapter 2, Volume I). However, nalorphine did not excite the Renshaw cells[56] though it is a potent inhibitor of rat brain cholinesterase, consequently this explanation must be rejected.

An enhancement by morphine of sensitivity of cholinoceptive neurones was suggested by Duggan et al.[56] If opioids had a direct postsynaptic action, the antagonism by naloxone of excitation caused either by morphine or acetylcholine would be easy to explain, but how to interpret the opposite effects of levorphanol and enkephalins?[62,151] Moreover, all the in vitro pharmacological data are against such an explanation.

If opioids release acetylcholine via an opiate specific presynaptic mechanism, the majority, of previously mentioned experimental data could be interpreted. Nevertheless, it is difficult to explain that why is the interaction of acetylcholine with morphine different from that with opioid peptides and other opiates such as levorphanol[62] and etorphine.[152] Morphine is known not to increase but to decrease the release of acetylcholine in the conventional in vitro models (Chapter 1, Volume I and Chapter 1, Volume II). This latter objection might be eliminated if a disinhibition by opioids of the presynaptic cells is supposedly due to inhibition of the liberation of another transmitter responsible for the activation of motoneurones.

If opioids cause a presynaptic depolarization as they do in the dorsal horn synapses (see later), an enhanced rate of spontaneous acetylcholine release may be expected, which explains the increase in Renshaw cell firing but not the increased sensitivity of these neurones to morphine in dependent animals.[153]

Finally if opiates hyperpolarize the presynaptic terminals, a receptor supersensitivity may be expected on the postsynaptic side due to the decreased rate of spontaneous acetylcholine liberation. Such an action could explain the increased sensitivity of Renshaw cells to all excitants and also the decreased potency of depressants, but not the curious finding that morphine excited a significantly greater proportion of Renshaw cells in dependent rats than in naive animals.[153]

None of the above-mentioned hypotheses may give account for the observed differences in the effects of morphine, other opiates, and enkephalins. No unitarian pharmacodynamic explanation can be found. Obviously some other ways of actions should also be taken into consideration.

Opiates, beside their own action, displace the endogenous ligands of opiate receptors. This action may result in an effect just opposite to those discussed above, but experimentally the netto result of the opposite actions is recorded. A part of the effects exerted by the opiates is certainly connected with their local anesthetic actions. Finally, all these data suggest the multiplicity of opiate receptors with differential sensitivity to opiates and opioid peptides. Therefore, many of these apparent contradictions can be eliminated if supposing that the working mechanism is similar to those observed in the hippocampus and in the brain stem.

If opiates act via disinhibition, i.e., by suppression of inhibitory interneurones, the observed facilitation of Renshaw cells would be the outcome explaining the enhanced activity of stimulants such as acetylcholine or homocysteate and the decreased inhibitory potency of glycine as well. The Renshaw cells receive a facilitatory input by the motor-axon collaterals, consequently if such interneurones exist at all, they should be located beyond the motoneurones. For this reason the existence of a neurone population should be supposed, which tonically inhibits the motoneurones, and opiates would facilitate the motoneurones and hence the Renshaw neurones by suppressing these inhibitory units. From this point of view the experiments of Jurna et al.[158] deserve special attention. These authors[158] observed the facilitation of spinal α-motoneurones upon morphine treatment in rats. In this study morphine reduced the latency of α-reflex discharges elicited by electrical stimulation of the dorsal root.[158] Surprisingly, this facilitation of α-motoneurone-discharges by morphine was not modified if naloxone was coadministered, and naloxone alone did not mimic this facilitatory effect of morphine at least at low dose level.[158] A facilitation by morphine of monosynaptic α-motoneurone discharges was observed also by Kuschinsky et al.[159] upon eliciting the reflex responses by stimulation of the gastrocnemius-soleus nerve.[159]

According to another explanation opiates, agonists, and antagonists as well, would displace certain endogenous ligands from their presynaptic receptors; just as supposed in the case of brain stem neurones. If these natural ligands of opiate receptors tonically inhibit the acetylcholine release as they do in many other areas of the nervous system, their displacement might result in increased synaptic conductance between the motor-axon collaterals and Renshaw cells leading to facilitation of the latter ones. If these receptors are of δ-type, both morphine and naloxone will behave like competitive antagonists against enkephalins, as the intrinsic δ-agonist activity of the latter is much stronger than those of either morphine or naloxone. Consequently, no wonder that naloxone inhibits the effect of morphine but the former has a certain morphine-like activity if given alone since both might be antagonists of the enkephalins on receptors where the latter is the selective and strong agonist.

Upon repeated morphine treatment not only tolerance but also cross-tolerance might develop between opiates and enkephalins. If such a process were really initiated, the enkephalins should lose, at least partially, their agonist potency. If so, a smaller amount of morphine might displace enough of the natural ligands to elicit a noticeable diminution of presynaptic inhibition of acetylcholine release, hence to facilitate the Renshaw cells. Since tolerance develops to morphine, giving naloxone, a pure antagonist, it will show an enhanced antagonism against morphine as well.

Obviously such a displacement of endogenous tonically released opioid peptides might take place between the motor-axon collaterals and Renshaw cells or between other interneurones and the motoneurones. In the latter case an enhanced firing of α-motoneurones could be expected, which has been demonstrated so far only by one group.[158]

K. Spinal Cord Dorsal Horn Neurones

Of the spinal neurones the role of the lamina V dorsal horn cells has been studied the most extensively. These neurones are preferentially activated by C-fibers, i.e., by noxious stimuli and also receive visceral afferents. Moreover, lamina V cells represent the site of origin of the spino-thalamic connections.[160] Consequently, if opioids exert a certain antinociceptive action also at spinal level, the transmission of nociceptive signals might be inhibited at these neurones, or at least the inhibition of these cells is expected to reflect the blockade of nociceptive information at spinal level.

First Kitahata et al.[161] reported that i.v. administered morphine inhibited the activation of interneurones in the dorsal horn upon noxious stimulation. Since then it has been repeatedly demonstrated that morphine and opioid peptides inhibit the spontaneous activity of dorsal horn neurones.[73,81,90,160-164] (The recently isolated new opioid peptide, Tyr-Arg called also "kyotorphin" showed similar action.)[92] Although several authors found no effect on the spontaneous activity,[46] the specific activation of these cells upon nociceptive stimulation was found suppressed in almost all relevant studies where opioids were applied[64,65,73,81,90,160,164-167] (Table 1). Such a statement is valid mainly for the nociceptive cells located in lamina V and sometimes in laminae VI and VII. Even more important to note that mainly those units can be inhibited by opioids which respond specifically to painful stimulation but not those driven by other kinds of sensory stimuli, such as tactile, non-noxious thermal, or other modalities.[54,64,160] As the nociceptive information is carried by Aδ and C fibers, it is not surprising that those neurones activated by Aα cutaneous afferents were not influenced by morphine.[167,168] The specificity of the action is shown by its antagonizability with naloxone or other opiate antagonists, though in some studies some cells displayed naloxone-insensitive depression upon morphine treatment. It might be asked whether these effects of enkephalins (and morphine) cannot be explained by simple excitation or depression of neuronal firing. If yes, they should antagonize the actions of phoretically applied excitatory amino acids and potentiate the inhibition of inhibitory ones. Reviewing the data available, no such correlation can be found. Morphine inhibited the majority of neurones excited by glutamate and aspartate in one study,[169] while in another work[65] variable effect has been recorded. The d,l-homocysteic acid elicited firing was not consistently blocked by either morphine[65] or Met-E.[73] Also the interaction of morphine with glycine or GABA-induced inhibition was found to be variable.[65,73,169,170]

Sometimes it seems to be difficult to separate the different kinds of sensory stimulation. The usually applied mechanical, thermal, or electrical stimulation of the skin may simultaneously activate receptors selectively sensitive to noxious and to non-noxious stimuli. Of the methods used, probably the intraarterial injection of bradykinin represents the most powerful activation of nociceptive receptors.[171] This mode of sensory stimulation was extensively used in experiments on the antinociceptive effects of opiates.[154,162,171,172] These authors surely have the right that intra-arterial bradykinin is the most powerful stimulus of the "pain-receptors". Nevertheless, this kind of their stimulation hardly can be regarded natural. Moreover, the exact localization of the receptors activated is not known. Only two groups reported on negative findings, i.e., on the inability of systemically administered morphine to suppress the facilitation by bradykinin of nociceptive neurones in the dorsal horn, in both cases in spinal preparations.[171,173] In one case[173] morphine did suppress the activation elicited by other kinds of nociceptive stimulation, only the effect of this modality of chemical stimulation could not be overcome by morphine. It is conceivable that intra-arterial bradykinin represents such an intense stimulation of pain mechanisms that they cannot be suppressed in spinal animals only in intact ones. Consequently, these data also show that the strictly intraspinal antinociceptive mechanisms might be too weak in themselves.

In the first such study of the Takagi group[173] spinal transection completely abolished the ability of morphine to suppress the unitary discharges elicited by intra-arterially administered bradykinin. However, in the experiments of Grossmann and Jurna,[174] morphine depressed the neuronal activity in the axons of the ventrolateral tract of the spinal cord evoked by stimulation of A-fibers of the sural nerve also in spinal cats. And in the studies of Le Bars et al.[160,167,168] spinal transection did not modify at all the ability of systemically applied morphine to suppress the dorsal horn neuronal activation to C-fiber activation. Thus, there are sites of direct spinal actions too. Moreover, under certain experimental conditions[160,167,168] the supraspinal actions seem to play no significant role in suppression of the spinal nociceptive response. Otherwise, as reported by Akaike et al.,[154] microinjecting morphine into the NRG in rat or rabbit, the bradykinin-induced activation of lamina V cells was depressed. Consequently, this caudal medullary area or the adjacent structures might be initiating sites of the descending pain inhibitory pathways.

Hence the question arises, where are the spinal opiate receptors responsible for inhibition of these lamina V cells located? This question can only be answered by localized microiontophoretic application of the drugs under study. The first guess was that the opiate receptors were located on the cell bodies of the lamina V neurones. As shown by Kitahata et al.,[161] morphine suppressed the unit activity only in those areas (laminae I and V) which were involved in the nociception. Those cells, however, which did not take part in processing of the painful information (laminae IV and VI) were not influenced.[161] In other studies beside laminae I and V also some units in laminae VI and VII were inhibited but never the cells in lamina IV which respond generally to non-noxious stimuli (Table 1). In one study,[164] however, even these cells could be depressed if more intense electrophoretic currents were used. Obviously the method of local microiontophoresis has certain disadvantages. The outcome of phoretic application depends to a large degree on unspecific factors, such as the current strength, concentration of the drug solution used, etc. Therefore, the local application may give precise information on the site of action, but sometimes it is rather difficult to decide whether the observed effects have any physiological significance or not. For this reason the studies on microiontophoretic application of opioids into lamina V cells gave conflicting results. Some workers found inhibition[73,90] while others sometimes observed no effect[64,171] and also excitation has been reported.[153] The inability to detect a clear-cut effect upon local morphine application might be attributed to the difficult localizability of these cells in several cases,[56] since the units in the adjacent regions were not expected to be inhibited. The rather long debate on the exact site of opiate action within the dorsal horn seems to be decided by Duggan et al.[64] and Davies and Dray.[46] In their experiments morphine ejected in the vicinity of the cell bodies in lamina V did not depress their responses to noxious stimulation. In marked contrast with these negative findings the cell responses were abolished if morphine or Met-E were ejected in the region of substantia gelatinosa (laminae II-III).[64]

The substantia gelatinosa, i.e., the dorsalmost tip of the compact part of the dorsal horn, contains the central endings of the primary afferents, i.e., there are the first synapses of the pathways carrying noxious information and also many interneurones modulating the pain transmission. Furthermore, this region is the richest in opiate receptors and enkephalins within the spinal cord (Chapters 1 and 3, Volume III). Consequently, the efficacy of opioids ejected specifically into this region is not surprising. More relevant is the question why did opiates act at all when they were injected into the lamina V, i.e., in the vicinity of the cell bodies.? The same group observed later, that Met-E and Met-E amide influenced the activity of these cells differently if they were microiontophoresed near to the cell bodies or administering them into the substantia gelatinosa.[73] In the former case enkephalins reduced the spontaneous activity and that induced by any kind of noxious or non-noxious stimuli; while ejecting them into the substantia gelatinosa, a more or less selective inhibition of nociceptive responses was observed.[73] Naloxone given intravenously readily reversed the effects of enkephalins ejected into the dorsal laminae but their direct effects on the cell bodies not only upon its local iontophoretic application.[73]

Otherwise in the experiments where the opiates and the opioid peptides were compared no qualitative differences have been found.[46,64] Therefore, the previously discussed experiments on the mode of action of opiates are more or less valid for the opioid peptides. Injecting Met-E or Met-E-amide into the substantia gelatinosa, the noxious stimulation induced firing of lamina V cells was inhibited.[64] In experiments of Davies and Dray,[46] however, Met-E contrary to morphine did not depress the response to noxious heat stimulation. The enzymatically stabilized and more active enkephalin analogue, (D-Ala^2, Leu^5)-E, however, selectively depressed the heat-evoked responses in the laminae IV and V. The apparent inability of the natural enkephalin to do the same was attributed to its rapid metabolic inactivation. The consequences of metabolic instability of natural enkephalins was confirmed by Duggan et al.[64] who described first this effect of the enkephalins. In their experiments more consistent effects were observed with Met-E-amide than with the natural Met-E. Otherwise, Zieglgänsberger and Tulloch[90] found the spontaneous or stimulation facilitated firing of the units in laminae IV, V and VI depressed by local microiontophoretic application of Met-E or Leu-E. A very surprising finding reported in this study was that local application of enkephalins into the lamina IV depressed the neuronal activation to light tactile stimuli. Thus either the previously mentioned contentions should be revoked according to opioids selectively depress the pathways carrying noxious information and the precise site of their action is the substantia gelatinosa, or the feasibility of data obtained by local microiontophoretic application in the vicinity of cell bodies under study must contested as mentioned above. A further danger of oversimplification arises if only the depressant effects of opioids are considered. A certain (sub)-population of dorsal horn neurones is actually facilitated by morphine[56,172,173,175] as seen previously in the cases of other brain structures. Again the problem emerges: do other opiates and opioid peptides also excite certain dorsal horn neurones or it is a specific property of morphine? Furthermore, is the effect naloxone sensitive or not. These questions were examined by Belcher and Ryall,[172] who reported on two surprising findings: while morphine depressed the majority of nociceptive cells of the dorsal horn, it excited the non-nociceptive neurones. Naloxone reversed the excitation of non-nociceptive units, but as for the depression of nociceptive neurones by morphine, naloxone reversed the effect only at a part of them.[172] It is important to add that also levorphanol depressed the nociceptive cells, but rarely excited the non-nociceptive ones.[172]

Consequently, among the dorsal horn neurones there are some which are excited by morphine, but not by levorphanol and enkephalins just as in the brain stem. On the other hand contrary to these brain stem units and similarly to the hippocampal pyramidal and ventral spinal Renshaw neurones, this excitation is naloxone reversible. As discussed above, the indirect nature of opioid-induced excitation of hippocampal pyramidal cells has sufficiently been documented, while in the case of Renshaw cells a similar way of action can only be suggested. Probably this is valid also for these dorsal neurones. Let us mention here, that in a recent study of Le Bars et al.[121] injecting morphine into the n. raphe magnus, an activation of the dorsal horn "convergent units" could be observed, while the same cells were depressed upon i.v. application of the drug. It is difficult to interpret those data, that opioids depress the dorsal horn neurones somewhere along the afferent sensory pathways, while some of them are activated via the descending brain stem-spinal pathways. Obviously the latter mediate the centrally induced analgesic effect of opiates.

The functions of these descending pain inhibitory pathways have been intensively studied by electrographic methods. The relative roles of local and supraspinal mechanisms in inhibition of lamina V neurones have been quantitated by Hanoaka et al.[176] Morphine inhibited both the spontaneous activity and that induced by high threshold peripheral cutaneous stimulation. This inhibitory action was dose-dependent and it could be detected in intact and in spinal cats as well. But comparing the effects of the same doses, the degree of and the duration of suppression were greater in intact than in spinal animals. The differences were larger at low than at high dose levels. Thus, opiates have a basic analgesic effect at spinal level, which is only further enhanced by supraspinal mechanisms.

Also other experiments render it very probable that the descending pain inhibitory pathways, facilitated by opiates, originate somewhere in the lower brain stem. Morphine was reported to inhibit the evoked potentials in the ventrolateral funiculus of the spinal cord induced by splanchnic nerve stimulation in decerebrated preparation but not in spinal animals.[177] More precisely, if the brain stem of cats was transected more cranially than the 11 mm posterior level, the action of morphine was not reduced, but transecting the brain at the posterior level of 14 to 16 mm, the drug-induced inhibition disappeared.[177]

NRG lies just between these lines of transections, therefore, also these data confirm its primary role in the downward control of pain reactions. Nevertheless, it is not the only site responsible for this action. Electrical stimulation of the central inferior nucleus of the raphe inhibited the activation by painful stimuli of lamina V units in the dorsal horn without affecting the lamina IV cells, i.e., those neurones which are activated by non-nociceptive stimulation.[141] Electrical stimulation of n. raphe magnus also inhibited the units located in laminae I, V, and VII, but only the high threshold neurones were suppressed.[140,178] Ipsilateral lesion of the dorsolateral funiculus markedly reduced the inhibition of neurones lying caudally to the lesion.[140] Not only the spinal nociceptive neurones were inhibited by these descending pathways but also the corresponding cells of medullary trigeminal nuclei. Recently Yokota and Hashimoto[179] reported that the unit discharges elicited by electrical tooth pulp stimulation were depressed by simultaneous stimulation of the PAG. These units were scattered in the trigeminal subnucleus caudalis, in the n. reticularis ventralis and dorsalis.[179] In the latter nuclei also antidromic potentials could be detected upon PAG stimulation,[179] showing that there is a direct fiber projection from the n. reticularis into the PAG.

Consequently, the spinal nociceptive neurones can be inhibited by electrical stimulation of those lower brain structures (NRG, raphe nuclei, PAG) where some units showed naloxone sensitive or insensitive excitation upon treatment with opioids. Moreover, these are the brain sites from where analgesia may be elicited by local application of minute amounts of opiates (Chapter 2, Volume I). Presumably these regions represent the starting points of the descending pain inhibitory pathways. The influence exerted by these pathways are obviously added to the actions of local opiate mechanisms. Of course a lot of questions emerge if such a general scheme is accepted.

Is the existence of these descending pathways also morphologically demonstrated?

There is no evidence for a direct PAG-spinal cord projection. The former projects, however, to the n. raphe magnus and also the projection of the latter to the caudal trigeminal nucleus and to the dorsal horn of the spinal cord has been demonstrated.[119,120,178] The long descending fibers end in laminae I, II, V, and VI, which might explain the inhibition of lamina V cells to noxious stimuli upon simultaneous stimulation of the raphe nuclei.[140,141,178] Interestingly no projection from the NRG to the dorsal horn has been found.[178] Thus, if accepting that PAG stimulation acts via the raphe nuclei, only the role NRG (NRPG) requires further clarification. *Á priori* a direct connection between the NRG (NRPG) and the spinal cord might be supposed, since electrical stimulation of this area elicits powerful analgesic effect.[146]

Are the brain stem and spinal opiate receptors mediating the inhibition of nociceptive neurones similar to those responsible for the behavioral effects in the forebrain?

As for these caudally located opiate receptors, no more can be told that their facilitation by opioids can be reversed by opiate antagonists just as the more cranially mediated behavioral effects. Nothing is known on their sensitivity to naloxone in quantitative aspect. Therefore, it is difficult to find out which type of opiate receptors mediate the above-discussed effects. Since the spinal cord contains exclusively enkephalins and even the medullary concentration of β-endorphin is very low (Chapter 3, Volume III) a preponderance of δ-receptors might be proposed. In a recent study Fields et al.[180] demonstrated that the dorsal root tissue and the dorsal horn contained both μ- and δ-receptors with a relatively high proportion to the former. Otherwise, Yaksh[181] reported that the pA_2 values of naloxone against morphine were practically the same for depression of the spinal nociceptive neurones and in different analgesic assays ranging from the tail-flick test to the shock titration paradigm. The similarity in the pA_2 values and in the relative proportion of μ- and δ-receptors suggests that no significant differences can be found if the opiate receptors located in the cranial and caudal parts of the central nervous system are compared. Nevertheless the problem requires further detailed experimental study.

A further question is that which neurotransmitters are released at the spinal nerve terminals of the opiate sensitive supraspinal neurones mediating the centrally induced analgesic effect of opioids.

From this point of view the data presented by Headley et al.[182] deserve special attention. These authors[182] showed that noradrenaline and serotonin ejected iontophoretically near to the cell bodies in lamina V or into the substantia gelatinosa inhibited the nociceptive responses of cat dorsal horn neurones. Both substances selectively inhibited the responses to noxious stimuli, leaving the inocuous stimuli more or less unaffected, while GABA iontophoresis reduced both types of responses.[182] It is important to note that serotonin is more effective in depressing the nociceptive responses if giving it into the substantia gelatinosa than into lamina V, i.e., close to the cell bodies.[183] Furthermore, the action of serotonin could be antagonized by locally administered methysergide, while administering the latter alone, the nociceptive responses either remained unaltered or they were only slightly increased.[183] Consequently, the receptors mediating the analgesic activity of serotonin are located in the substantia gelatinosa, i.e., in the region of the first synapses of nociceptive pathways. The lack of significant effect upon methysergide administration in itself shows that the serotoninergic pathways are not tonically active. Nevertheless, these data explain the analgesic action of serotonin if giving it directly into the intrathecal space.[124] Probably, these neurotransmitters might mediate the spinal effects of centrally administered opioids. Actually, Takagi et al.[162] found that the analgesic effect of electrical stimulation could be prevented by tetrabenazine, i.e., by depletion of the catecholamines, and this effect was reversed by l-DOPA. In their further experiments the antinociception produced by microinjecting morphine into the NRG or NRPG was suppressed by pretreatment with phenoxybenzamine giving it either i.p. or intrathecally.[184] Propranolol and methysergide, however, had no effect.[184] The direct proof of this working mechanism has been given by Yaksh[124] who could reverse the analgesic action of morphine injected into the PAG by intrathecally administered methysergide or phentolamine. Giving the two agents together, the analgesia was totally blocked, showing that PAG elicits a joint activation of noradrenergic and serotoninergic descending pathways.[124] The analgesia elicited by electrical stimulation of NRGP seems to be mediated only by noradrenaline releasing axons, since methysergide does not antagonize this effect.[184]

Probably not all the spinal nociceptive neurones are influenced (inhibited) by the descending serotoninergic (and noradrenergic) pathways. Le Bars et al.[168] reported that in intact cats two thirds of the dorsal horn nociceptive neurones showed normal dose-dependent antinociceptive responses even after depletion by *p*-chlorophenylalanine of brain and spinal cord serotonin, however, one third of the cells could not suppressed by morphine after previous depletion of the serotonin. It is easy to suppose that the former cells are directly inhibited by opiates, while the latter are depressed via the serotonin releasing descending pathways. However, of the endogenous neurotransmitters, first of all, substance P has been related to the analgesic effect of opioids, so far the latter have been suggested to be the natural antagonists of the former (Chapter 2, Volume II). Hence the next question arises.

Do the endogenous opioids exert their analgesic effect by antagonism of the substance P?

Substance P is supposed by many authors to be a neurotransmitter released by the primary pain afferents[185-187] Its distribution within the spinal cord is similar to that of the opiate receptors (Chapter 3, Volume III), it is located in sensory axons synapsing with the dorsal horn neurones in the substantia gelatinosa, and the depolarization of these terminals results in its release.[185,187,188] It has repeatedly been demonstrated that substance P (applied by microintophoresis) selectively stimulates the nociceptive neurones.[188-190] The liberation of substance P upon "painful" chemical stimulation by capsaicin has been recently demonstrated in the spinal cord.[191] However, contrary to the expectation morphine did not antagonize the substance P elicited excitation of nociceptive units in the dorsal horn.[188,192] Of course these data exclude only the direct interaction between enkephalins and substance P on the postsynaptic side but do not rule out the possibility that opiates act via inhibition of the substance P release, which has been demonstrated in vivo[191] and in vitro.[185] If so, the site of interaction is presynaptic as it will be discussed later (Chapter 2, Volume II). All these interesting experiments, however, do not answer the crucial question.

Have these opiate receptor mediated antinociceptive mechanisms any physiological significance?

The effects of local opiate application or those of the electrical stimulation only show that certain possibilities exist to elicit antinociception via opiate receptors. But it is a further question whether these opiate receptor mediated mechanisms exert a tonic or phasic effect on the pain reflexes under physiological conditions.

In their first related study made in cats Duggan et al.[193] found practically no effect on the dorsal horn neuronal response to noxious heat upon naloxone treatment in intact cat under chloralose anesthesia. However, if the spinal nociceptive pathways were tonically inhibited by supraspinal mechanisms maintained by endogenous opioids, a facilitation should have been obtained. The same group reported that in spinal cats naloxone did not merely antagonize the depression of nociceptive response to morphine[65] or to enkephalins[73] but enhanced these responses above the control values. Others observed a facilitation by small doses of naloxone of dorsal horn unit discharges upon supramaximal transcutaneous electrical stimulation. These experiments were carried out also in intact rats anesthetized by chloralose. These latter studies[65,73,91] prove the existence of tonic intraspinal pain inhibitory mechanisms maintained by endogenous opioids. Finally it is worth stressing that in all these studies, reporting on facilitation by naloxone of nociceptive neuronal responses, the phenomenon was observed only at a part of the units examined.[65,73,91] Consequently, only a subpopulation of the nociceptive units are tonically controlled by endogenous opioids. Moreover, it is obviously easier to record the inhibition of nociceptive responses and the suppression of the latter by naloxone than facilitation of a response which is almost maximal under normal conditions. Furthermore, even the general anesthesia might distort the results, but these spinal electrographic studies require either anesthesia or spinal (supraspinal) transection.

Consequently, more informative might be the experiments on the effect of naloxone on spinal nociceptive reflexes made in awake animals or in humans.

V. EFFECTS OF OPIOIDS ON SPINAL MONO- AND POLYSYNAPTIC REFLEXES

Morphine has been reported by numerous investigators to depress the polysynaptic reflexes in spinal cats and dogs as well.[1,194] Since the polysynaptic reflexes are most frequently activated by nociceptive stimuli, this is the effect that had been expected. The data concerning the monosynaptic reflexes (not related to the pain transmission) are rather contradictory. In the experiments of Wikler[1,195] made in spinal cats the monosynaptic reflexes were either unaltered or even stimulated upon morphine treatment, which otherwise resulted in definite depression of the polysynaptic responses. The inhibition of flexor reflex in spinal dogs was found dose-dependent and the relative potencies of different opiates in this relation were found well correlated with their analgesic activities in animal experiments and in the clinical practice as well.[196] Only one study contended that morphine did not modify either the monosynaptic, or the polysynaptic reflexes in low spinal cats.[197] Morphine inhibited the nociceptive flexion reflexes (RIII) but did not modify the monosynaptic ones (H reflex) in humans with paraplegia due to high spinal injury, and the effect could be reversed by naloxone.[198,199] Therefore, it seems that opiates selectively inhibit the nociceptive reflexes also in humans.

Considering that opiates inhibit specifically the spinal input of nociceptive information, and the spinal reflexes elicited by painful stimulation are polysynaptically mediated, the above-mentioned selectivity seems to be justified. Nevertheless, the previously reviewed facilitation of Renshaw cells also may result in modification of spinal reflexes. Since no data indicate that opioids had direct effects on spinal motoneurones, all changes observed in the spinal reflexes might reflect primarily the altered function of spinal interneurones. Of course, in the sense according to all elements are called interneurones which are interposed between cells of the ganglion spinale and the ventral horn motoneurones.

Thus, opiates inhibit the ipsilateral flexor and crossed extensor reflexes in intact and in spinal animals as well.[1] These effects meet the main criteria of opiate-specific actions. First, the inhibition of spinal polysynaptic reflexes may be reversed by naloxone[200] and levallorphan,[194] i.e., by pure opiate antagonists. The weak morphine-like effects of cyclazocine and nalorphine[196] have obviously no significance as in their cases it is difficult to separate the agonist and antagonist actions. Characteristically, nalorphine and cyclazocine display much shallower dose-response curve than morphine just as in all other in vivo assays specific for opiate agonists.[196] Upon repeated application of morphine or methadone to chronic spinal dogs tolerance developed to the depression of ipsilateral flexor and the crossed extensor reflexes but not to the excitant effect on the ipsilateral extensor thrust.[201]

These data obtained in spinal preparations prove that the basic effects of opioids on spinal reflex activity is brought about by intraspinal mechanisms, which of course does not exclude the participation of supraspinal centers.

The almost classical experiments on depression of spinal reflexes by opiates were repeated during the last decade by applying electrophysiological methods, i.e., using electrical stimulation of the afferent fibers instead of natural sensory stimulation and by recording the evoked response of motoneurones instead of mechanical registration of the muscle movements. These more sophisticated methods have helped to clarify some previous contradictions.

Stimulating the dorsal roots and recording the reflex responses (evoked potentials) on the corresponding ventral roots, morphine inhibited both certain monosynaptic and polysynaptic reflexes in spinal and in decerebrated cats as well.[200] A very important finding of this study was that the influence of morphine on monosynaptic reflex was more evident at higher stimulation frequency.[200] This observation seems to be at variance with the in vitro experiments (in isolated organ preparations), where the inhibitory effect of opioids is stronger at low frequency (Chapter 1, Volume II). But just such a relationship is predicted if trying to explain the data based on the "gate-theory" of Melczak and Wall[202] (see later).

First Grossmann et al.[203] reported that morphine inhibited the α-flexor motoneurone discharges elicited by conditioning stimulation of the dorsal root both in intact and in spinal cats. This effect was mimicked by levorphanol and meperidine[203] but not by dextrorphan.[204] The specificity of this effect is proven by its reversal upon naloxone treatment.[204] The inhibition of α-flexor motoneurones by morphine is mediated both by spinal and supraspinal mechanisms. In the experiments of Grossmann et al.[203] the inhibitory action of morphine was preserved after spinal transection but in another study[204] the same effect was abolished upon spinalization. In the latter experiment[204] the peroneal nerve was activated by stimuli of low intensity which excited Group II but not Group III and C afferents. In the experiments of Grossmann et al.,[203] however, presumably also the afferents mediating pain impulses were activated.

However, not all α-motoneurones are inhibited by opiates. The extensor α-motoneurones activated by tetanic stimulation of ipsilateral gastrocnemius soleus nerve were further facilitated by morphine and levorphanol as shown by Kuschinsky et al.[159] Thus, the inhibition of α-flexor motoneurones is accompanied by facilitation of the α-extensor units, which is not surprising considering the reciprocal innervation of flexor and extensor muscles.[159,204] (The disinhibition of γ-motoneurones may be a secondary consequence of increased extensor tone.) Also this facilitation of α-extensor motoneurones may be reversed by naloxone.[159]

Consequently, the spinal motoneurones are affected by opiates in two basically different ways. The α-motoneurones activated by painful stimuli are inhibited by opiates via intraspinal mechanisms. Opiates exert a reciprocal effect on the spinal flexor and extensor α-motoneurones via supraspinal centres. The supraspinal activation of extensor α-motoneurones and inhibition of flexor α-motoneurones are probably interdependent events, unrelated to the antinociceptive actions of opiates. These supraspinal actions, however, explain the well-known actions of opiates on muscular tone and posture or more exactly their cateleptogenic effect. All these data give indirect evidence of a contingent physiological role of enkephalins and/or endorphins in regulation of the muscular tone.

Only few studies are known which directly demonstrate their participation. In rats DALA given i.c.v. was found to elicit myoclonic twitches, which were blocked by naloxone or GABA or ergometrine but potentiated by haloperidol[37] as discussed previously. Therefore, the increase in muscular tone in intact animals might be at least partially related to inhibition of striatal or extrastriatal dopaminergic functions, just as in the cases of opiates.

In some other experiments of Jurna et al.,[194] also performed in spinal cats, morphine did not depress the monosynaptic activation of motoneurones, whereas the polysynaptic evoked responses were markedly attentuated. Thus the apparent contradictions in relation of the monosynaptic reflex might probably be explained so, that contrary to the study of Krivoy et al.,[200] single shocks have been applied in the first study.[194] Also Goldfarb et al.[205,206] observed that morphine and oxymorphone depressed both the mono- and polysynaptic reflexes in spinal cats. Also in these studies[205,206] the afferent peripheral nerves were stimulated by repeated shocks. But repeated stimulation of the afferent fibers results in temporal facilitation and opiates might inhibit this temporal facilitation. Jurna et al.[194] actually demonstrated the abolition of temporal facilitation of monosynaptic excitatory postsynaptic potentials in gastrocnemius-soleus motoneurones in spinal cats. The same group demonstrated the inhibition by morphine of interneurones, i.e., depression of their discharges to activation by single pulses.[194]

Recording by intracellular microelectrodes the activation of spinal motoneurones, the excitatory action potentials generated by antidromic or by monosynaptic activation were not modified upon morphine treatment. The polysynaptic excitatory potentials induced by dorsal root stimulation were depressed.[194] Morphine also inhibits the disynaptic antagonist inhibition, namely it was found to abolish the inhibition of gastrocnemius soleus monosynaptic reflex upon stimulation of tibialis anterior nerve.[206]

Hence, opiates seem to inhibit the spinal reflexes both by abolition of temporal facilitation and by inhibition of interneurones. The former mechanism explains the depression of monosynaptic reflexes if eliciting them by a train of electrical stimuli, whereas the other effect accounts for the inhibition of polysynaptic reflexes even if they are elicited by temporally separated single pulses.

The next question is whether these effects could not be explained by the facilitation of Renshaw cells upon opiate treatment. Neither mecamylamine nor atropine had any effect on the reflex depression produced by morphine.[206] For this reason the previously discussed effects of opiates on the recurrent collateral-Renshaw neurone synapses cannot explain these effects. The facilitation by opiates of Renshaw neurones is accompanied by depression of the motoneurones as seen above.[194,203]

Though a considerable amount of information has been obtained on the contingent role of endogenous opioid peptides in regulation of the spinal reflexes by examining the action of opiate antagonists(see later), only one study is known on the effect of opioid peptides. As reported by Polc,[85] administering FK 33-824, a very active enkephalin analogue, to spinal cats, depression of the polysynaptic reflexes and motoneurone activity could be observed.[85]

Considering that upon opioid treatment the absence of nociceptive reactions is accompanied by increased muscular tone, catalepsy, and rigidity, just these changes in the activity of motoneurones may be supposed. The inhibition of reflex discharges of flexor α-motoneurones reflects the lack of pain reflexes, whereas the facilitation of extensor α-motoneurones might obviously be related to the increased muscular tone.

Now the question emerges that to what degree are these effects mediated by direct intraspinal mechanisms. The effects of opioids on spinal reflexes were studied mostly in spinal animals, thus the previously discussed alterations in spinal reflexes might be brought about by more or less local mechanisms. As for the activation of extensor α-motoneurones, this effect of opiates is likely to be mediated by supraspinal mechanisms. Their activation can be abolished by spinalization,[159] but the facilitation of extensor α-motoneurones can be inhibited also by apomorphine in intact animals.[159] In another study the α-motoneurone activity was facilitated upon depletion of brain catecholamines by tetrabenazine, α-methyl-*p*-tyrosine, or reserpine also only in intact animals.[203] In prenigrally decerebrated rats nigral stimulation facilitated the monosynaptic α-reflex activity, while inhibited the polysynaptic antinociceptive reflexes.[158] Therefore, it is reasonable to suppose, that the facilitation of extensor α-motoneurones is due to the central antidopaminergic effect of opiates.

However, it is not sure that all monosynaptic reflexes are facilitated by opiates via supraspinal mechanisms. Satoh and Takagi[207] reported that morphine facilitated the postsynaptic inhibition of monosynaptic reflex discharges evoked by electrical stimulation of Ia fibers. This effect of morphine was diminished but not abolished by low spinal transection.[207]

Accordingly, not all flexor α-motoneurones are depressed by opiates and the inhibition is not mediated exclusively by intraspinal mechanisms. Conversely the extensor α-motoneurones are not always facilitated by opiates and not exclusively via supraspinal pathways.

Returning to the question of physiological role of endogenous opioids, first of all data on the effects of opiate antagonist deserve special attention. Obviously, displacement by naloxone-like agents of the endogenous opioids can reveal the role of the latter in normal functioning of the spinal cord. The relevant experimental data were obtained mostly during the last years, which might be surprising considering how voluminous the literature is on the opiates. However, the first really "pure" opiate antagonist, naloxone, was developed relatively recently[208] and usage of partial agonists instead of the pure ones led to much confusion.

First McClane and Martin[196] reported that naloxone, contrary to narcotic analgesics and some mixed agonist-antagonists, (nalorphine, cyclazocine) did not depress the flexor reflex elicited by nociceptive stimuli in chronic spinal dogs but rather enhanced it. Since then it has been repeatedly demonstrated that the spinal mono- and polysynaptic reflexes are facilitated by naloxone,[75,198,206,209] naltrexone,[75] and levallorphan[194] The only conflicting data have been reported by Krivoy et al.,[200] who observed no increase in the amplitude of the ventral root polysynaptic potential evoked by dorsal root stimulation upon naloxone treatment in acutely spinalized cats. The failure to detect facilitation upon naloxone treatment in this experiment[200] might be due to the application of too strong electrical stimuli directly to the dorsal root, while in the other studies generally milder stimuli were applied and with one exception[194] not to the dorsal root but on the peripheral nerves.

Moreover, Goldfarb and Hu[205] found a negative correlation between the stimulus strength and the enhancement of nociceptive reflexes produced by naloxone.

The decreased efficacy of opiate antagonists to enhance the spinal reflexes elicited by stronger stimuli may be explained in two different ways.

- Stronger (generally repetitive) stimulation elicits enhanced release of endogenous opioids, which is more difficult to overcome by naloxone.
- The stronger stimulation results in enhanced activation of pain reflexes, which can be inhibited to smaller degree by the endogenous opioids. If the inhibitory effect of endogenous opioids is decreased, there is less room for disinhibition by opiate antagonists.

It must, however, be taken into consideration that very different doses have been applied in the different laboratories. Mostly doses between 0.1 to 1.0 to 2.0 mg/kg were applied, but facilitation of C-fiber reflexes were observed already applying 0.005

mg/kg naloxone or 0.0025 mg/kg naltrexone.[75] If the facilitatory effect of opiate antagonists is simply due to inhibition of endogenous opioid peptides, an initially steep then at the higher doses flattening dose-response curve could be expected. The ceiling effect had actually been reported,[75] but nobody found clear dose-response relationship. A further explanation for the lack of regular dose-response relationship may be the existence of a "hidden" agonist potency of these otherwise relatively pure opiate antagonists.[210] As the intrinsic agonist potency of naloxone could be detected at relatively high dose level,[210] the only negative finding in relation of flexor reflex potentiation[200] also might be attributed to the usage of a too high dose (1 mg/kg).

In healthy awake humans naloxone facilitated the monosynaptic extensor reflex elicited by stimulation of the tibial nerve, i.e., it increased the excitability of α-extensor motoneurones, while the polysynaptic tactile and nociceptive reflexes were unaltered.[209]

However, in humans suffering in chronic paraplegia of traumatic origin just the opposite effect has been observed, i.e., the polysynaptic nociceptive flexor reflex was facilitated, while the monosynaptic extensor reflex remained unaltered.[198,199]

As seen above the data obtained in spinal animals and those recorded in the corresponding neurologic cases are consistent with each other. Consequently, the existence of a tonic inhibitory control of pain reflexes mediated by naloxone-sensitive mechanisms may be regarded as proven. The only contradictory experiments[209] were made in "intact" humans, showing that these local spinal mechanisms might be confounded by supraspinal effects.

Of course, it can be asked whether naloxone does not merely act as a general excitatory agent. A very important point as in the case of such an unspecific action the facilitation by naloxone of nociceptive reactions would not necessarily prove the existence of endogenous pain relieving substances. The effect of very low dose of opiate antagonists, the lack of regular dose-response relationships, and the inability of naloxone to cause further reflex facilitation in the presence of nalorphine[75] contradict such as unspecific action. From this point of view the data presented Fry et al.[211] deserve special attention, who compared the effects of (+) and (−) isomers of naloxone on the unit activity. In naive animals no effect or sometimes a nonstereospecific depression of the cortical and striatal neurones was observed, while the excitatory action in dependent animals was found to be strictly stereospecific.[211]

Thus, just the occasionally observed depressant effect of naloxone might be a nonspecific action. It is tempting to speculate that apart from the controversies concerning the effect of naloxone in drug naive animals, the previously mentioned lack of dose-response relationship might be attributed to a nonspecific depressant action, confounding the stereospecific excitatory effects. (Of course, in interaction experiments, especially upon opiate withdrawal, the antagonism of opiate-induced depression might conceal the weak unspecific depressant effect.)

A further possibility is that naloxone blocks certain unspecified tonic inhibitory input into the motoneurones. Also this mode of action can be excluded. The inhibition of monosynaptic extensor reflex upon stimulation of the nerve to biceps muscle was not modified by naloxone[205] and neither morphine facilitated the postsynaptic inhibition of a monosynaptic reflex in spinal cats.[207] Morphine, however, did increase the postsynaptic inhibition in intact animals.[207] From these data it follows that in spinal animals the postsynaptic inhibition cannot be attributed to endogenous substances. Certain descending pathways, however, may facilitate the reflex inhibitory processes at sites located beyond the first sensory synapses.

Hence, the next question emerges. Where does naloxone exert its action within the spinal cord? The reflex enhancement cannot be explained by excitation of otherwise silent afferent fibers as the mono- and polysynaptic reflexes were facilitated by naloxone even if the corresponding peripheral nerves were activated by supramaximal stimuli. As mentioned above, even an interaction at the recurrent collateral-Renshaw cell synapses can be excluded.[206]

But Jurna et al.[194] convincingly demonstrated the facilitation by levallorphan of interneurones and excitatory postsynaptic potentials of motoneurones. In the same work[194] also the temporal facilitation of excitatory postsynaptic potentials of motoneurones was enhanced. The same group[212] demonstrated previously that opiates inhibited the post-tetanic potentiation of monosynaptic reflexes. As the post-tetanic potentiation is a characteristic manifestation of temporal facilitation, these observations[194,212] confirm each other.

Therefore, opiate antagonists seem to enhance the spinal transmission from the sensory input to the interneurones. The main site of such transmission is the substantia gelatinosa, the area richest in opiate receptors and enkephalins (Chapters 1 and 3, Volume II) in the spinal cord. Consequently, all the data suggest that the endogenous opioids tonically inhibit the processing of nociceptive sensory information at the first synapses of spino-thalamic pathways.

Let us mention here, that local microiontophoretic application of naloxone into the dorsal horn (mostly into the substantia gelatinosa) resulted in a slight activation of dorsal horn nociceptive neurones.[46,90,91,213] Thus, endogenous pain transmission inhibiting substances must be tonically released in laminae II and III. Considering the high density of opiate receptors and the high concentrations of Met-E and Leu-E just in these layers of the spinal cord (Chapters 1 and 3, Volume III), these opioid peptides might be the hypothetical pain transmission inhibiting substances. Accordingly, the pain transmission were under a permanent enkephalinergic control in the dorsal horn. As the above-discussed facilitating actions of naloxone also could be observed in spinal cats,[213] this enkephalin mediated inhibitory sytem is at least in part independent of the brain stem-spinal descending pathways.

The strength of the naloxone's effect depends, among others, on the previously applied treatment with opioid agonists and antagonists. Tachyphylaxis to the reflex antagonizing effect has repeatedly been demonstrated. After the first application of naloxone the second dose caused a much weaker facilitation of spinal reflexes.[75,205,206] This loss of efficacy seems to be fairly significant[75] and long lasting.[205,206] Some authors observed an enhancement of the naloxone's reflex facilitatory effect if it was given after morphine treatment,[205,206] though others did not detect such a reflex overshoot.[196,200] Also these observations confirm the existence of tonically active opiatergic intraspinal antinociceptive mechanisms. Obviously, if the endogenous opioids are displaced from their binding sites, receptor supersensitivity develops, which attenuates the effect of the second dose. Conversely supersensitivity may develop upon morphine treatment resulting in enhancement of the action of succeeding naloxone treatment.

Finally, in experiments of Bell and Martin,[75] naloxone was much more effective in reversing the depressant effect of morphine (the prototype of μ-receptor agonists) on C fiber and radiant heat evoked reflexes than against the similar action of ethylketocyclazocine (a κ-receptor agonist). Consequently, even if the spinal opiate receptors are not uniform,[180] the tonic inhibitory control should be attributed to both μ- and κ-receptors, concluded these authors.[75]

As the naturally occurring opioid peptides of the spinal cord are enkephalins (Chapter 3, Volume III) this is a rather surprising conclusion. The enkephalins are much more active agonists of the δ-receptors, than of the μ-receptors, as shown by Lord et al.,[214] Rónai et al.[215] (see also Chapter 5, Volume III). But how could the μ-receptors, which are relatively insensitive just to the locally present opioids, i.e., to enkephalins, mediate the tonic inhibitory effects of the latter. Upon trying to explain this apparent contradiction, probably the forthcoming experimental findings should be taken into consideration.

In the classical experiment of the Martin's group[216] only the effects of opiates were studied in spinal dogs. But for the time being no similar data are available on the antinociceptive effects of opioid peptides and on their antagonizability by different opiate antagonists in spinal animals.

The existence of two separate subtypes of opiate receptors, one preferentially interacting with the natural opioids (δ-type) and the other preferentially binding the classical morphine congeners (μ-type) has been proven in three different ways; in receptor binding assay opioid peptides displace more easily each other than opiates and vice versa[216] (see also Chapter 1, Volume I and Chapters 1 and 5, Volume III). The relative potencies of opioid peptides and morphine congeners in different opiate sensitive isolated organ preparations also are different.[214,215] (see also Chapter 1, Volume III). And both in the receptor binding assay and in isolated organ preparations the antagonizability by naloxone of opiates and opioid peptides are different (Chapter 1, Volume III). However, all these experiments were made in vitro and these in vitro data do not necessarily relate to the in vivo conditions. In experiments of Yaksh and Rudy[217] the pA_2 values of naloxone against intrathecally administered morphine and β-endorphin were the same in relation of analgesia. In our experiments[218] made in intact mice, the pA_2 values of naloxone were practically the same against morphine, β-endorphin and a potent enkephalin analogue D-Met, Pro-EA, while in rats the corresponding pA_2 values against morphine and the opioid peptides were different (see also Chapter 3, Volume II). For these reasons probably no conclusions can be drawn from the in vitro data in relation of differential antagonizability by antagonists in vivo. But even such a negative conclusion is disputable, since β-endorphin and the active enkephalin analogues are more or less equiactive agonists of the μ- and δ-receptors, respectively (Chapter 5, Volume III). On the other hand the natural enkephalins and their analogues, acting preferentially on δ-receptors are in vivo practically inactive or very short-acting (Chapter 2, Volume II and Chapter 5, Volume III), rendering the quantitative measurement of pA_2 values very difficult. In the above reviewed electrophysiological experiments the effects of iontophoretically applied enkephalins and their analogues were invariably found antagonized by naloxone. Such experiments are not suitable for the quantitative measurement of antagonizability, as in the case of iontophoretic application the real tissue concentrations of the drugs under study can not be measured. This question could only be answered in possession of systemically active selective δ-receptor agonist opioid peptides which might be examined not only by local iontophoretic application but also by systemic route. Unfortunately, no such compounds are available or examined yet. (Of the systemically active enkephalin analogues, (D-Ala^2, D-Leu^5)-E has been reported to be fairly selective agonist of δ-receptors (Chapter 5, Volume III). The spinal reflex inhibitory action of this compound and its antagonizability by naloxone has not been examined nor compared to the effects of morphine in similar experiments.)

Finally it should not be left out of consideration that Fields et al.[180] proved the presence of both μ- and δ-receptors in the spinal cord.

VI. THE "GATE CONTROL SYSTEM" AND THE OPIOIDS (TOWARD A WORKING HYPOTHESIS ON SPINAL MODULATION BY ENKEPHALINS OF PAIN REFLEXES)

Upon trying to interpret the data on the spinal action of opioids, first of all the so-called "gate control system", postulated by Melzack and Wall,[202] deserves serious consideration. Reading their almost classical work[202] the impression arises that all recent investigations on opiate receptors and opioid peptides have not done more but provided a neuropharmacological basis for their hypothesis formulated in pure physiological terms. Figure 1 summarizes their hypothesis. According to it ongoing afferent fibers before synapsing with cells, having thalamic projection, give collaterals to a neurone population located within the substantia gelatinosa. These cells of the substantia gelatinosa tonically inhibit the pain transmission in the first synapses of the ascending sensory pathways, i.e., they produce a permanent negative feedback control of pain transmission. Large diameter fibers carrying non-noxious information activate this negative feedback system. The small diameter (mostly C) fibers not only convey the nociceptive stimuli but also inhibit these interneurones. Consequently, by inhibiting the negative feedback system they significantly facilitate the transmission of sensory impulses. Accordingly, non-noxious sensory modalities inhibit the transmission of pain signals. Unfortunately, this ingenious hypothesis[202] does not refer to the opposite phenomenon, i.e., to the inhibition of non-nociceptive sensory functions during pain perception. Of course, also the central pain inhibitory mechanisms could exert their action via these neurones; their activation obviously would result in "closing of the gate," i.e., in inhibition of impulse transmission in the first synapses of the ascending sensory pathways.[202]

The most convincing evidence of this hypothesis has probably been given by Belcher and Ryall.[172] In their experiments, made in intact cats, morphine excited the non-nociceptive cells of the dorsal horn, while depressed the nociceptive ones.[172] Moreover, naloxone readily reversed the excitatory action of morphine, but only one third of the cells depressed by morphine were disinhibited by naloxone.[172] Obviously, the naloxone-sensitive activation of the non-nociceptive neurones might reflect the activity of those neurones which tonically inhibit the pain pathways. Naloxone-sensitive facilitation of certain units in the dorsal horn has been observed upon microinjecting opiates into the n. raphe magnus.[120-122]

The other experimental data are far from being so unequivocal. First, levorphanol, also a strong opiate agonist, was found only rarely excite the non-nociceptive neurones.[172] Examining the data summarized in Table 1 (Chapter 2, Volume II) it can be seen that the dorsal horn sensory neurones were found mostly inhibited upon treatment with opioids. Of course, mostly the big sensory neurones located deeply in the dorsal horn were studied, i.e., such cells which lay already beyond the "gate". And the inhibitory interneurones should obviously be found in laminae I through III. The cells located in these laminae were selectively examined only in several studies[81,175] but mostly inhibition has been found. Nevertheless, Randič and Miletič[81] reported that even in laminae I and II the majority of nociceptive neurones were depressed by Met-E, while those activated by mechano-receptors were not influenced or rather facilitated by Met-E. The latter effect, however, could not be antagonized by naloxone.[81] Thus, the data of Belcher and Ryall[172] are not consistent with many other experiments, though many observations made in connection of endorphin research seem to confirm the "gate" theory.

Having a look at the schematic diagram on the hypothetical enkephalinergic control of dorsal horn sensory neurones by the interneurones located in the substantia gelatinosa as suggested by Zieglgänsberger and Tulloch,[90] the proposed interrelationships seem to be very familiar (Figure 2). According to the proposition of these authors[90] enkephalin releasing cells would tonically inhibit the sensory neurones in the dorsal horn. Activation or suppression of these cells of substantia gelatinosa would mean inhibition or facilitation of the sensory transmission. Also according to the latter scheme[90] (Figure 2) the small diameter fibers would inhibit these enkephalin releasing cells via intercalated interneurones while the large diameter afferent fibers and the descending pathways activate them. (The latter should activate this inhibitory system

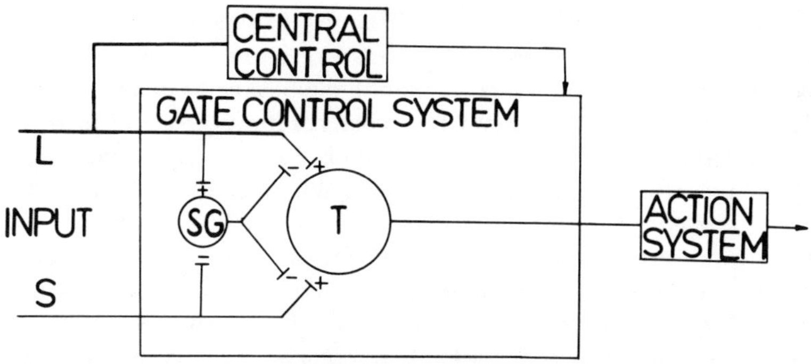

FIGURE 1. Schematic diagram of the gate control theory of pain mechanisms: L, the large-diameter fibers; S, the small-diameter fibers. The fibers project to the substantia gelatinosa (SG) and first central transmission (T) cells. The inhibitory effect exerted by SG on the afferent fiber terminals is increased by activity in L fibers and decreased by activity in S fibers. The central control trigger is represented by a line running from the large-fiber system to the central control mechanisms; these mechanisms, in turn, project back to the gate control system. The T cells project to the entry cells of the action system. +, excitation; −, inhibition. (From Melzack, R. and Wall, P. D., *Science*, 150, 971, 1965. With permission.)

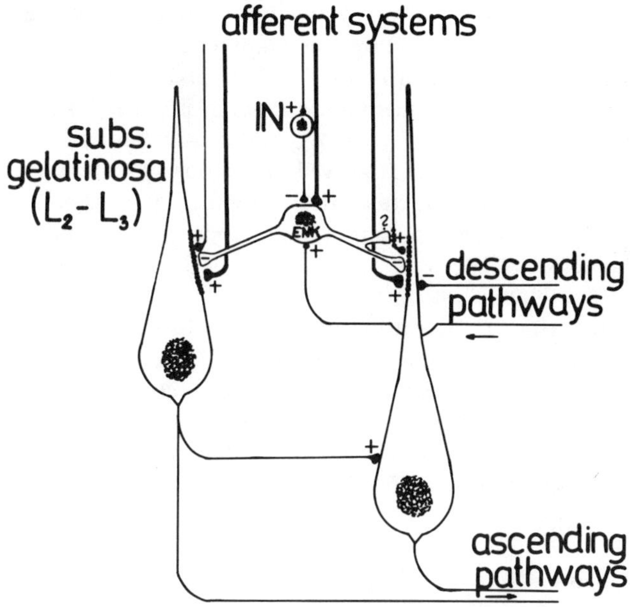

FIGURE 2. Schematic diagram illustrating a hypothesized (enkephalinergic) control mechanism of neurones in the substantia gelatinosa (L_2- L_3) over neurones involved in somatosensory perception. The small, bipolar (ENK) neurone impinges together with afferent C-fibers carrying opiate receptors (dots). Inhibitory interneurones (IN) have to be postulated to make C-fiber input to these inhibitory neurones. Descending pathways can act directly, e.g., via noradrenergic or serotoninergic pathways, on neurones of the spino-thalamic system (L_4 and L_5) or via the enkephalinergic neurones in the substantia gelatinosa; only this part of the descending influence would be naloxone sensitive. The presynaptic contact is marked with a question mark; although recent experiments on the excitability of primary afferent fiber terminals have shown a selective increase in C-fiber threshold following systematic administration of analgesic doses of morphine. Excitatory/inhibitory influences (+/−). (From Zieglgänsberger, W. and Tulloch, I. F., *Brain Res.*, 167, 53, 1979. With permission.)

via adrenaline or serotonin.)[182] The main difficulty of this theory[90] is connected with the possible site of opiate receptors. Zieglgänsberger and Tulloch[90] suppose that the opiate receptors are located on the postsynaptic membrane of dendrites of laminae IV and V cells, which contradicts the much discussed supposition on the presynaptic localization of opiate receptors. A further possibility also included into this schema is that the opiate receptors are located on the presynaptic endings of small diameter primary afferents (see question mark in Figure 2). Accordingly, enkephalins would inhibit the release of transmitters responsible for the activation of small neurones exerting a tonic inhibition of sensory transmission. Such a hypothesis is obviously in agreement with the general opinion. If the opiate receptors are located presynaptically, the gating cells probably ought not to be either enkephalinergic or enkephalin releasing.

Of course, also the supposed tonically active inhibitory cells in the substantia gelatinosa are presynaptic in relation to the dorsal horn sensory neurones. It is questionable, however whether these cells act directly on the sensory neurones as put forward by Zieglgänsberger and Tulloch[90] or enkephalins are released on the afferent input of primary sensory cells as proposed by Melzack and Wall.[202] On the other hand also a postsynaptic site of action could be suggested since both Met-E and Leu-E strongly inhibited the activation by glutamate of dorsal horn neurones (Chapter 2, Volume II), and glutamate is known to act postsynaptically via depolarization.

Nevertheless, there is a good deal of compelling evidence for a presynaptic site of action. Opiates inhibited the activation by acetylcholine of laminae IV and V neurones, which points to a postsynaptic site of action.[164] The opiates did not modify either the resting membrane potential or the membrane input resistance.[90] Furthermore, enkephalins had no effect on spike initiation or on the overshoot of action potential[90] though they inhibited the Na^+ influx triggered at the postsynaptic membrane.[164] Furthermore opioids slowed down the rate of rise of excitatory postsynaptic potentials at a dose level, which did not alter the amplitude or slope of action potentials and their effect was naloxone sensitive.[164]

The presynaptic site of action of opioids has been confirmed also in other ways. Le Bars et al.[167] reported, that the activation of the same dorsal horn units was differentially modified by morphine if the discharges were elicited by stimulation of Aα, Aδ, or C-fibers. Stimulating the corresponding peripheral nerve (sural nerve), morphine did not alter the first component of evoked response, related to Aα input, but inhibited the latter discharges corresponding to the Aδ or C inputs.[167] The differences in the effects exerted on the same cells can not be explained otherwise, but by a presynaptic site of action.

A. The Role of Primary Afferent Hyperpolarization

First Krivoy and Huggins[219] reported that morphine, methadone, meperidine decreased the fifth component of dorsal root potential (DRP) in spinal cats. ("Dorsal root potential" is the compound evoked potential elicited by peripheral nerve stimulation and recorded near to the dorsal root entry zone.) Nalorphine antagonized this effect of the narcotics, while giving it alone it induced similar effect.[219] These findings have been later confirmed.[207,220,221] It is not surprising that examining the potent enkephalin analogue, FK 33-824 also a depression of the late component has been observed just in the case of another active analogue (D-Ala^{2}Pro^5)-E-amide.[93] This effect of opioid peptides was readily reversed by naloxone,[85,93] while naloxone alone did not modify the DRP.[93] Thus, the action is strictly opiate specific, the observed similar effect of nalorphine[93] was obviously due to its partial agonist character. The failure of naloxone to elicit any action in this test if given alone,[84] shows that endogenous opioids do not tonically depress the DRP.

As this phenomenon could be demonstrated also in spinal cats,[93,207,219] primarily intraspinal mechanisms might be involved; though in one study[207] made in intact animals a small dose of morphine had no noticeable effect on the ipsilateral DRP but markedly depressed it on the contralateral side and this depression was abolished by acute spinal transection.

Of course, the question emerges what is the physiological meaning of this attenuation of evoked potentials? The DRP consists of at least six components. The first three to four components (DRP I, II, III, IV) are associated with the intramedullary invasion of action potentials,[219] thus their alteration would suggest a certain local anesthetic effect. In all the above-mentioned works the late negative (DRP V) and the subsequent positive (DRP VI) components were examined, first of all the DRP V.

This negativity might be related either to the depolarization of secondary neurones[219] or to that of the primary afferents.[93] In all the recent studies reviewed here the second explanation has been accepted, i.e., DRP V is regarded an indicator of primary afferent depolarization. The primary afferent depolarization is, however, a reliable indicator of presynaptic inhibition.[221] Eccles et al.[222] proposed that certain interneurones made axo-axonal connections with the primary afferents. The depolarization of afferents brought about by activation of these synapses results in inhibition of transmitter release in the first sensory synapses. This phenomenon is called "presynaptic inhibition". The similarity of concepts called "presynaptic inhibition"[222] and "gate control system"[202] does not require much explanation. Consequently, based on these considerations opioids are expected to elicit presynaptic depolarization, hence an augmentation of DRP or at least of its fifth component. As reviewed above, it was reduced, i.e., instead of afferent depolarization hyperpolarization took place. This hyperpolarization has been demonstrated also by direct microintophoretic application of the above-mentioned potent enkephalin analogue, FK 33-824.[85]

Upon trying to resolve these apparent contradictions it should be taken into consideration that the hyperpolarization of primary afferents increases the magnitude of the depolarization of these fibers produced by stimulation of other fibers.[223] Hence the possibility arises that the excitabilities of different afferents are differentially modified by the opioids.

Recently Sastry[82,83] demonstrated that morphine and Met-E, applying them iontophoretically into the substantia gelatinosa, produced an increase in the threshold of antidromic activation of sural C and A afferent fibers. This increase in the threshold indicates that opioids hyperpolarize the cutaneous afferent terminals. Such an hyperpolarization explains the repeatedly observed diminution of DRP V.

As postulated by Eccles[223] prior hyperpolarization of primary afferents potentiates the magnitude of the depolarization of these fibers produced by other fibers, this phenomenon has been observed by Sastry.[82,83] Met-E and morphine not only decreased the excitability of afferent terminals as detected by the increase in the threshold for antidromic activation but also potentiated the decrease in the threshold brought about by simultaneous conditioning stimulation of another afferent (superficial peroneal nerve).[82] Naloxone antagonized both the increase in the threshold for antidromic activation and the decrease in the threshold produced by stimulation of the superficial peroneus nerve.[82,83]

All these data are compatible with the gate control theory,[202] if supposing that the decrement of DRP V reflects the hyperpolarization of non-nociceptive afferents and the nociceptive afferents themselves are depolarized, but this depolarization is concealed by the hyperpolarization of non-nociceptive afferents. Actually Chin et al.[221] reported that small doses of morphine increased the DRP V if the high threshold afferents were stimulated.

Consequently, primary afferent depolarization of non-nociceptive afferents might be the mode of action of opioids in the spinal cord. Otherwise, the lack of any effect upon naloxone administration alone,[84] indicates that this mechanism is not tonically active. Finally, it is questionable that these changes in the activation threshold in the magnitude of 10 to 20% can explain the dramatic increase in pain threshold in intact animals. Only further experiments will give appropriate answers to this question.

B. Substance P and Enkephalins

Finally a third model deserves special attention, since all the above-mentioned data are compatible with the mechanism proposed by Jessel and Iversen.[185] These authors[185] demonstrated in vitro the inhibition of substance P liberation from superfused slices of spinal trigeminal nucleus upon adding morhpine, β-EP, or DALA to the incubation medium. The same process has recently been demonstrated also in vivo.[191] As shown in Figure 3, these authors suppose that the opioid peptides, liberated from interneurones located in the substantia gelatinosa inhibit the release of substance P from small-diameter primary afferents.[185] As mentioned previously, excitation by substance P of neurones in the various parts of the brain, especially in the dorsal horn is a well-established fact.[185,189,190,224] Substance P is particularly concentrated in the terminals of sensory neurones synapsing in the dorsal horn,[187] and its distribution especially in the spinal cord and in the lower brain stem is rather similar to that of the enkephalins (Chapter 3, Volume III). Thus, substance P seems to be the specific excitant of the dorsal horn cells which respond specifically to noxious stimuli.[188]

As morphine does not antagonize the firing induced by substance P,[188,192] they cannot act on the same receptors. Supposing, however, that substance P is one of the physiological transmitters in the first synapses of the nociceptive pathways,[185] the opiates might act via inhibition of its release. The previously discussed hyperpolarization of primary afferents by opioids and the high level of substance P in just those layers of the spinal cord which are rich also in opioid peptides confirm such a hypothesis.

Let us mention here another theoretically possible explanation according to enkephalins should inhibit while substance P facilitate the liberation of a third transmitter in the sensory pathways. In mouse vas deferens the inhibition by Met-E of potassium induced release of noradrenaline was found counteracted by substance P, while giving the latter alone, the noradrenaline release was facilitated.[225]

Such an explanation might be attractive for those who regard all the recently discovered polypeptides only neuromodulators, which contrary to the neurotransmitters of much simpler structure do not act postsynaptically but modify the liberation of the latter. Otherwise, according to this hypothesis substance P should act as the "regular" classical neurotransmitters on the postsynaptic membrane.

The significance of interaction between substance P and opioid peptides has been further substantiated by studying the effects of *capsaicin.* Capsaicin is the active ingredient of the pungent Hungarian red pepper called "paprika", and the experiments connected with it represent a spicy subchapter in the story of pain regulating peptides, written mostly by Hungarian researchers.

Capsaicin has been known for a long time to produce irritation of the skin and mucous membranes if applied topically in acute experiments, and generally it is one of the strongest chemical irritants of natural origin used in pharmacological experiments.[186] Acute capsaicin treatment increases the firing of certain sensory afferents, but only those fibers are affected which participate in the transmission of chemically irritating or thermal stimuli, while the fibers mediating mechanical information are not involved.[226] Surprisingly, chronic capsaicin treatment (topically or parenterally) has been found to result in an insensitivity to different types of chemical or thermal irritation, while the responsivity to other types of noxious stimuli or low threshold

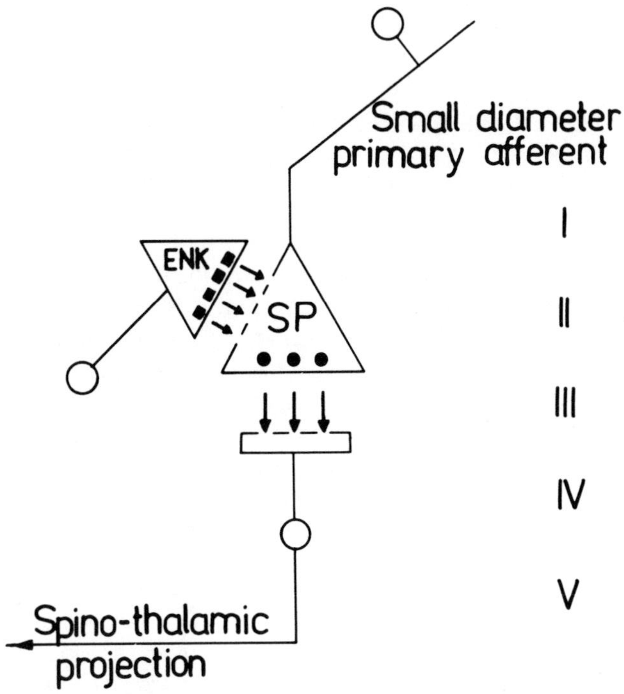

FIGURE 3. Schematic representation of a possible mechanism for opiate-induced suppression of SP release. SP is shown localized within the terminal of a small diameter afferent fiber which forms an excitatory axodendritic synapse with process of a spinal cord neurone originating in lamina IV or V and projecting rostrally. A local enkephalin-containing inhibtory interneurone (ENK), confined to laminae II and III forms a presynaptic contact on the terminal of the primary afferent. Opiate receptor sites are depicted presynaptically on the primary afferent terminal. Numbers on the right refer to the laminae of Rexed. (From Jessell, T. M. and Iverson, L. L., *Nature (London)*, 268, 549, 1977. With permission.)

mechanical stimulation remained the same.[226,227] This selective insensitivity persists for months. In these animals no inflammatory reaction could be induced by either capsaicin or other substances (e.g., by mustard oil), known to elicit neurogenic inflammation.[288]

For a long time these surprising data of Pórszász and Jancsó,[226] Szolcsányi et al.,[227] Jancsó et al.[228,229] were difficult to interpret, therefore, they did not arouse peculiar interest. However, when it turned out that capsaicin treatment resulted in depletion of the substance P in the substantia gelatinosa Rolandi,[230] the significance of these observations became suddenly obvious. According to a recent study of Yaksh et al.,[231] capsaicin produced its effect by an initial liberation of substance P from the primary afferent terminals followed by prolonged depletion of this sensory transmitter. Decrease in the concentration of substance P upon intrathecal capsaicin treatment was accompanied by analgesia.[231] This analgesia was apparent in those pharmacological assays where chemical irritation (phenylquinone writhing test) or thermal stimulation (hot plate test) were applied, but no effect was observed in assays involving noxious mechanical stimulation.[231]

Thus, the validity of the model proposed by Jessel and Iversen[185] seems to be confirmed in many ways. However, it is questionable whether this working mechanism is valid for all kinds of noxious stimulation or not. Since depletion by capsaicin of substance P results in selective elevation of pain threshold to chemically irritant and thermal stimuli but not to high threshold mechanical stimulation, also this schema might be valid only for certain kinds of noxious stimuli. Such a conclusion would, however, mean that pain is not a homogeneous sensory modality, though all kinds of noxious stimulation are thought to act via an uniform mechanism, i.e., by lesioning of the intact tissue hence by liberating endogenous algogen substances which directly stimulate the peripheral endings of C-fibers. However, it must be taken into consideration that the most reliable pharmacological methods used in measuring the analgesic activity (Chapter 2, Volume I), such as the tail-flick, hot plate, phenylquinone, or acetic acid writhing test use thermal or chemical stimuli to excite the nociceptive receptors. Consequently, those data referring to mechanical stimulation as pain stimulus may probably be regarded as irrelevant. In connection with this problem it is probably worth recalling the old clinical experience according to the superficial cutaneous sharp pain sensation and the deep dull pain are differentially affected by opiate treatment. Morphine treatment does not significantly suppress the pain sensation caused by a pinprick or pinching of the skin, while it relieves the patients of their unsupportable deep pain. As the superficial sharp pain is mediated mainly by large diameter fibers, while the pathological pain sensation is mediated mostly (or exclusively) by small diameter C-fibers, it is tempting to relate only the latter ones to the mechanisms working with substance P.

Finally let us mention here, that according to Stone and Perkins[232,233] the inhibitory effect of morphine on single unit discharges is mediated by *adenosine* liberation. These authors demonstrated that the striatal and cortical units inhibited by morphine were inhibited also by adenosine, an agent known to suppress the Ca^{++} dependent transmitter release. And recently Phillis et al.[234] demonstrated the naloxone sensitive potentiation adenosine release from the cerebral cortex in vivo upon morphine treatment.

VII. EFFECTS OF OPIOIDS ON NEURONAL ACTIVITY IN VITRO

Upon trying to clarify the intracellular mode of action of opioid peptides probably the in vitro examinations seem to be the most reliable. Studying their effects in isolated organ preparations provides certain advantages. First of all, considering the basically simple and sufficiently clarified organization of peripheral nervous elements, the direct effects on neurones under study can easily be differentiated from those mediated transsynaptically. Furthermore general anesthesia does not distort the neuronal activity, the drug concentration in the bathing medium can be held constantly, etc. Nevertheless, it cannot be contended that the previously discussed controversy concerning the mechanism of action of opioids has been solved by the in vitro experiments. On the contrary, some contradictions became even more conspicuous, as the apparent discrepancies in the experimental data cannot be ascribed to uncontrollable factors due to the in vivo experimental conditions.

The first in vitro studies were made with single neurones of *myenteric plexus of guinea pig ileum*. This plexus contains a class of neurones sensitive to narcotic drugs as proven by naloxone reversible inhibition of muscle contractions and that of the concomitant release of acetylcholine upon treatment with opioids (Chapter 1, Volume I). The neuronal organization of this plexus is more or less clarified. In addition to the acetylcholine releasing cells directly activating the cholinergic muscle fibers via neuromuscular synapses also the presence of adrenergic and serotoninergic fibers has been proven.[235] As in this preparation the nerve terminals impinging upon muscle cells act exclusively via acetylcholine liberation, the serotoninergic and adrenergic neurones may act only by modulating the release of the former transmitter. Thus, this preparation provides an excellent model for the in vitro study of opiate action.[235]

Sato et al.[236] examined the effect of opiates on single myenteric neurones. Using glass suction electrodes they found the spontaneous activity of these neurones suppressed upon morphine treatment. These observations were confirmed and extended by Dingledine and Goldstein,[235,237,238] who demonstrated the reversal by naloxone of this action and also the stereospecificity by demonstrating the effectivity of levorphanol and the inactivity of dextrorphan under the same experimental conditions.

Otherwise, these neurones also may be inhibited by norepinephrine[235] and stimulated by serotonin acetylcholine, and nicotine.[235,237,238] Consequently, these neurones may be activated by nicotinic-cholinergic and serotoninergic receptors. Morphine prevents the increase in firing elicited by serotonin, and the units not stimulated by serotonin were insensitive also to morphine and levorphanol.[237] These observations are consistent with the general view according to serotonin and norepinephrine act indirectly via facilitation and inhibition of the final acetylcholine releasing neurones supplied with opiate receptors.

The firing of these myenteric neurones is depressed by natural enkephalins and β-EP as well.[239-242] In this respect Met-E and morphine seem to be roughly equipotent, namely morphine inhibited these cells in certain studies in concentrations about 10^{-7} *M*[235,237,238] and the similar effect of Met-E was demonstrated also in the concentration of 10^{-7} *M* in another laboratory.[239] Nevertheless North and Williams[239] found Met-E about five to ten times more potent than morhine or normorphine. Also the dose-dependency of this action has been demonstrated.[235,241] Comparing the different opioid peptides, Met-E proved to be three to five times more active than Leu-E.[239,241] In the only study in which β-EP was also examined, its potency seemed to be similar to that of the Met-E.[241]

To date it is rather difficult to guess which type of receptors is responsible for these electrographic actions. Considering that Met-E and morphine are roughly equiactive in inhibition of electrical stimulation induced contractions of guinea pig ileum (Chapter 5, Volume III) and this preparation is supposed to contain primarily μ-receptors (Chapter 1, Volume III), a similar parallelism might be supposed in relation to the electrographic effects. Since Met-E is more potent suppressant of neuronal firing than morphine in this system,[239] at least partially different receptors should mediate the two kinds of actions. Consequently, these observations point to the heterogeneity of opiate receptors in the guinea pig ileum. From this point of view it would be very informative to learn about the quantitative aspects of the antagonizability by naloxone of this effect. Unfortunately no such study has been made yet. Dingledine et al.[237] used naloxone in concentrations of $3X10^{-7}$-10^{-8} *M* to reverse the effect of morphine, while North et al.[239,241,242] used even smaller concentrations against enkephalins. As the agonist actions of opioid peptides is more difficult to antagonize by naloxone on the δ-receptors than that of the opiates on μ-receptors (Chapter 1, Volume III), even this comparison of data published by different groups seems to confirm the role of μ-receptors in this effect.

However, it must be noted that all these data refer only to a subpopulation of myenteric plexus neurones, called Type 1 cells by North and Henderson.[243] The activity of neurones having no synaptic connection within the plexus, i.e., the Type 2 cells are not significantly affected by the opioids.[243]

Tolerance may develop also to this inhibitory action. In acute experiments no sign of tachyphlaxis was found in a study,[239] i.e., repeated application of Met-E did not lead to attenuation of its inhibitory potency. In another work of the same group[242] previous application to the bathing fluid of Met-E in high concentration resulted in a reduction of its hyperpolarizing action. Thus, the contingent development of acute tolerance remains to be elucidated.

It has, however, been unambiguously proven, that myenteric neurones prepared from guinea pigs repeatedly pretreated with morphine in vivo are much less sensitive to the suppressant action of opiates in vitro.[244] Adding naloxone to these preparations, a marked enhancement of firing could be observed.[244] More or less similar phenomena may be observed if the guinea pig ileum is incubated in vitro in morphine solution for periods lasting at least 24 hr.[245] These data are in good agreement with the findings of Waterfield et al.,[246] who found the inhibitory action on longitudinal muscle strip contractions caused by enkephalins or morphine attenuated if the animals were treated in vivo with morphine pellets. Accordingly, the in vitro electrographic experiments prove that the tolerance development is an intracellular process, which does not require a highly organized neronal network like the intact brain.

The next question is what is the mechanism of this effect. According to the simplest theoretical possibility the inhibition of myenteric neurones might be due to the local anesthetic effect of opioids. However, in the experiments of Kosterlitz and Wallis,[247] Dingledine and Goldstein[235] morphine did not change the size of action potentials, conduction velocity, or excitability of isolated vagus and hypogastric nerves in concentrations 10 to 100 times as high as those required to depress the transmission at the above-mentioned autonomic nerve-muscle cell junctions. Only meperidine and methadone caused a moderate decrease in the size of action potentials and the conduction velocity without modifying the resting membrane potential and excitability in the heroic dose of 0.1 mg/mℓ.[247] Though the opioid peptides have not been examined in this direction, based on the above-mentioned findings the role of a local anesthetic action may probably be excluded.

According to a further theoretical possibility, opioids might inhibit the firing of myenteric plexus neurones by preventing their transsynaptic excitation via the above-mentioned serotoninergic or nicotinergic receptors. The longitudinal muscle strip of guinea pig ileum was found to retain its antagonizability by opiates even after blocking or desensitizing these receptors by hexamethonium or large doses of serotonin, respectively.[238] Moreover, the spontaneous activity of these cells could be suppressed by opiates also in Ca^{++} free Ringer solution.[238] As elimination of Ca^{++} ions from the bathing medium results in a complete blockade of synaptic transmission, this finding proves that the site of action of opioids must be sought at the myenteric cells and not proximally to them, i.e., neither an increase in the release of inhibitory transmitters nor a decrease that of the excitatory transmitters may explain this finding.

As opioids seem to exert their depressant effect directly on the myenteric neurones, more reliable information may be hoped upon recording their action by intracellular microelectrodes.

The first such study was carried out by North and Henderson.[243] In their experiments morphine did not modify either the resting membrane potential or the excitatory postsynaptic potentials elicited by single focal stimulus via microelectrodes located close to the cells examined. In the same study,[243] contrary to the previously mentioned data obtained by extracellular electrodes,[235,237] serotonin depressed the excitatory postsynaptic potentials and the acetylcholine release and these actions could not be reversed by morphine.[243] For these reasons the authors concluded that the inhibitory action of opioids (and the opposite effect of serotonin) occurred at the varicose cell process, from which acetylcholine was released toward the muscle cell.[243] Of course such a postulation includes a further one, namely that the conventional extracellular electrodes record the spike activity of these varicose cell processes and not the soma activity. In a short time the same group[248,249] partially changed its previous postulate, as using normorphine instead of morphine in much smaller concentration, a dose-dependent hyperpolarization has been found. Of course, not only (nor) morphine but also Met-E and Leu-E[240,242,249] were found to hyperpolarize these neurones, thus all considerations based primarily on experiments using morphine as agonist also are valid in relation of the opioid peptides. This membrane hyperpolarization was accompanied by a concomitant fall in membrane resistance and the changes in the two parameters were strictly proportional.[248] A decrease in membrane potential means decrease in the steady-state membrane potential displacement upon passing current of known intensity through electrodes, one of them located within the cells, the other extracellularly. Consequently, hyperpolarization and decrease in input resistance results *per definitionem* in suppression of spike generation. The hyperpolarizing action, which was repeatedly contested in the subsequent studies, would provide a satisfactory explanation for many electrographic effects of opioid peptides. The decrease both in the spontaneous firing and activation in response to electrical or noxious stimulation or to iontophoretic application of acetylcholine, glutamate, or aspartate is easy to interpret if there is a general decrease in the excitability of neurones due to hyperpolarization. Moreover, if opioids are supposed to act directly by suppression of acetylcholine release, it is rather difficult to explain why the inhibitory potency of opioids depends on the frequency and intensity of electrical stimulation (Chapter 1, Volume I). However, if they act by hyperpolarization, it is easy to understand that the decrease in excitability upon opioid treatment can be more easily overcome by high frequency and intensity stimulation.

Interestingly, enkephalins and β-EP[241] and morphine[238] maintained their inhibitory potencies also in medium completely free of Ca^{++} ions, though the spontaneous activity of myenteric neurones increased upon omitting from the bathing solution these bivalent cations which play such an enormous role in transmitter release. On the other hand, lowering the K^+ or Cl^- concentration in the bathing solution decreased the ability of opiates to depress the spontaneous firing of myenteric neurones.[250] Hence the possibility arises that opiates act via facilitation of Na^+/K^+ exchange, which results in hyperpolarization and decreased excitability of the cells. In the same study,[250] however, perfusion of the tissue with ouabain did not alter the inhibitory potency of morphine. Thus, the blockade by ouabain of ATPase did not modify the opiate action, which shows that changes in the conductance of monovalent cations can not be the primary event in their mechanism of action.

The previous observation on the role of Ca^{++} ions is, however, not surprising if regarding the well-known increase in neuronal excitability upon lowering the extracellular Ca^{++} concentration. The finding that enkephalin is active also in the absence of extracellular Ca^{++}[241] has, however, certain implications. According to Williams and North[241] also this fact proves that the inhibition of firing is not mediated by virtue of altered (decreased) transmitter release. The inhibition of neuronal activity cannot be ascribed to a decrease or increase of transmembranal Ca^{++} flux. These authors[241] are certainly correct upon stating that the presence of extracellular Ca^{++} ions is not necessary for the inhibitory effects of opiates. The validity of this statement is, however, even broader. As discussed in Chapter 2, Volume I; Chapter 1, Volume II; and Chapter I, Volume III, Ca^{++} ions antagonized the characteristic actions of opioids both in vitro and in vivo. Considering that Ca^{++} influx is a very important step in the depolarization of the excitable membrane of neurones or muscle cells, the increased potency of opioids to elicit hyperpolarization upon omitting these ions, seems to be self-evident.

The question is, however, whether this finding may be regarded as experimental proof of the supposition that hyperpolarization *per se* is the central event in the action of opioids.

Apparently these questions have been more or less sufficiently answered by the same group. In a recent study North et al.[242] reported that addition of Met-E or Leu-E to the perfusing solution hyperpolarized the plexus myentericus cells, whereas applying them by microiontophoresis, very frequently a depolarization of soma membrane was observed. This depolarization was not modified by naloxone, consequently the phenomenon has nothing to do with the opiate receptors.[242] However, by electrophoretic application of enkephalins to sites away from the soma, i.e., close to the varicose cell processes, hyperpolarization was induced.[242] Consequently, the primary site of action within the neurones cannot be the soma itself. The authors[157,242] suppose that opioids inhibit the acetylcholine release via hyperpolarization of those sites on the cellular processes from which acetylcholine is released. This hyperpolarization results in propagation block or an increase in conductance (i.e. a decrease in input resistance), which gives rise to shunting of the action potentials.[157,242]

As seen above, the detailed microelectrophysiological analysis only confirmed some previous suggestions[238] according to opioids block the impulse invasion into the terminal varicosities of the cholinergic neurones.

The plexus myentericus of guinea pig ileum is not the only organ preparation suitable for studying the effects of opioids in vitro. Also the isolated *mouse vas deferens* preparation may give valuable information, where noradrenaline is the transmitter in the neuromuscular synapses. So far no data are known on the electrographic effects of opioids in this preparation. Nevertheless, the data on the effects of morphine suggest that the working mechanism of opiates in this organ is similar to that followed up in the guinea pig ileum. Opiates were repeatedly found to diminish dose-dependently the excitatory junction potentials recorded by intracellular microelectrodes placed into the effector muscle cells.[243,251,252] The action is naloxone sensitive,[243] stereospecific,[252] and prone to tolerance development.[251] As opiates did not change the resting membrane potentials and input resistance of muscle cells[251,252] nor their sensitivity to exogenous noradrenaline,[252] a presynaptic site of action may be suggested. Thus, the reduction by opiates of excitatory junction potentials might be due to the inhibition of noradrenaline release from the nerve varicosities just in the case of myenteric cells, since recently Ca^{++} was reported to reverse to morphine-induced inhibition of excitatory junction potentials.[253] Since the guinea pig ileum contains opiate receptors of predominantly μ-type, while those of the vas deferens are mostly of δ-type, it would be very interesting to learn about the action of opioids peptides too.

Pepper and Henderson[254] working with *brain slice preparation* observed dose-dependent and naloxone reversible hyperpolarization of neuronal membrane of *locus coeruleus* cells upon normorphine treatment. This hyperpolarization was associated with an increase in membrane conductance.[254]

Another group[255] observed inhibition by FK 33-824 of the spontaneous firing of *paraventricular* neurones in hypothalamic slices. This inhibitory action presumably explains the suppression of vasopressin release from the posterior pituitary upon treatment with opioids (Chapter 5, Volume II).

Isolated *frog sympathetic ganglion* represents a further preparation where the mechanism of action of opioids has been studied. Wouters and van den Bercken[256,257] observed the hyperpolarization of the postsynaptic neurones in this ganglion and a decrease in the amplitude of slow inhibitory postsynaptic potentials (IPSP) upon treatment with Met-E, DALA, and morphine. According to these authors[256,257] opioids hyperpolarized both sides of the synapses examined; the hyperpolarization of postsynaptic membrane has been recorded directly, while the similar process on the presynaptic side explained the attenuation of transmitter release and hence the diminution of the IPSP.[256,257]

Beside the myenteric plexus preparation, the most valuable information has probably been obtained upon studying the effect of opioids on *cultured mouse spinal neurones*. Explanting spinal cord cross-sections with attached dorsal root ganglia from fetal mice onto collagene-coated glass slides and incubating them in standard culture medium, a very useful in vitro preparation may be obtained.[258,259] This tissue culture contains the dorsal root ganglia, i.e., the primary cells of the sensory pathways and the dorsal horn region of spinal cord, i.e., the second sensory neurones and their synapses. As shown by Hiller et al.,[260] homogenates made of such cultures exhibit stereospecific binding of labeled diprenorphine, i.e., they contain opiate receptors. These opiate receptors are located mostly on the neuritic outgrowth of dorsal root ganglion cells, i.e., on the presynaptic terminals of primary afferents.[260]

Electrical stimuli applied to the dorsal root ganglia in this preparation evoked negative slow-wave potentials with a latency of 2 to 3 msec in the dorsal horn region of the spinal cord explant.[258] This sensory evoked potential could be antagonized by morphine, etorphine, and levorphanol but not by dextrorphan, the formers' effect being naloxone sensitive.[258] Thus the preparation meets the main criteria of specific effects mediated by opiate receptors, since development of tolerance has also been observed upon chronic exposure to morphine.[258] The attenuation of opioid-induced inhibition of evoked potentials upon raising the Ca^{++} level in the bathing medium is a further evidence of the reliability of this preparation in measurement of typical opiate-like activity.[259] On the other hand, the sensory evoked potentials were enhanced upon adding naloxone proving that in this system not only opiate receptors but also their endogenous ligands must be present.

The inhibitory potencies of different opioid peptides were compared in this preparation by Crain et al.[259] As shown in Table 2 morphine, α-endorphin, and β-EP were found about 30, 10, and 100 times more potent than the natural enkephalins. The potencies of the two known strongest enkephalin analogues, i.e., those of D-Met,Pro-EA, and FK 33-824 exceeded 2000 and 5000 times the activity of their natural ancestors. These quantitative data are in good agreement with their in vivo analgesic potencies upon i.c.v. application, and the relative in vitro potencies measured in guinea pig ileum but not with those found in the mouse vas deferens preparation (Chapter 5, Volume III).

From these correlations summarized in Table 2, some interesting conclusions may be drawn.

- Spinal cord tissue culture is the most reliable known in vitro system in prediction of the in vivo analgesic efficacy (see also Chapter 5, Volume III).
- The δ-receptors present in high concentration in mouse vas deferens have not much to do with the analgesic effect of opioids.
- If the "classical" μ-receptors are supposed to mediate the analgesic effects of opiates and opioid peptides as well, the same receptors must be located on the efferent terminals of dorsal root ganglion cells.

The data obtained in this tissue culture preparations are "irregular" only in two cases; the two natural enkephalins were found equiactive, though[259] Leu-E is generally weaker in those preparations which contain mainly μ-receptors (Chapter 5, Volume III). And α-endorphin was found only 10 times less potent than β-EP,[259] though the analgesic activity of the formers is weaker at least three to four orders of magnitude (Chapter 3, Volume II).

MacDonald and Nelson[261] using a very similar preparation, stimulated by microelectrodes the dorsal root ganglions cells and recorded the excitatory postsynaptic potentials (EPSP) in the spinal cord neurones. Etorphine applied iontophoretically into regions of contact between dorsal root ganglion and spinal cord cells, depressed the monosynaptic EPSPs. The degree of depression was proportional to the current strength and could easily be reversed by naloxone.[261] Supposing that the transmitter release was quantal and the distribution of Poisson type, based on their data the authors supposed that only the number of released quanta decreased but the mean quan-

Table 2
COMPARISON OF DEPRESSANT POTENCIES OF OPIOID PEPTIDES IN SPINAL CORD TISSUE CULTURE WITH THEIR ANALGESIC EFFECTS[a]

Structure	Relative potencies in: Suppression of evoked response in spinal cord tissue culture[b]	Relative potencies in: Analgesia testing upon i.c.v. treatment[c]
Met-E	1	<0.01
Leu-E	1	<0.01
(D-Ala²)-Met-E	10—100	3—23
(D-Ala²D-Leu⁵)-E	500	≈30
(D-Ala¹)-Met-E	<<1	<<0.01
D-Met,Pro-EA	1000—2000	60—80
FK 33-824	2000—5000	1000
β-EP	100	7—33
α-endorphin	10	<0.01
Morphine	33	1

[a] Reproduced with small corrections from the paper of Crain et al.[259] The courtesy of Dr. Crain is gratefully acknowledged.

[b] Data published by Crain et al.[259]

[c] For data see Chapter 3, Volume II and Crain et al.[259]

tal content did not.[261] Consequently, etorphine should act via decreasing the number of synaptosomes released, i.e., presynaptically.

The same group demonstrated that Leu-E attenuated the glutamate-induced depolarization of spinal cord neurones in the same preparation.[262] As glutamate exerts its effect postsynaptically, any interference by enkephalins with the glutamate-induced depolarization might indicate a postsynaptic action. Nevertheless, a detailed data analysis revealed that Leu-E contrary to the conventional neurotransmitters did not compete for the glutamate receptors, nor did it modify the membrane conductance, i.e, iontophoretic application of enkephalin attenuated the glutamate response but did not alter the ion-current measured during voltage clamping. A decrease in excitability of postsynaptic membrane also could be demonstrated when the ability of neurones to generate action potentials either by suprathreshold constant current or by abrupt cessation of hyperpolarizing current pulses was measured.[263]

These experiments do not exclude the possibility of a postsynaptic mode of action, but the mechanism of this effect of enkephalin on the postsynaptic membrane is different from that of the classical neurotransmitters.

Of course the in vitro recording may be useful sometimes also in the analysis of certain supracellular phenomena, i.e., in elucidation of hidden intercellular connections. Working on *hippocampal slices* Lee et al.[264] demonstrated that the inhibitory interneurones were depressed by potent enkephalin derivatives, while an inactive analogue had no effect. Thus, the previously mentioned hypotheses,[77,94] according to excitation of the hippocampal pyramidal cells is due to inhibition of otherwise tonically active inhibitory units, obtained final confirmation by an in vitro study.[264]

And recently Nicoll et al.[114] demonstrated that DALA also in *hippocampal slices* facilitated the pyramidal cells via suppression of the inhibitory postsynaptic potentials, while contrary to another observation[265] the excitatory postsynaptic potentials were not found affected. A similar suppression of dendrodendritic inhibition and blockade of presynaptic inhibition have been observed in the isolated turtle olfactory bulb preparation and in the isolated spinal cord of the frog, respectively.[114] Thus just these in vitro preparations helped to clarify the basic mode of action of opiates, i.e., the *presynaptic inhibition of transmitter release,* which may result in inhibition, disinhibition (hippocampus), or blockade of transmission (spinal cord) depending on the local neuronal organization (see also Table 3).

Also the interaction of opioid peptides with other neurotransmitters or neuromodulators may be studied in vitro. Culturing *chicken dorsal root ganglia,* Mudge et al.[267] confirmed and extended the previously mentioned in vivo data on the inhibition by opioids of substance P liberation. In these experiments Met-E and DALA inhibited the depolarization of cultured dorsal root ganglion cells elicited by adding of K^+ solution.[267] Analyzing the shape of action potentials upon enkephalin treatment, a decrease in the duration of the portion due to inward Ca^{++} current was observed.[267] As neither the resting membrane potential nor the membrane conductance were found modified, the specific action of opioids was obviously related to the blockade of Ca^{++} movement.[267] Needless to stress that this conclusion is in good agreement with the data on the antagonism by Ca^{++} of in vivo (Chapter 2, Volume I) and in vitro (Chapter 1, Volume II; Chapter 1, Volume III) effects of opiates. The question remains whether this decrease in Ca^{++} conductance is the cause or the consequence of opiate specific actions.

Finally let us mention here that the inhibitory influence on firing by opioids was related to their well-known inhibitory action on adenylate cyclase stimulated by prostaglandins (Chapter 2, Volume I). As mentioned above, adenosine was also reported to inhibit the firing of striatal and cortical units.[232,233] In a recent study[269] on myenteric plexus neurones, elevation of the intracellular content of cyclic nucleotides by PGE_2, dibutyryl-cAMP, or by a phosphodiesterase inhibitor did not modify the ability of morphine to attenuate the unit firing. Therefore, solely a reduction of intracellular cAMP content can not explain the effects of opioids.

VIII. CONCLUSIONS

In spite of the above-discussed controversies, some conclusions are probably justified.

1. The opiate-sensitive neurones are generally inhibited by the opioid peptides. This effect explains the general depressant action of opioids. Presumably μ-receptors mediate this effect. The occasionally observed facilitation of unit activity is due to suppression of inhibitory interneurones.
2. Opioids induce seizure activity in high dosage. This effect is presumably mediated by δ-receptors and appears only at pharmacological dose level. This phenomenon has scarcely any physiological significance.
3. Endogenous opioid peptides are involved in the physiological regulation (suppression) of sensory transmission especially in relation of noxious stimuli.
4. Their sites of action is multiple, but the first synapses of sensory pathways may have critical importance. At these sites opiates may take part in "gating" of the noxious sensory impulses.
5. Opioids exert their action primarily by inhibition of transmitter release, i.e., by presynaptic inhibition.
6. This presynaptic inhibition may be brought about by a direct action on the primary sensory afferent fibers and indirectly by facilitation of intraspinal and brain stem-spinal descending pathways which activate inhibitory interneurones connected to the primary afferent neurones.
7. The presynaptic inhibitory action is realized by hyperpolarization and/or decreased conductance in fine nerve terminals. Opioids probably also directly inhibit the postsynaptic depolarization, but the significance of this latter action is questionable, and its mechanism is different from that of the classical neurotransmitters.

Table 3
EFFECTS OF OPIOID PEPTIDES AND NALOXONE ON NEURONAL ACTIVITY IN VITRO

Treatment	Species	Structure	Mode of recording	Effect	Reversal by opiate antagonists	Ref.
Met-E, Leu-E	Guinea pig	Myenteric plexus neurones	Extracellularly	Decrease in spontaneous activity	Yes	239
Met-E, Leu-E	Guinea pig	Myenteric plexus neurones	Intracellularly	Increase in resting membrane potential and fall in membrane resistance	Yes	240
Met-E, Leu-E, β-EP	Guinea pig	Myenteric plexus neurones	Extracellularly	Inhibition of spontaneous firing in Ca^{++} free medium	Yes	241
Met-E, Leu-E	Guinea pig	Myenteric plexus neurones	Intracellularly	Hyperpolarization and fall in input resistance if adding the enkephalins to the bathing medium, but no effect if applying them iontophoretically into the cell soma	Yes	242
Met-E	Mouse	Vas deferens	Intracellularly	Decrease in the amplitude of the excitatory junction potentials	Yes	266
Met-E, DALA	Chicken embryo	Dorsal root ganglion	Intracellularly	Decrease in spike duration of the action potentials; no change interesting membrane potential or membrane conductance	Yes	267
Leu-E	Fetal mouse	Spinal tissue culture	Intracellularly	Inhibition of depolarization induced by glutamate, glycine or GABA; blockade of action potentials induced by electrical stimulation; no change in membrane potential	Yes	262 263
Met-E, Leu-E, DALA, (D-Ala², D-Leu⁵)-E, D-Met,Pro-EA, FK 33-824, β-EP, α-endorphin	Fetal mouse	Spinal cord tissue culture (dorsal horn region)	Extracellularly	Depression of evoked potentials elicited by stimulation of the dorsal root ganglia	Yes	259

DALA	Frog	Isolated spinal cord	Extracellularly	Inhibition of polysynaptic reflex response to dorsal root stimulation	Yes	268
Naloxone	Frog	Isolated spinal cord	Extracellularly	Facilitation of polysynaptic reflex response to dorsal root stimulation	—	268
Met-E, DALA	Frog	Paravertebral sympathetic ganglion	Sucrose gap technique	Hyperpolarization of ganglionic neurones; depression of inhibitory postsynaptic potentials	Yes	256 257
DALA, Met-E	Frog	Isolated spinal cord	Sucrose gap technique	Inhibition of dorsal root potentials elicited by ventral root stimulation; hyperpolarization of the primary afferents; blockade of presynaptic inhibition	Yes	114
DALA	Turtle	Isolated olfactory bulb	Intracellularly	Suppression of inhibitory postsynaptic potentials evoked ortho-or antidromically	Yes	114
DALA, Met-E	Rat	Hippocampal slice preparation (pyramidal cell layer)	Extracellularly	Augmentation of the orthodromically induced pyramidal population spike	Yes	114
DALA, Met-E	Rat	Hippocampal slice preparation (pyramidal cell layer)	Intracellularly	Multiplication of orthodromically induced action potentials; suppression of inhibitory postsynaptic potentials;	Yes	114
FK 33-824	Rat	Hippocampal slice	Extracellularly	Potentiation of field potentials evoked by stimulation of Schaffer collaterals	Not mentioned	265
FK 33-824	Rat	Hippocampal slice	Intracellularly	Increase in the amplitude of the excitatory postsynaptic potentials, no change in the membrane potential	Not mentioned	265
(D-Ala^2)-Met-E, Met-E	Rat	Hippocampal slice (pyramidal cells)	Extracellularly	Increase in the amplitude of the population spikes, no change in the excitatory postsynaptic potentials	Yes	94
N-acetyl-(D-Ala^2, Lys^3)-Met-E-amide	Rat	Hippocampal slice (pyramidal cells)	Extracellularly	No effect	—	94

Table 3 (continued)
EFFECTS OF OPIOID PEPTIDES AND NALOXONE ON NEURONAL ACTIVITY IN VITRO

Treatment	Species	Structure	Mode of recording	Effect	Reversal by opiate antagonists	Ref.
(D-Ala^2)-Met-E	Rat	Hippocampal slice (inhibitory interneurones)	Extracellularly	Depression of excitability	Yes	264
N-acetyl-D-Ala^2, Lys^3)-Met-E-amide	Rat	Hippocampal slice (inhibitory interneurones)	Extracellularly	No effect	—	264
(D-Ala^2)-Met-E (D-Ala^2,D-Leu^5)-E	Guinea pig	Pontin slices (locus coeruleus neurones)	Intracellularly	Hyperpolarization and increase in membrane conductance	Yes	254
FK 33-824	Rat	Hypothalamic slices (paraventicular neurones)	Extracellularly	Decrease in spontaneous firing	Yes	255

REFERENCES

1. **Wikler, A.**, Sites and mechanisms of action of morphine and related drugs in the central nervous system, *Pharm. Rev.*, 2, 435, 1950.
2. **Havlicek, V., LaBella, F., Pinsky, C., Childiaeva, R., and Friesen, H.**, Behavioral and EEG responses to α-endorphin intracerebroventricularly in rats, in *Proc. Int. Narcotic Res. Conf.*, van Ree, J. M. and Terenius, L., Eds., Elsevier/North-Holland, Amsterdam, 1978, 423.
3. **Albus, K. and Herz, A.**, Inhibition of behavioral and EEG activation induced by morphine acting on lower brain-stem structures, *Electroenceph. Clin. Neurophysiol.*, 33, 579, 1972.
4. **Corrado, A. P. and Longo, V. G.**, An electrophysiological analysis of the convulsant action of morphine, codeine and thebaine, *Arch. Int. Pharmacodyn.*, 132, 255, 1961.
5. **Navarro, G. and Elliott, H. W.**, The effects of morphine, morphinone and thebaine on the EEG and behavior of rabbits and cats, *Neuropharmacology*, 10, 367, 1971.
6. **Gangloff, H. and Monnier, M.**, The topical action of morphine, levorphanol (levorphan) and the morphine antagonist levallorphan on the unanesthetized rabbit's brain, *J. Pharmacol. Exp. Ther.*, 121, 78, 1957.
7. **Leimdorfer, A.**, An electroencephalographic analysis of the action of amidone, morphine and strychnine on the central nervous system, *Arch. Int. Pharmacodyn. Ther.*, 76, 153, 1948.
8. **Colasanti, B. and Khazan, N.**, Agonistic properties of narcotic analgesics and antagonists on the electroencephalogram and behavior in the rat and their reversal by naloxone, *Neuropharmacology*, 12, 619, 1973.
9. **Tortella, F. C., Cowan, A., and Adler, M. W.**, EEG and behavioral effects of ethylketocyclazocine, morphine and cyclazocine in rats; differential sensitivities towards naloxone, *Neuropharmacology*, 19, 845, 1980.
10. **Colasanti, B. and Khazan, N.**, Antagonism of the acute electroencephalographic and behavioral effects of morphine in the rat by depletion of brain biogenic amines, *Neuropharmacology*, 12, 463, 1973.
11. **Roubicek, J., Zaks, A., and Freedman, A. M.**, EEG changes produced by heroin and methadone, *Electroenceph. Clin. Neurophysiol.*, 27, 667, 1969.
12. **Colasanti, B. K.**, Involvement of brain biogenic amines in the electroencephalographic and behavioral effects of morphine in post-addict rats, *Neuropharmacology*, 16, 235, 1977.
13. **Székely, J. I., Dunai-Kovács, Z., Miglécz, E., and Tarnawa, I.**, The *in vivo* pharmacology of (D-Met^2,Pro^5)-enkephalinamide, in *Endorphins '78* Gráf, L., Palkovits, M., and Rónai, A. Z., Eds., Publishing House of the Hungarian Academy of Science, Budapest, Hungary, 1973, 319.
14. **Sawyer, C. H., Critchlow, B. V., and Barraclough, C. A.**, Mechanism of blockade of pituitary activation in the rat by morphine, atropine and barbiturates, *Endocrinology*, 57, 345, 1955.
15. **Wikler, A. and Altschul, S.**, Effects of methadone and morphine on the electroencephalogram of the dog, *J. Pharmacol. Exp. Ther.*, 98, 437, 1950.
16. **Teitelbaum, H., Blosser, J., and Catravas, G.**, Bilateral electroencephalographic response and unilateral tolerance to unilateral intracerebral morphine injections, *Nature (London)*, 260, 158, 1976.
17. **Green, J. D. and Arduini, A. A.**, Hippocampal electrical activity in arousal, *J. Neurophysiol.*, 17, 533, 1954.
18. **Gibbs, F. A. and Maltby, G. L.**, Effects on the electrical activity of the cortex of certain depressant and stimulant drugs-barbiturates, morphine, caffeine, benzedrine and adrenaline, *J. Pharmacol.*, 78, 1, 1943.
19. **Andrews, H. L.**, Brain potentials and morphine addiction, *Psychosom. Med.*, 3, 399, 1941.
20. **Andrews, H. L.**, Changes in the electroencephalogram during a cycle of morphine addiction, *Psychosom. Med.*, 5, 143, 1943.
21. **Silvestrini, B. and Longo, V. G.**, Selective activity of morphine on the 'EEG arousal reaction' to painful stimuli, *Experientia*, 12, 436, 1956.
22. **Chin, J. H. and Domino, E. F.**, Effects of morphine on brain potentials evoked by stimulation of the tooth pulp of the dog, *J. Pharmacol. Exp. Ther.*, 132, 74, 1961.
23. **Freye, E. and Arndt, J. O.**, Perfusion of the fourth cerebral ventricle with fentanyl induces naloxone-reversible bradycardia, hypotension, and EEG synchronisation in conscious dogs, *Naunyn-Schmiedeb. Arch. Pharmacol.*, 307, 123, 1979.
24. **Yeung, J. C., Yaksh, T. L., and Rudy, T. A.**, Effect on the nociceptive threshold and EEG activity in the rat of morphine injected into the medial thalamus and the periaqueductal gray, *Neuropharmacology*, 17, 525, 1978.
25. **Pickworth, W. B. and Sharpe, L. G.**, EEG-behavioral dissociation after morphine- and cyclazocine-like drugs in the dog: further evidence for two opiate receptors, *Neuropharmacology*, 18, 617, 1979.
26. **Urca, G., Frenk, H., Liebeskind, J. C., and Taylor, A. N.**, Morphine and enkephalin: analgesic and epileptic properties, *Science*, 197, 83, 1977.

27. **Frenk, H., Urca, G., and Liebeskind, J. C.**, Epileptic properties of leucine- and methionine-enkephalin: comparision with morphine and reversibility by naloxone, *Brain Res.*, 147, 327, 1977.
28. **Tortella, F. C., Moreton, J. E., and Khazan, N.**, Electroencephalographic and behavioral effects of D-Ala²-methionine-enkephalinamide and morphine in the rat, *J. Pharmacol. Exp. Ther.*, 206, 636, 1978.
29. **Aloisi, F., de Carolis, S. A., and Longo, V.**, EEG and behavioral effects of morphine, enkephalins and derivatives administered into the lateral cerebral ventricles of rats and rabbits, *Pharm. Res. Commun.*, 12, 467, 1980.
30. **Henriksen, S. J., Bloom, F. E., McCoy, F., Ling, N., and Guillemin, R.**, β-Endorphin induces nonconvulsive limbic seizures, *Proc. Natl. Acad. Sci. USA*, 75, 5221, 1978.
31. **Teitelbaum, H.**, Cross-tolerance between morphine and cholinergic blocking agents microinjected into anterior amygdala, *Nature (London)*, 267, 452, 1977.
32. **Frenk, H., McCarty, B. C., and Liebeskind, J. C.**, Different brain areas mediate the analgesic and epileptic properties of enkephalin, *Science*, 200, 335, 1978.
33. **Reynolds, D. V.**, Surgery in the rat during electrical analgesia induced by focal brain stimulation, *Science*, 164, 444, 1969.
34. **Mayer, D. J., Wolfle, T. L., Akil, H., Carder, B., and Liebeskind, J. C.**, Analgesia from electrical stimulation in the brainstem of the rat, *Science*, 174, 1351, 1971.
35. **Taylor, D., Hoffer, B., Zieglgänsberger, W., Siggins, G., Ling, N., Seiger, A., and Olson, L.**, Opioid peptides excite pyramidal neurons and evoke epileptiform activity in hippocampal transplants in oculo, *Brain Res.*, 176, 135, 1979.
36. **Elazar, Z., Motles, E., Ely, Y., and Simantov, R.**, Acute tolerance to the excitatory effect of enkephalin microinjections into hippocampus, *Life Sci.*, 24, 541, 1979.
37. **Dzoljic, M. R., Lely v.d., A. J., and Mourik, v., J. B. A.**, Enkephalin-induced myoclonic twitches blocked by ergometrine and potentiated by haloperidol, *Psychopharmacology*, 66, 111, 1979.
38. **Schreiber, R. A.**, The effect of naloxone on audiogenic seizures, *Psychopharmacology*, 66, 205, 1979.
39. **George, R., French, E. D., and Vasquez, S.**, Naloxone-induced seizures and its reversal by morphine in rats, rabbits and cats, *Pharmacologist*, 21, 157, 1979.
40. **De Wied, D., Bohus, B., van Ree, J. M., and Urban, I.**, Behavioral and electrophysiological effects of peptides related to lipotropin (β-LPH), *J. Pharmacol. Exp. Ther.*, 204, 570, 1978.
41. **Volavka, J., James, B., Reker, D., Mallya, A., Cho, D., and Pevnick, J.**, EEG and other effects of naltrexone and heroin in man, *Pharmakopsychiat, Neuro-Psychopharmacol.*, 12, 79, 1979.
42. **Frederickson, R. C. A. and Norris, F. H.**, Enkephalin-induced depression of single neurons in brain areas with opiate receptors-antagonism by naloxone, *Science*, 194, 440, 1976.
43. **Satoh, M., Zeiglgänsberger, W., and Herz, A.**, Interaction between morphine and putative excitatory neurotransmitters in cortical neurones in naive and tolerant rats, *Life Sci.*, 17, 75, 1975.
44. **Palmer, M. R., Morris, D. H., Taylor, D. A., Stewart, J. M., and Hoffer, B. J.**, Electrophysiological effects of enkephalin analogs in rat cortex, *Life Sci.*, 23, 851, 1978.
45. **Korf, J., Bunney, B. S., and Aghajanian, G. K.**, Noradrenergic neurons: morphine inhibition of spontaneous activity, *Eur. J. Pharmacol.*, 25, 165, 1974.
46. **Davies, J. and Dray, A.**, Pharmacological and electrophysiological studies of morphine and enkephalin on rat supraspinal neurones and cat spinal neurones, *Br. J. Pharmacol.*, 63, 87, 1978.
47. **Hill, R. G., Pepper, C. M., and Mitchell, J. F.**, The action of iontophoretically applied met-enkephalin on single neurones in rat brain, in *Proc. Int. Narcotic Res. Club Meeting*, Kosterlitz, H. W., Ed., North-Holland, Amsterdam, 1976, 225.
48. **Zieglgänsberger, W. and Fry, J. P.**, Actions of enkephalin on cortical and striatal neurones of naive and morphine tolerant/dependent rats, in *Proc. Int. Narcotic Res. Club Meeting*, Kosterlitz, H. W., Ed., North-Holland, Amsterdam, 1976, 231.
49. **Frederickson, R. C. A., Nickander, R., Smithwick, E. L., Shuman, R., and Norris, F. H.**, Pharmacological activity of met-enkephalin and analogues: *in vitro* and *in vivo*-depression of single neuronal activity in specified brain regions, in *Proc. Int. Narcotic Res. Club Meeting*, Kosterlitz, H. W., Ed., North-Holland, Amsterdam, 1976, 239.
50. **Mallari, C. G. and Klemm, W. R.**, Morphine-induced regional and dose-response differences on unit impulse activity in decerebrate rats, *Psychopharmacology*, 56, 261, 1978.
51. **Zieglgänsberger, W., Fry, J. P., Herz, A., Moroder, L., and Wünsch, E.**, Enkephalin-induced inhibition of cortical neurones and the lack of this effect in morphine tolerant/dependent rats, *Brain Res.*, 115, 160, 1976.
52. **Nicoll, R. A., Siggins, G. R., Ling, N., Bloom, F. E., and Guillemin, R.**, Neuronal actions of endorphins and enkephalins among brain regions: a comparative microiontophoretic study, *Proc. Natl. Acad. Sci. USA*, 74, 2584, 1977.
53. **Satoh, M., Zieglgänsberger, W., Fries, W., and Herz, A.**, Opiate-agonist interaction at cortical neurones of naive and tolerant/dependent rats, *Brain Res.*, 82, 378, 1974.

54. **Hill, R. G., Pepper, C. M., and Mitchell, J. F.,** Depression of nociceptive and other neurones in the brain by iontophoretically applied met-enkephalin, *Nature (London),* 262, 604, 1976.
55. **Palmer, M. R. and Hoffer, B. J.,** Catecholamine modulation of enkephalin-induced electrophysiological responses on cerebral cortex, *J. Pharmacol. Exp. Ther.,* 213, 205, 1980.
56. **Duggan, A. W., Davies, J., and Hall, J. G.,** Effects of opiate agonists and antagonists on central neurons of the cat, *J. Pharmacol. Exp. Ther.,* 196, 107, 1976.
57. **Eidelberg, E. and Bond, M. L.,** Effects of morphine and antagonists on hypothalamic cell activity, *Arch. Int. Pharmacodyn.,* 196, 16, 1972.
58. **Frederickson, R. C. A., Norris, F. H., and Hewes, C. R.,** Effects of naloxone and acetylcholine on medial thalamic and cortical units in naive and morphine dependent rats, *Life Sci.,* 17, 81, 1975.
59. **Hill, R. G. and Pepper, C. M.,** The effects of morphine and metenkephalin on nociceptive neurones in the rat thalamus, *Br. J. Pharmacol.,* 58, 459P, 1976.
60. **Bradley, P. B., Briggs, I., Gayton, R. J., and Lambert, L. A.,** Effects of microiontophoretically applied methionine-enkephalin on single neurones in rat brainstem, *Nature (London),* 261, 425, 1976.
61. **Gent, J. P. and Wolstencroft, J. H.,** Effects of methionine-enkephalin and leucine-enkephalin compared with those of morphine on brainstem neurones in cat, *Nature (London),* 261, 426, 1976.
62. **Davies, J. and Dray, A.,** Effects of enkephalin and morphine on Renshaw cells in feline spinal cord, *Nature (London),* 262, 603, 1976.
63. **Davies, J. and Dray, A.,** Actions of enkephalin and morphine on spinal cord and brain stem neurones, *Br. J. Pharmacol.,* 58, 458, 1976.
64. **Duggan, A. W., Hall, J. G., and Headley, P. M.,** Morphine, enkephalin and the substantia gelatinosa, *Nature (London),* 264, 456, 1976.
65. **Duggan, A. W., Hall, J. G., and Headley, P. M.,** Suppression of transmission of nociceptive impulses by morphine: selective effects of morphine administered in the region of the substancia gelatinosa, *Br. J. Pharmacol.,* 61, 65, 1977.
66. **Gent, J. P. and Wolstencroft, J. H,** Actions of morphine, enkephalin and endorphin on single neurones in the brainstem, including the raphe and the periaqueductal gray, of the cat, in *Proc. Int. Narcotic Res. Club Meeting,* Kosterlitz, H. W., Ed., North-Holland, Amsterdam, 1976, 217.
67. **Forbes, J. E. and Dewey, W. L.,** The effects of narcotic antagonists on ganglionic transmission in the rat, *Life Sci.,* 19, 401, 1976.
68. **Segal, M.,** Morphine and enkephalin interactions with putative neurotransmitters in rat hippocampus, *Neuropharmacology,* 16, 587, 1977.
69. **McCarthy, P. S., Walker, R. J., and Woodruff, G. N.,** Depressant actions of enkephalins on neurones in the nucleus accumbens, *J. Physiol. (London),* 267, 40, 1977.
70. **Young, W. S. III, Bird, S. J., and Kuhar, M. J.,** Iontophoresis of methionine-enkephalin in the locus coeruleus area, *Brain Res.,* 129, 366, 1977.
71. **Bradley, P. B., Gayton, R. J., and Lambert, L. A.,** A microiontophoretic study of enkephalin and enkephalin analogues on brain stem neurones in the rat, *Br. J. Pharmacol.,* 61, 147, 1977.
72. **Bashir, A. A. and Bramwell, G.J.,** Effect of etorphine on brain stem neurones in the rat: a microiontophoretic study, *Br. J. Pharmacol.,* 61, 480p, 1977.
73. **Duggan, A.W., Hall, J. G., and Headley, P. M.,** Enkephalins and dorsal horn neurones of the cat: effects on responses to noxious and innocuous skin stimuli, *Br. J. Pharmac.,* 61, 399, 1977.
74. **Guyenet, P. G. and Aghajanian, G. K.,** Excitation of neurons in the nucleus locus coeruleus by substance P and related peptides, *Brain Res.,* 136, 178, 1977.
75. **Bell, J. A. and Martin, W. R.** The effect of the narcotic antagonists naloxone, naltrexone and nalorphine on spinal cord C-fiber reflexes evoked by electrical stimulation or radiant heat, *Eur. J. Pharmacol.,* 42, 147, 1977.
76. **Gayton, R. J., Lambert, L. A., and Bradley, P. B.,** Failure of (+)-naloxone to antagonise responses to opioid peptides, *Neuropharmacology,* 17, 549, 1978.
77. **Zieglgänsberger, W., French, E. D., Siggins, C. R., and Bloom, F. E.,** Opioid peptides may excite hippocampal pyramidal neurons by inhibiting adjacent inhibitory interneurons, *Science,* 205, 415, 1979.
78. **Woodruff, G. N. and McCarthy, P. S.,** Actions of (D-Ser^2,Met^5)-enkephalin methyl ester on neurones in rat brain, *7th Int. Congr. Pharm. Paris, 1978, Abstract,* 173.
79. **Aghajanian, G. K.,** Tolerance of locus coeruleus neurones to morphine and suppression of withdrawal response by clonidine, *Nature (London),* 276, 186, 1978.
80. **Hill, R. G. and Pepper, C. M.,** The depression of thalamic nociceptive neurones by D-ala^2,D-leu^5-enkephalin, *Eur. J. Pharmacol.,* 47, 223, 1978.
81. **Randič, M. and Miletič, V.,** Depressant actions of methionine-enkephalin and somatostatin in cat dorsal horn neurones activated by noxious stimuli, *Brain Res.,* 152, 196, 1978.
82. **Sastry, B. R.,** Morphine and Met-enkephalin effects on sural A δ afferent terminal excitability, *Eur. J. Pharmacol.,* 50, 269, 1978.

83. **Sastry, B. R.**, Presynaptic effects of morphine and methionine-enkephalin in feline spinal cord, *Neuropharmacology*, 18, 367, 1979.
84. **Hu, J. W., Dostovsky, J. O., and Sessle, B. J.**, Primary afferent depolarization of tooth pulp afferents is not affected by naloxone, *Nature (London)*, 276, 283, 1978.
85. **Polc, P.**, Effects of a synthetic enkephalin analogue on the cat spinal cord, *Experientia*, 34, 928, 1978.
86. **Guyenet, P. G., and Aghajanian, G. K.**, ACh, substance P and Met-enkephalin in the locus coeruleus: pharmacological evidence for independent sites of action, *Eur. J. Pharmacol.*, 53, 319, 1979.
87. **Fry, J. P., Zieglgänsberger, W., and Herz, A.**, Specific versus non-specific actions of opioids on hippocampal neurones in the rat brain, *Brain Res.*, 163, 295, 1979.
88. **Fry, J. P., Herz, A., and Zieglgänsberger, W.**, Non-specific actions of opioids in the rat hippocampus, *J. Physiol. (London)*, 280, 15P, 1978.
89. **Satoh, M., Akaike, A., and Takagi, H.**, Excitation by morphine and enkephalin of single neurons of nucleus reticularis paragigantocellularis in the rat: a probable mechanism of analgesic action of opioids, *Brain Res.*, 169, 406, 1979.
90. **Zieglgänsberger, W. and Tulloch, I. F.**, The effect of methionine- and leucine-enkephalin on spinal neurones of the cat, *Brain Res.*, 167, 53, 1979.
91. **Rivot, J. P., Chaouch, A., and Besson, J. M,.** The influence of naloxone on the C fiber response of dorsal horn neurons and their inhibitory control by raphe magnus stimulation, *Brain Res.*, 176, 355, 1979.
92. **Satoh, M., Kawajiri, S., Ikai, Y., and Takagi, H.**, Effects of a new opioid peptide (Tyr-Arg) on lamina V type neurons of the rabbit spinal dorsal horn, *Jpn. J. Pharmacol.*, 29, 191 p, 1979.
93. **Pomeranz, B. and Gurevich, N.**, Effects of enkephalin analogue and naloxone on cat spinal cord dorsal root potentials, *Eur. J. Pharmacol.*, 60, 307, 1979.
94. **Dunwiddie, T., Mueller, A., Palmer, M., Stewart, J., and Hoffer, B.**, Electrophysiological interactions of enkephalins with neuronal circuitry in the rat hippocampus. I. Effects on pyramidal cell activity, *Brain Res.*, 184, 311, 1980.
95. **Ono, T., Oomura, Y., Nishino, H., Sasaki, K., Muramoto, K., and Yano, I.**, Morphine and enkephalin effects on hypothalamic glucoresponsive neurons, *Brain Res.*, 185, 208, 1980.
96. **Hosford, D. A. and Haigler, H. J.**, Morphine and methionine-enkephalin: different effects on spontaneous and evoked neuronal firing in the mesencephalic reticular formation of the rat, *J. Pharmacol. Exp. Ther.*, 213, 355, 1980.
97. **Bell, J. A., Sharpe, L. G., and Berry, J. N.**, Depressant and excitant effects of intraspinal microinjections of morphine and methionine-enkephalin in the cat, *Brain Res.*, 196, 455, 1980.
98. **Biscoe, T. J., Duggan, A. W., and Lodge, D.**, Effect of etorphine, morphine and diprenorphine on neurones of the cerebral cortex and spinal cord of the rat, *Br. J. Pharmacol.*, 46, 201, 1972.
99. **Bramwell, G. J. and Bradley, P. B.**, Actions and interactions of narcotic agonists and antagonists on brain stem neurones, *Brain Res.*, 73, 167, 1974.
100. **Hill, R. G. and Pepper, C. M.**, Selective effects of morphine on the nociceptive responses of thalamic neurones in the rat, *Br. J. Pharmacol.*, 64, 137, 1978.
101. **Duggan, A. W. and Hall, J. G.**, Morphine, naloxone and the responses of medial thalamic neurones of the cat, *Brain Res.*, 122, 49, 1977.
102. **Chou, D. T. and Wang, S. C.**, Unit activity of amygdala and hippocampal neurons: effects of morphine and benzodiazepines, *Brain Res.*, 126, 427, 1977.
103. **Miyagawa, T., Shima, K., Sakurada, S., and Kisara, K.**, Effects of morphine on single unit activity of the amygdala in cats, *Jpn. J. Pharmacol.*, 28, 58P, 1978.
104. **Iwatsubo, K. and Clouet, D. H.**, Effects of morphine and haloperidol on the electrical activity of rat nigrostriatal neurons *J. Pharmacol. Exp. Ther.*, 202, 429, 1977.
105. **Lee, C. M., Wong, P. C. L., and Chan, S. H. H.**, The involvement of dopaminergic neurotransmission in the inhibitory effect of morphine on caudate neurone activities, *Neuropharmacology*, 16, 571, 1977.
106. **Kondo, Y., and Iwatsubo, K.**, Diminished responses of nigral dopaminergic neurons to haloperidol and morphine following lesions in the striatum, *Brain Res*, 181, 237, 1980.
107. **Gayton, R. J., and Bradley, P. B.**, Comparison of the effects of morphine and related substances with those of dopamine in single neurones in the rat caudate nucleus, in *Proc. Int. Narcotic Res. Club Meeting*, Kosterlitz, H. W., Ed., North-Holland, Amsterdam, 1976, 213.
108. **Iwamoto, E. T., Loh, H. H., and Way, E. L.**, Circling behavior after narcotic drugs and during naloxone-precipitated abstinence in rats with unilateral nigral lesions, *J. Pharmacol. Exp. Ther.*, 197, 503, 1976.
109. **Iwamoto, E. T. and Way, E. L.**, Circling behavior and stereotypy induced by intranigral opiate microinjections, *J. Pharmacol. Exp. Ther.*, 203, 347, 1977.

110. **Scheel-Krüger, J., Golembiowska, K., and Mogilnicka, E.,** Evidence for increased apomorphine-sensitive dopaminergic effects after acute treatment with morphine, *Psychopharmacology,* 53, 55, 1977.
111. **De la Baume, S., Patey, G., Marcais, H., Protais, P., Costentin, J., and Schwartz, J. C.,** Changes in dopamine receptors in mouse striatum following morphine treatments, *Life Sci.,* 24, 2333, 1979.
112. **Dafny, N. and Burks, T. F.,** 5-HT and morphine interaction, effects on sensory input in caudate nucleus and substantia nigra, *Neuropharmacology,* 16, 577, 1977.
113. **Kerr, F. W. L., Triplett, J. N., Jr., and Beeler, G. W.,** Reciprocal (push-pull) effects of morphine on single units in the ventromedian and lateral hypothalamus and influences on other nuclei: with a comment on methadone effects during withdrawal from morphine, *Brain Res.,* 74, 81, 1974.
114. **Nicoll, R. A., Alger, B. E., and Jahr, C. E.,** Enkephalin blocks inhibitory pathways in the vertebrate CNS, *Nature (London),* 287, 22, 1980.
115. **Hong, J. S., Wood, P. L., Gillin, J. C., Yang, H. Y. T., and Costa, E.,** Changes in hippocampal Met-enkephalin content after recurrent motor seizures, *Nature (London),* 285, 231, 1980.
116. **Urca, G. and Liebeskind, J. C.,** Electrophysiological indices of opiate action in awake and anesthetized rats, *Brain Res.,* 161, 162, 1979.
117. **Shima, K., Anezaki, K., Sakurada, S., Kisara, K., and Nakahama, H.,** Effects of morphine on single unit activity of central gray in cats, *Jpn. J. Pharmacol.,* 27, 51P, 1977.
118. **Haigler, H. J.,** Morphine: ability to block neural activity evoked by nociceptive stimulus, *Life Sci.,* 19, 841, 1976.
119. **Anderson, S. D., Basbaum, A. I., and Fields, H. L.,** Response of medullary raphe neurons to peripheral stimulation and to systemic opiates, *Brain Res.,* 123, 363, 1977.
120. **Dostrovsky, J. O.,** Raphe and periaqueductal gray induced suppression of non-nociceptive neuronal responses in the dorsal column nuclei and trigeminal subnucleus caudalis, *Brain Res.,* 200, 184, 1980.
121. **Le Bars, D., Dickenson, A. H., and Besson, J. M.,** Microinjection of morphine within nucleus raphe magnus and dorsal horn neurone activities related to nociception in the rat, *Brain Res.,* 189, 467, 1980.
122. **Dickenson, A. H., Oliveras, J. L., and Besson, J. M.,** Role du noyau raphe magnus dans l'analgesie morphinique: Etudes par microinjections intracerebrales chez le rat, *C. R. Acad. Sci. Ser. D.,* 287, 955, 1978.
123. **Proudfit, H. K.,** Reversible inactivation of raphe magnus neurones: effects on nociceptive threshold and morphine-induced analgesia, *Brain Res.,* 201, 459, 1980.
124. **Yaksh, T. L.,** Direct evidence that spinal serotonin and noradrenaline terminals mediate the spinal antinociceptive effects of morphine in the periaqueductal gray, *Brain Res.,* 160, 180, 1979.
125. **Oleson, T. D., Twombly, D. A., and Liebeskind, J. C.,** Effects of pain-attenuating brain stimulation and morphine on electrical activity in the raphe nuclei of the awake rat, *Pain,* 4, 211, 1978.
126. **Bradley, P. B. and Bramwell, G. J.,** Stereospecific actions of morphine on single neurones in the brain stem of the rat, *Neuropharmacology,* 16, 519, 1977.
127. **Bradley, P. B. and Dray, A.,** Actions and interactions of microiontophoretically applied morphine with transmitter substances on brain stem neurones, *Br. J. Pharmacol.,* 47, 642P, 1973.
128. **Bramwell, G. J. and Bradley, P. B.,** Effects of morphine on brainstem neurones in naive and chronic morphine-treated rats and effects of PCPA, *Neuropharmacology,* 17, 975, 1978.
129. **Bradley, P. B. and Dray, A.,** The effects of microiontophoretically applied morphine and transmitter substances in rats during chronic treatment and after withdrawal from morphine, *Br. J. Pharmacol.,* 51, 104, 1974.
130. **Bradley, P. B. and Dray, A.,** Morphine and neurotransmitter substances: micro-iontophoretic study in the rat brain stem, *Br. J. Pharmacol.,* 50, 47, 1974.
131. **Bradley, P. B. and Bramwell, G. J.,** A stereospecific action of morphine on brain stem neuronal activity: a microiontophoretic study, *Br. J. Pharmacol.,* 53, 462P, 1975.
132. **Goldinger, A., Müller, W. E., and Wollert, U.,** The interaction of several excitatory drugs with specific ^{3}H-strychnine binding to synaptosomal membranes of rat spinal cord, *Naunyn-Schmiedeb. Arch. Pharmacol.,* 311, R73, 1980.
133. **Segal, M.,** Serotonergic innervation of the locus coeruleus from the dorsal raphe and its action on responses to noxious stimuli, *J. Physiol. (London),* 286, 401, 1979.
134. **Bird, S. J., Atweh, S. F., and Kuhar, M. J.,** Microiontophoretic study of the effects of opiates on autoradiographically localized opiate receptors, in *Proc. Int. Narcotic Res. Club Meeting,* Kosterlitz, H. W., Ed., North-Holland, Amsterdam, 1976, 199.
135. **Bird, S. J. and Kuhar, M. J.,** Iontophoretic application of opiates to the locus coeruleus, *Brain Res.,* 122, 523, 1977.
136. **Strahlendorf, H. K., Strahlendorf, J. C., and Barnes, C. D.,** Endorphin-mediated inhibition of locus coeruleus neurons, *Brain Res.,* 191, 284, 1980.

137. **Cedarbaum, J. M. and Aghajanian, G. K.**, Activation of locus coeruleus neurons by peripheral stimuli:modulation by a collateral inhibitory mechanism, *Life Sci.*, 23, 1383, 1978.
138. **Sasa, M., Munekiyo, K., and Takaori, S.**, Morphine interference with noradrenaline-mediated inhibition from the locus coeruleus, *Life Sci.*, 17, 1373, 1975.
139. **Takagi, H., Doi, T., and Akaike, A.**, Microinjection of morphine into the medial part of the bulbar reticular formation in rabbit and rat: inhibitory effects on lamina V cells of spinal dorsal horn and behavioral analgesia, in, *Proc. Int. Narcotic Res. Club Meeting,* Kosterlitz, H. W., Ed., North-Holland, Amsterdam, 1976, 191.
140. **Fields, H. L., Basbaum, A. I., Clanton, C. H., and Anderson, S. D.**, Nucleus raphe magnus inhibition of spinal cord dorsal horn neurons, *Brain Res.*, 126, 441, 1977.
141. **Guilbaud, G., Oliveras, J. L., Giesler, G., Jr., and Besson, J. M.**, Effects induced by stimulation of the centralis inferior nucleus of the raphe on dorsal horn interneurons in cat's spinal cord, *Brain Res.*, 126, 355, 1977.
142. **Mayer, M. L. and Hill, R. G.**, The effects of intravenous fentanyl, morphine and naloxone on nociceptive responses of neurones in the rat caudal medulla, *Neuropharmacology,* 17, 533, 1978.
143. **Ayliffe, S. J. and Hill, R. G.**, The actions of opiates on the nociciceptive responses of neurones in the trigeminal nucleus caudalis of the rat, *J. Physiol. (London),* 293, 65P, 1979.
144. **Gatipon, G. B.**, Morphine and brain stem neurons: microiontophoretic study of responses to noxious stimulation in the rat, *Pharmacologist,* 18, 177, 1976.
145. **Azami, J., Roberts, M. H. T., and Wright, D. M.**, Effects of iontophoretic application of morphine and naloxone on responses of nucleus reticularis paragigantocellularis neurones to noxious stimulation in the rat, *J. Physiol. (London),* 293, 63P, 1979.
146. **Akaike, A., Shibata, T., Satoh, M., and Takagi, H.**, Analgesia induced by microinjection of morphine into, and electrical stimulation of, the nucleus reticularis paragigantocellularis of rat medulla oblongata, *Neuropharmacology,* 17, 775, 1978.
147. **Scott Mohrland, J. and Gebhart, G. E.**, Medullary reticular formation activity after PAG or microiontophoretic morphine, *Pharmacologist,* 21, 158, 1979.
148. **Kruglov, N. A.**, Effect of the morphine-group analgesics on the central inhibitory mechanisms, *Int. J. Neuropharmacol.*, 3, 197, 1964.
149. **Felpel, L. P., Sinclair, J. G., and Yim, G. K. W.**, Effects of morphine on Renshaw cell activity, *Neuropharmacology,* 9, 203, 1970.
150. **Duggan, A. W. and Curtis, D. R.**, Morphine and the synaptic activation of Renshaw cells, *Neuropharmacology,* 11, 189, 1972.
151. **Davies, J. and Duggan, A. W.**, Opiate agonist-antagonist effects on Renshaw cells and spinal interneurones, *Nature (London),* 250, 70, 1974.
152. **Lodge, D., Headley, M. P., Duggan, A. W., and Biscoe, T. J.**, The effects of morphine, etorphine and sinomenine on the chemical sensitivity and synaptic responses of Renshaw cells and other spinal neurones in the rat, *Eur. J. Pharmacol.*, 26, 277, 1974.
153. **Davies, J.**, Effects of morphine and naloxone on Renshaw cells and spinal interneurones in morphine dependent and nondependent rats, *Brain Res.*, 113, 311, 1976.
154. **Akaike, A., Kawasaki, K., Doi, T., and Takagi, H.**, Effect of morphine microinjection into the medulla oblongata on the spinal dorsal horn neuron, *Jpn. J. Pharmacol.*, 26, 119P, 1976.
155. **Davies, J. and Dray, A.**, Substance P and opiate receptors, *Nature (London),* 268, 351, 1977.
156. **Ryall, R. W. and Belcher, G.**, Substance P selectively blocks nicotinic receptors on Renshaw cells: a possible synaptic inhibitory mechanism, *Brain Res.*, 137, 376, 1977.
157. **North, R. A.**, Opiates, opioid peptides and single neurones, *Life Sci.*, 24, 1527, 1979.
158. **Jurna, I., Heinz, G., Blinn, G., and Nell, T.**, The effect of substantia nigra stimulation and morphine on α-motoneurones and the tail flick response, *Eur. J. Pharmacol.*, 51, 239, 1978.
159. **Kuschinsky, K., Ropte, H., Meseke, R., Cremer, H., and Sontag, K. H.**, Specific action of narcotics on reflex activation of rat alpha-motoneurones, *Naunyn-Schmiedeb. Arch. Pharmacol.*, 296, 249, 1977.
160. **Le Bars, D., Menétrey, D., Conseiller, C., and Besson, J. M,**. Depressive effects of morphine upon lamina V cells activities in the dorsal horn of the spinal cat, *Brain Res,*. 98, 261, 1975.
161. **Kitahata, L. M., Kosaka, A., Taub, A., Bonikos, K., and Hoffert, M.**, Laminaspecific suppression of dorsal horn unit activity by morphine sulphate, *Anesthesiology,* 41, 39, 1974.
162. **Takagi, H., Doi, T., and Kawasaki, K.**, Effects of morphine, 1-DOPA and tetrabenazine on the lamina V cells of spinal dorsal horn, *Life Sci.*, 17, 67, 1975.
163. **Toyooka, H., Hanaoka, K., Ohtani, M., Yamashita, M., Taub, A., and Kitahata, L. M.**, Suppressive effect of morphine on single-unit activity of cells in rexed lamina VII, *Anesthesiology,* 47, 513, 1977.
164. **Zieglgänsberger, W. and Bayerl, H.**, The mechanism of inhibition of neuronal activity by opiates in the spinal cord of cat, *Brain Res.*, 115, 111, 1976.

165. **Bella, D. D., Benelli, G., and Beeson, J. M.,** Viminol stereoisomers and lamina V interneurons activity: preliminary results, *Life Sci.*, 17, 73, 1975.
166. **Besson, J. M., Wyon-Maillard, M. C., Benoist, J. M., Conseiller, C., and Hamann, K. F.,** Effects of phenoperidine on lamina V cells in the cat dorsal horn, *J. Pharmacol. Exp. Ther.*, 187, 239, 1973.
167. **Le Bars, D., Guilbaud, G., Jurna, I., and Besson, J. M.,** Differential effects of morphine on responses of dorsal horn lamina V type cells elicited by A and C fiber stimulation in the spinal cat, *Brain Res.*, 115, 518, 1976.
168. **Le Bars, D., Rivot, J. P., Guilbaud, G., Menétrey, D., and Besson, J. M.,** The depressive effect of morphine on the C fibre response of dorsal horn neurones in the spinal rat pretreated or not by pCPA, *Brain Res.*, 176, 337, 1979.
169. **Dostrovsky, J. and Pomeranz, B.,** Morphine blockade of amino acid putative transmitters on cat spinal cord sensory interneurones, *Nature (London)*, 246, 222, 1973.
170. **Curtis, D. R. and Duggan, A. W.,** The depression of spinal inhibition of morphine, *Ag. Actions*, 1, 14, 1969.
171. **Piercey, M. F. and Hollister, R. P.,** Morphine fails to block the discharges evoked by intra-arterial bradykinin in dorsal horn neurones of spinal cats, *Neuropharmacology*, 16, 425, 1977.
172. **Belcher, G. and Ryall, R. W.,** Differential excitatory and inhibitory effects of opiates on non-nociceptive and nociceptive neurones in the spinal cord of the cat, *Brain Res.*, 145, 303, 1978.
173. **Satoh, M., Nakamura, N., and Takagi H.,** Effect of morphine on bradykinin-induced unitary discharges in the spinal cord of the rabbit, *Eur. J. Pharmacol.*, 16, 245, 1971.
174. **Grossmann, W. and Jurna, I.,** Depression by morphine of activity in the ventrolateral tract evoked from cutaneous A-fibers, *Eur. J. Pharmacol.*, 29, 171, 1974.
175. **Dostrovsky, J. and Pomeranz, B.,** Interaction of iontophoreticaly applied morphine with responses of interneurones in cat spinal cord, *Exp. Neurol.*, 52, 325, 1976.
176. **Hanaoka, K., Ohtani, M., Toyooka, H., Dohi, S., Ghazdi-Saidi, K., Taub, A., and Kitahata, L. M.,** The relative contribution of direct and supraspinal descending effects upon spinal mechanisms of morphine analgesia, *J. Pharmacol. Exp. Ther.*, 207, 476, 1978.
177. **Satoh, M. and Takagi, H.,** Enhancement by morphine of the central descending inhibitory influence on spinal sensory transmission, *Eur. J. Pharmacol.*, 14, 60, 1971.
178. **Basbaum, A. I., Clanton, C. H., and Fields, H. L.,** Opiate and stimulus-produced analgesia: functional anatomy of a medullospinal pathway, *Proc. Natl. Acad. Sci. USA*, 73, 4685, 1976.
179. **Yokota, T., and Hashimoto, S.,** Periaqueductal gray and tooth pulp afferent interaction on units in caudal medulla oblongata, *Brain Res.*, 117, 508, 1976.
180. **Fields, H. L., Emson, P. C., Leigh, B. K., Gilbert, R. F. T., and Iversen, L. L.,** Multiple opiate receptor sites on primary afferent fibres, *Nature (London)*, 284, 352, 1980.
181. **Yaksh, T. L.,** Opiate receptors for behavioral analgesia resemble those related to the depression of spinal nociceptive neurons, *Science*, 199, 1231, 1978.
182. **Headley, P. M., Duggan, A. W., and Griersmith, B. T.,** Selective reduction by noradrenaline and 5-hydroxytryptamine of nociceptive responses of cat dorsal horn neurones, *Brain Res.*, 145, 185, 1978.
183. **Griersmith, B. T. and Duggan, A. W.,** Prolonged depression of spinal transmission of nociceptive information by 5-HT administered in the substantia gelatinosa: antagonism by methysergide, *Brain Res.*, 187, 231, 1980.
184. **Kuraishi, Y., Harada, Y., Satoh, M., and Takagi, H.,** Antagonism by phenoxybenzamine of the analgesic effect of morphine injected into the nucleus reticularis gigantocellularis of the rat, *Neuropharmacology*, 18, 107, 1979.
185. **Jessell, T. M. and Iversen, L. L.,** Opiate analgesics inhibit substance P release from rat trigeminal nucleus, *Nature (London)*, 268, 549, 1977.
186. **Virus, R. M. and Gebhart, G. F.,** Pharmacologic actions of capsaicin: apparent involvement of substance P and serotonin, *Life Sci.*, 25, 1273, 1979.
187. **Hökfelt, T., Kellerth, J. O., Nilsson, G., and Pernow, P.,** Experimental immunohistochemical studies on the localization and distribution of substance P in cat primary sensory neurones, *Brain Res.*, 100, 235, 1975.
188. **Piercey, M. F., Einspahr, F. J., Dobry, P. J. K., Schroeder, L. A., and Hollister, R. P.,** Morphine does not antagonize the substance P mediated excitation of dorsal horn neurons, *Brain Res.*, 186, 421, 1980.
189. **Henry, J. L., Sessle, B. J., Lucier, G. E., and Hu, J. W.,** Effects of substance P on nociceptive and non-nociceptive trigeminal brain stem neurons, *Pain*, 8, 33, 1980.
190. **Andersen, R. K., Lund, J. P., and Puil, E.,** Enkephalin and substance P effects related to trigeminal pain, *Can. J. Physiol. Pharmacol.*, 56, 216, 1978.
191. **Yaksh, T. L., Jessell, T. M., Gamse, R., Mudge, A. W., and Leeman, S. E.,** Intrathecal morphine inhibits substance P release from mammalian spinal cord in vivo, *Nature (London)*, 286, 155, 1980.

192. **Henry, J. L.,** Effects of substance P on functionally identified units in cat spinal cord, *Brain Res.,* 114, 439, 1976.
193. **Duggan, A. W., Hall, J. G., Headley, P. M., and Griersmith, B. T.,** The effect of naloxone on the excitation of dorsal horn neurones of the cat by noxious and non-noxious cutaneous stimuli, *Brain Res.,* 138, 185, 1977.
194. **Jurna, I., Grossmann, W., and Theres, C.,** Inhibition by morphine of repetitive activation of cat spinal motoneurones, *Neuropharmacology,* 12, 983, 1973.
195. **Wikler, A.,** Studies on the actions of morphine on the central nervous system of the cat, *J. Pharmacol.,* 80, 176, 1944.
196. **McClane, T. K. and Martin, W. R.,** Effects of morphine, nalorphine, cyclazocine, and naloxone on the flexor reflex, *Int. J. Neuropharmacol.,* 6, 89, 1967.
197. **Takagi, H., Matsumura, M., Yanai, A., and Ogiu, K.,** The effect of analgesics on the spinal reflex activity of the cat, *Jpn. J. Pharmacol.,* 4, 176, 1955.
198. **Willer, J. C. and Bussel, B.,** Evidence for a direct spinal mechanism in morphine-induced inhibition of nociceptive reflexes in humans, *Brain Res.,* 187, 212, 1980.
199. **Willer, J. C. and Bussel, B.,** Possible explanation for analgesia mediated by direct spinal effect of morphine, *Lancet,* 1, 158, 1980.
200. **Krivoy, W., Kroeger, D., and Zimmermann, E.,** Actions of morphine on the segmental reflex of the decerebrate-spinal cat, *Br. J. Pharmacol.,* 47, 457, 1973.
201. **Wikler, A. and Frank, K.,** Hindlimb reflexes of chronic spinal dogs during cycles of addiction to morphine and methadone, *J. Pharmacol.,* 94, 382, 1948.
202. **Melzack, R. and Wall, P. D.,** Pain mechanisms: a new theory, *Science,* 150, 971, 1965.
203. **Grossmann, W., Jurna, I., Nell, T., and Theres, C.,** The dependence of the antinociceptive effect of morphine and other analgesic agents on spinal motor activity after central monoamine depletion, *Eur. J. Pharmacol.,* 24, 67, 1973.
204. **Seeber, U., Kuschinsky, K., and Sontag, K. H.,** Inhibition by opiate narcotics of rat flexor α-motoneurones, *Naunyn-Schmiedeb. Arch. Pharmacol.,* 301, 181, 1978.
205. **Goldfarb, J. and Hu, J. W.,** Enhancement of reflexes by naloxone in spinal cats, *Neuropharmacology,* 15, 785, 1976.
206. **Goldfarb, J., Kaplan, E. I., and Jenkins, H. R.,** Interaction of morphine and naloxone in acute spinal cats, *Neuropharmacology,* 17, 569, 1978.
207. **Satoh, M. and Takagi, H.,** Effect of morphine on the pre- and postsynaptic inhibitions in the spinal cord, *Eur. J. Pharmacol.,* 14, 150, 1971.
208. **Foldes, F. F., Lunn, J. N., Moore, J., and Brown, J. M.,** N-allyl-noroxymorphone: a new potent narcotic antagonist, *Am. J. Med. Sci.,* 245, 23, 1963.
209. **Boureau, F., Willer, J. C., and Dauthier, C.,** Study of naloxone in normal awake man: effects on spinal reflexes, *Neuropharmacology,* 17, 565, 1978.
210. **Sewell, R. D. E. and Spencer, S. J.,** Antinociceptive activity of narcotic agonist and partial agonist analgesics and other agents in the tail-immersion test in mice and rats, *Neuropharmacology,* 15, 683, 1976.
211. **Fry, J. P., Herz, A., and Zieglgänsberger, W.,** A demonstration of naloxone-precipitated opiate withdrawal on single neurones in the morphine-tolerant/dependent rat brain, *Br. J. Pharmacol.,* 68, 585, 1980.
212. **Jurna, I. and Schäfer, H.,** Depression of post-tetanic potentiation in the spinal cord by morphine and pethidine, *Experientia,* 21, 226, 1965.
213. **Henry, J. L.,** Naloxone excitation of dorsal horn units in the spinal cats, cit. post 66.
214. **Lord, J. A. H., Waterfield, A. A., Hughes, J., and Kosterlitz, H. W.,** Endogenous opioid peptides: multiple agonists and receptors, *Nature (London),* 267, 495, 1977.
215. **Rőnai, A. Z., Gráf, L., Székely, J. I., Dunai-Kovács, Z., and Bajusz, S.,** Differential behavior of LPH-(61-91)-peptide in different model systems: comparison of the opioid activities of LPH-(61-91)-peptide and its fragments, *FEBS Lett.,* 74, 182, 1977.
216. **Martin, W. R., Eades, C. G., Thompson, J. A., Huppler, R. E., and Gilbert, P. E.,** The effects of morphine- and nalorphine-like drugs in the nondependent and morphine dependent chronic spinal dog, *J. Pharmacol. Exp. Ther.,* 197, 517, 1976.
217. **Yaksh, T. L. and Rudy, T. A.,** Studies on the direct spinal action of narcotics in the production of analgesia in the rat, *J. Pharmacol. Exp. Ther.,* 202, 411, 1977.
218. **Székely, J. I., Dunai-Kovács, Z., Miglécz, E., Rőnai, A. Z., and Bajusz, S.,** *In vivo* antagonism by naloxone of morphine, β-endorphin and a synthetic enkephalin analog, *J. Pharmacol. Exp. Ther.,* 207, 878, 1978.
219. **Krivoy, W. A. and Huggins, R. A.,** The action of morphine, methadone, meperidine and nalorphine on dorsal root potentials of cat spinal cord, *J. Pharmacol. Exp. Ther.,* 134, 210, 1961.

220. **Carstens, E., Tulloch, I., Zieglgänsberger, W., and Zimmermann, M., II.**, Does morphine produce presynaptic inhibition in the spinal cord, *J. Physiol. (London)*, 284, 137P, 1978.
221. **Chin, J. H., Crankshaw, D. P., and Kendig, J. J.**, Changes in the dorsal root potential with diazepam and with the analgesics aspirin, nitrous oxide, morphine and meperidine, *Neuropharmacology*, 13, 305, 1974.
222. **Eccles, J. C., Schmidt, R. F., and Willis, W. D.**, The mode of operation of synaptic mechanism producing presynaptic inhibition, *J. Neurophysiol.*, 26, 523, 1963.
223. **Eccles, J. C.**, Presynaptic inhibition in the spinal cord; physiology of spinal neurones, in *Progress in Brain Research*, Vol. 12, Eccles, J. C. and Schade, J. P., Eds., Elsevier, Amsterdam, 1964. 65.
224. **Piercey, M. F. and Einspahr, F. J.**, Use of substance P partial fragments to characterize substance P receptors of cat dorsal horn neurons, *Brain Res.*, 187, 481, 1980.
225. **Segawa, T., Murakami, H., Ogawa, H., and Yajima, H.**, Effect of enkephalin and substance P on sympathetic nerve transmission in mouse vas deferens, *Jpn. J. Pharmacol.*, 28, 13, 1978.
226. **Pórszász, J. and Jancsó, N.**, Studies on the action potentials of sensory nerves in animals desensitized with capsaicin, *Acta Physiol. Acad. Sci. Hung.*, 16, 299, 1959.
227. **Szolcsányi J. and Jancsó-Gábor, A.**, Sensory effects of capsaicin congeners, *Arzneim-Forsch. (Drug Res.)*, 25, 1877, 1975.
228. **Jancsó, G., Király, E., and Jancsó-Gábor, A.**, Pharmacologically induced selective degeneration of chemosensitive primary sensory neurones, *Nature (London)*, 270, 741, 1977.
229. **Jancsó, N., Jancsó-Gábor, A., and Szolcsányi, I.**, Direct evidence for neurogenic inflammation and its prevention by denervation and by pretreatment with capsaicin, *Br. J. Pharmacol.*, 31, 138, 1967.
230. **Jessel, T. M., Iversen, L. L., and Cuello, A. C.**, Capsaicin-induced depletion of substance P from primary sensory neurones, *Brain Res.*, 152, 183, 1978.
231. **Yaksh, T. L., Farb, D. H., Leeman, S. E., and Jessell, T. M.**, Intrathecal capsaicin depletes substance P in the rat spinal cord and produces prolonged thermal analgesia, *Science*, 206, 481, 1979.
232. **Perkins, M. N. and Stone, T. W.**, Blockade of striatal neurone responses to morphine by aminophylline: evidence for adenosine mediation of opiate action, *Br. J. Pharmacol.*, 69, 131, 1980.
233. **Stone, T. W., and Perkins, M. N.**, Is adenosine the mediator of opiate action on neuronal firing rate? *Nature (London)*, 281, 227, 1979.
234. **Phillis, J. W., Jiang, Z. G., Chelack, B. J., and Wu, P. H.**, Morphine enhances adenosine release from the in vivo rat cerebral cortex, *Eur. J. Pharmacol.*, 65, 97, 1980.
235. **Dingledine, R. and Goldstein, A.**, Single neuron studies of opiate action in the guinea pig myenteric plexus, *Life Sci.*, 17, 57, 1975.
236. **Sato, T., Takayanagi, I., and Takagi, K.**, Pharmacological properties of electrical activities obtained from nuerones in Auerbach's plexus, *Jpn. J. Pharmacol.*, 23, 665, 1973.
237. **Dingledine, R., Goldstein, A., and Kendig, J.**, Effects of narcotic opiates and serotonin on the electrical behavior of neurons in the guinea pig myenteric plexus, *Life Sci.*, 14, 2299, 1974.
238. **Dingledine, R. and Goldstein, A.**, Effect of synaptic transmission blockade on morphine action in the guinea pig myenteric plexus, *J. Pharmacol. Exp. Ther.*, 196, 97, 1976.
239. **North, R. A. and Williams, J. T.**, Enkephalin inhibits firing of myenteric neurones, *Nature (London)*, 264, 460, 1976.
240. **North, R. A.**, Hyperpolarization of myenteric neurons by enkephalin, *Br. J. Pharmacol.*, 59, 504P, 1977.
241. **Williams, J. T. and North, R. A.**, Effects of endorphins on single myenteric neurons, *Brain Res.*, 165, 57, 1979.
242. **North, R. A., Katayama, Y., and Williams, J. T.**, On the mechanism and site of action of enkephalin on single myenteric neurons, *Brain Res.*, 165, 67, 1979.
243. **North, R. A. and Henderson, G.** Action of morphine on guinea-pig myenteric plexus and mouse vas deferens studied by intracellular recording, *Life Sci.*, 17, 63, 1975.
244. **North, R. A. and Zieglgänsberger, W.**, Opiate withdrawal signs in single myenteric neurones, *Brain Res.*, 144, 208, 1978.
245. **North, R. A. and Karras, P. J.**, Opiate tolerance and dependence induced in vitro in single myenteric neurones, *Nature (London)*, 272, 73, 1978.
246. **Waterfield, A. A., Hughes, J., and Kosterlitz, H. W.**, Cross tolerance between morphine and methionine-enkephalin, *Nature (London)*, 260, 624, 1976.
247. **Kosterlitz, H. W. and Wallis, D. I.**, The action of morphine-like drugs on impulse transmission in mammalian nerve fibres, *Br. J. Pharmacol.*, 22, 499, 1964.
248. **North, R. A.**, Effects of morphine on myenteric plexus neurones, *Neuropharmacology*, 15, 719, 1976.
249. **North, R. A. and Tonini, M.**, Hyperpolarization by morphine of myenteric neurones, in *Proc. Int. Narcotic Res. Club Meeting*, Kosterlitz, H. W. Ed., North-Holland, Amsterdam, 1976, 205.

250. **North, R. A. and Williams, J. T.**, Estracellular recording from the guinea pig myenteric plexus and the action of morphine, *Eur. J. Pharmacol.*, 45, 23, 1977.
251. **North, R. A. and Vitek, L. V.**, The effect of chronic morphine treatment on excitatory junction potentials in the mouse vas deferens, *Br. J. Pharmacol.*, 68, 399, 1980.
252. **Henderson, G. and North, R. A.**, Depression by morphine of excitatory junction potentials in the vas deferens of the mouse, *Br. J. Pharmacol.*, 57, 341, 1976.
253. **Illes, P., Zieglgänsberger, W., and Herz, A.**, Calcium reverses in inhibitory action of morphine on neuroeffector transmission in the mouse vas deferens, *Brain Res.*, 191, 511, 1980.
254. **Pepper, C. M. and Henderson, G.**, Opiates and opioid peptides hyperpolarize locus coeruleus neurones in vitro, *Science*, 209, 394, 1980.
255. **Mühlethaler, M., Gehwiler, B. H., and Dreifuss, J. J.**, Enkephalin-induced inhibition of hypothalamic paraventricular neurons, *Brain Res.*, 197, 264, 1980.
256. **Wouters, W. and van den Bercken, J.**, Hyperpolarisation and depression of slow synaptic inhibition by enkephalin in frog sympathetic ganglion, *Nature (London)*, 277, 53, 1979.
257. **Wouters, W. and van den Bercken, J.**, Effects of met-enkephalin on slow synaptic inhibition in frog sympathetic ganglion, *Neuropharmacology*, 19, 237, 1980.
258. **Crain, S. M., Pererson, E. R., Crain, B., and Simon, E. J.**, Selective opiate depression of sensory-evoked synaptic networks in dorsal horn regions of spinal cord cultures, *Brain Res.*, 133, 162, 1977.
259. **Crain, S. M., Crain, B., Peterson, E. R., and Simon, E. J.**, Selective depression by opioid peptides of sensory-evoked dorsal-horn net-work responses in organized spinal cord cultures, *Brain Res.*, 157, 196, 1978.
260. **Hiller, J. M., Simon, E. J., Crain, S. M., and Peterson, E. R.**, Opiate receptors in cultures of fetal mouse dorsal root ganglia (DRG) and spinal cord: predominance in DRG neurites, *Brain Res.*, 145, 396, 1978.
261. **MacDonald, R. L. and Nelson, P. G.**, Specific-opiate-induced depression of transmitter release from dorsal root ganglion cells in culture, *Science*, 199, 1449, 1978.
262. **Barker, J. L., Neale, J. H., Smith, T. G., Jr., and MacDonald, R. L.**, Opiate peptide modulation of amino acid responses suggest novel form of neuronal communication, *Science*, 199, 1451, 1978.
263. **Barker, J. L., Smith, T. G., Jr., and Neale, J. H.**, Multiple membrane actions of enkephalin revealed using cultured spinal neurons, *Brain Res.*, 154, 153, 1978.
264. **Lee, H. K., Dunwiddie, T., and Hoffer, B.**, Electrophysiological interactions of enkephalins with neuronal circuitry in the rat hippocampus. II. Effects on interneuron excitability, *Brain Res.*, 184, 331, 1980.
265. **Haas, H. L. and Ryall, R. W.**, Presynaptic facilitation by enkephalin in the hippocampus, *J. Physiol. (London)*, 301, 37P, 1980.
266. **Waterfield, A. A., Smokcum, R. W. J., Hughes, J., Kosterlitz, H. W., and Henderson, G.**, In vitro pharmacology of the opioid peptides, enkephalins and endorphins, *Eur. J. Pharmacol.*, 43, 107, 1977.
267. **Mudge, A. W., Leeman, S. E., and Fischbach, G. D.**, Enkephalin inhibits release of substance P from sensory neurons in culture and decreases action potential duration, *Proc. Natl. Acad. Sci. USA*, 76, 526, 1979.
268. **Wouters, W., Ermes-Busio M. C. E., van den Bercken, J.**, Potentiation by naloxone of reflex activity in the isolated spinal cord of Xenopus, *Eur. J. Pharmacol.*, 55, 431, 1979.
269. **Karras, P. J. and North, R. A.**, Inhibition of neuronal firing by opiates: evidence against the involvement of cyclic nucleotides, *Br. J. Pharmacol.*, 65, 647, 1979.

Chapter 3

ANALGESIC AND BEHAVIORAL EFFECTS OF OPIOIDS

J. I. Székely

TABLE OF CONTENTS

I. Opioid Peptides and Analgesia 110
- A. Natural Enkephalins 110
- B. β-Lipotropin and Its Fragments 116
- C. Do the Opiates Act Via Mobilization of the Endogenous Opioids? 121
- D. Do the Endogenous Opioids Also Act Peripherally? 122

II. Effects of Opiate Antagonists on the Pain Threshold 123

III. Behavioral Effects of Opioid Peptides 131
- A. Behavioral Effects Upon Central Administration 131
- B. Hypermotility and Stereotypy 135
- C. Behavioral Effects Upon Peripheral Administration 137
  1. Analgesia 137
  2. Other Behavioral Effects 137

IV. Effects of Opioid Peptides on Conditioned Reactions 137

V. Euphorigenic Effects and Stimulus Properties of Opioid Peptides 140

VI. Antagonism by Naloxone of the Analgesic and Behavioral Effects of Opioid Peptides 142

VII. Modulation by Opioid Peptides (and Opiate Antagonists) of Food and Water Intake 145

References 148

I. OPIOID PEPTIDES AND ANALGESIA

A. Natural Enkephalins

The discovery of analgesic effects of natural opioid peptides reminds one of a fierce race where some competitors had obviously missed the right track. In their pioneering report Hughes et al.[1] described the amino acid sequence and morphine-like in vitro actions of natural enkephalins, but did not refer to their contingent analgesic effect in vivo. As in the case of classical morphine derivatives the antinociceptive effect correlates very well with the activities in the classical opiate-sensitive isolated organ preparations (Chapter 1, Volume I), also the natural enkephalins were supposed to have analgesic activity. Thus, within several months some reports appeared on the analgesic activity of enkephalins. First Belluzzi et al.[2] reported that Met-E and Leu-E injected i.c.v. to rats in the doses of 100 to 200 μg resulted in a moderate but statistically significant lengthening of latency time in the tail flick test. This effect ensued rapidly after treatment and lasted for several minutes as detected by measuring the pain threshold in every 2 min.[2] In a subsequent but independent study Büscher et al.[3] reported on similar action in mice. Both groups also demonstrated the reversal of this analgesic effect by naloxone.[2,3] Accordingly, they concluded that "enkephalin may serve as a natural opiate-like neurohormone in a central mechanism for the inhibition of pain";[2] a very ardent statement which is still debated even today. Although in the meantime many data have been compiled proving the existence of enkephalin mediated pain-controlling mechanism in the brain, the findings presented by these groups did not seem to confirm such a conclusion. Belluzzi et al.[2] observed a lengthening of latency time by 37 to 78%, which is actually a very weak effect. According to our own experiments "saline treatment" alone may elicit an "analgesic effect" of nearly similar magnitude (Figure 1). In this case, obviously the stress effect of handling, the frequent measurement of the pain threshold, and the i.c.v. treatment itself resulted in a moderate analgesia-like effect. Probably it is not accidental that in most relevant studies, transforming the numerical values (latency times) into nonparametric data, only those animals were regarded positive for analgesia, at which the control latency period was at least doubled. However, the first group[2] did not calculate ED_{50} values at all, the other defined the ED_{50} as the dose prolonging the reaction time by more than 75% in half of the animals. Even according to such a mild criterium, in mice the ED_{50} values of morphine, Met-E, and Leu-E were found 1.1; 76 and 240 μg/animal, respectively.[3] In rats 100 to 200 μg of Met-E or Leu-E gave weaker response than 10 μg of morphine.[2] Moreover, both groups observed only a very short-lasting effect, which dissipated in a few minutes.[2,3] A further factor which rendered these positive observations debatable was that the antinociceptive effects of enkephalins appeared almost instantly upon injection.

Administering, however, morphine or β-endorphin or potent enkephalin analogues into the lateral ventricles, the maximal analgesic effect appeared only 15 to 45 min[4-10] (see also Figure 2). The relatively slow onset of analgesic action can be attributed to the anatomical localization of opiate receptors responsible for the analgesic action (Chapter 2, Volume I). If substances are administered into the lateral ventricles, they have to travel down into the brain stem and traverse the wall of caudal parts of ventricular system and even penetrate into the brain tissue. Consequently the instant effect of natural enkephalins recorded by these[2,3] and several other groups (Table 1) might have been mediated by receptors situated relatively close to the lateral ventricles and not via the caudal brain stem.

Consequently, it is probably not surprising that the subsequent studies rather contested than confirmed these initial observations. In our experiments,[5] measuring the latency time in the rat tail flick test in every 2 min after i.c.v. treatment, only a very

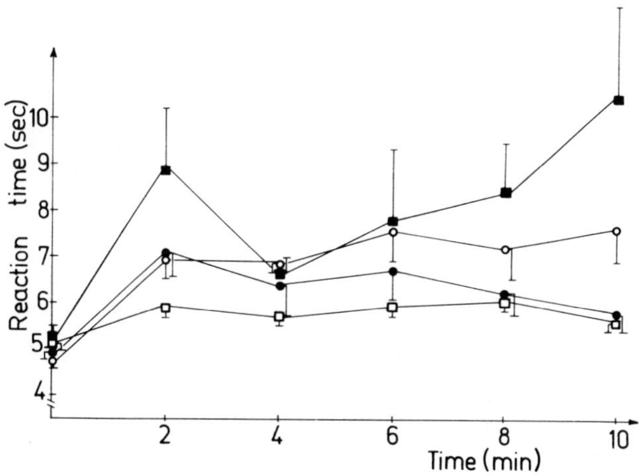

FIGURE 1. Effects of 30 nmole morphine (○), 670 nmole Met-E (●) 0.8 namole β-EP (■) and 20 μℓ saline (□) (given i.c.v.) in the rat tail flick test measuring the tail withdrawal latency time in every 2 min. On the abscissa the time elapsed since i.c.v. treatment, on the ordinate the reaction time (Mean ± S.E.).

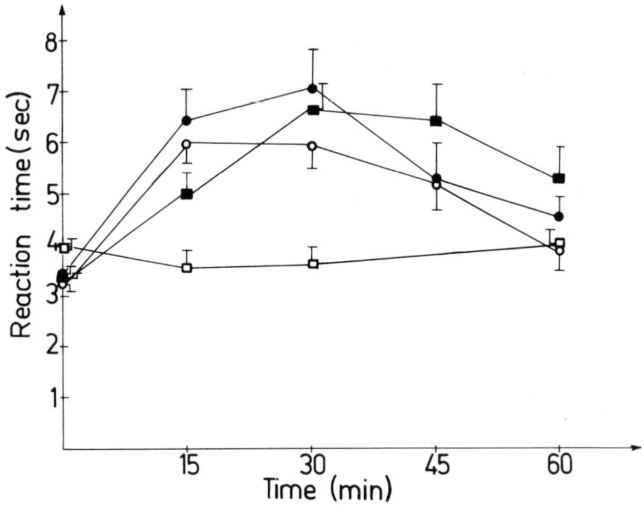

FIGURE 2. Effects of 4 nmole morphine (○), 0.1 nmole of D-Met, Pro-EA (●), 0.8 nmole β-EP (■) and 20 μℓ saline (□) (given i.c.v.) in the rat tail flick test measuring the tail withdrawal latency time in every 15 min. On the abscissa the time elapsed since i.c.v. treatment, on the ordinate the reaction time (Mean ± S.E.).

slight prolongation of latency time could be observed giving either Met-E, morphine or β-endorphin (Figure 1). Thus under the experimental conditions used by Belluzzi et al.[2] only a very slight effect was detected even upon treatment with such a powerful analgesic agent as morphine. However, following up the events for at least an hour, the morphine-induced analgesia was characterized by a slow onset of action, reaching its maximum at 15 to 45 min after treatment and lasting for 60 to 120 min depending on the dose applied[5] (Figures 2 and 3). Under this protocol, i.e., measuring the tail-

Table 1
"ANALGESIC" ACTION OF MET-E AND LEU-E

Treatment	Route of application	Dose	Species	Test	Effect	Reversal by opiate antagonists	Ref.
Met-E	i.c.v.	100-200 μg	Rat	Tail flick	Weak transient analgesia	Yes	2
Leu-E	i.c.v.	100-200 μg	Rat	Tail flick	Weak transient analgesia	Yes	2
Met-E	i.c.v.	ED_{50} = 75 μg	Mouse	Tail flick	Transient analgesia	Yes	3
Leu-E	i.c.v.	ED_{50} = 240 μg	Mouse	Tail flick	Transient analgesia	Not examined	3
Met-E	PAG	30-360 μg	Rat	Tail flick	Transient analgesia	Yes	11
Leu-E	PAG	30-360 μg	Rat	Tail flick	No effect	—	11
Met-E	i.c.v.	670 nmoles	Rat	Tail flick	No significant effect	—	5
Met-E	i.c.v.	400 μg	Cat	Tail pinch	No effect	—	6
Met-E	i.c.v.	100 μg	Rat	Hot plate	Hyperalgesia	—	12
Leu-E	i.c.v.	10-100 μg	Rat	Hot plate	Analgesia	Not examined	12
Met-E	PAG	1030 nmoles	Rat	Behavioral assessment	No effect	—	10 13
Leu-E	PAG	54-100 μg	Rat	Behavioral assessment	No effect	—	10 13
Met-E	i.c.v. or PAG	1-1000 μg	Rat	Tail flick	No effect	—	14
Leu-E	i.c.v. or PAG	1-1000 μg	Rat	Tail flick	No effect	—	14
Met-E	i.c.v.	125-500 μg	Rat	Writhing	Transient analgesia	Yes	14
Leu-E	i.c.v.	125-500 μg	Rat	Writhing	Marginal analgesic effect	Not examined	14
Met-E	i.c.v.	0.1-100.0 μg	Mouse	Tail flick	No effect	—	15
Leu-E	i.c.v.	1.0-100.0 μg	Mouse	Tail flick	No effect	—	15
Met-E	i.c.v.	30-400 μg	Cat	Tail pinch	Occasionally weak transient analgesia	Not examined	7

Met-E	PAG	$ED_{50} = 15.5\ \mu g$	Rat	Tail flick	Transient analgesia	Yes	16
Leu-E	PAG	$ED_{50} = 14.8\ \mu g$	Rat	Tail flick	Transient analgesia	Yes	16
Met-E	i.c.v.	100-400 μg	Cat	Jaw opening reflex	No effect	—	17
Met-E		200-400 μg	Rat	Tail flick hot plate	Transient analgesia	Yes	18
Met-E	Into n. ret. gigantocellularis	$ED_{50} = 1.4\ \mu g$	Rat	Tail pinch	Analgesia	Not examined	19
Leu-E	Into n. ret. gigantocellularis	$ED_{50} = 4.8\ \mu g$	Rat	Tail pinch	Analgesia	Not examined	19
Met-E	i.c.v.	$ED_{50} = 35\text{-}111\ \mu g$	Mouse	Hot plate	Strain dependent analgesia	Yes	20
Met-E	i.v.	50 mg/kg	Mouse	Hot plate	No effect	—	21
Leu-E	i.v.	50 mg/kg	Mouse	Hot plate	No effect	—	21
Met-E	Intracisternally	$ED_{50} = 146$ nmoles	Rat	Tail clip	analgesia	Yes	22
Met-E	Intrathecally	150-300 μg	Rat	Tail immersion	Analgesia lasting for 10 min	Not examined	23
Met-E	Intrathecally	$ED_{50} = 142\ \mu g$	Rat	Tail flick	Analgesia	Not examined	24

Note: PAG = periaqueductal gray.

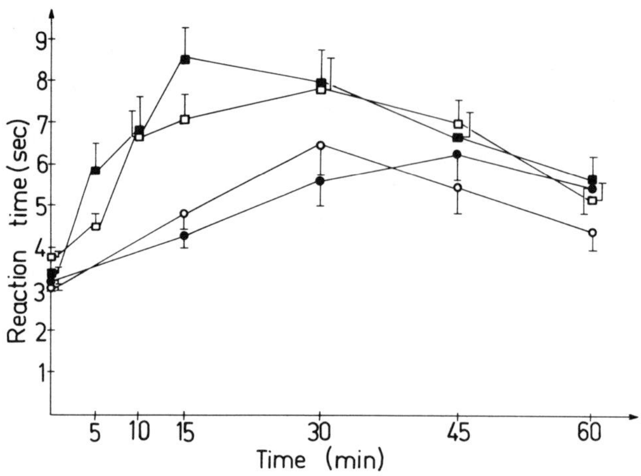

FIGURE 3. Effects of 1 mg/kg morphine (□) or D-Met, Pro-EA (■) given i.v. or 1 mg/kg morphine (○) or 2 mg/kg D-Met, Pro-EA (●) given s.c. in the rat tail flick test. See Figures 1 and 2 for further explanations.

flick latency in every 15 min, Met-E proved to be totally inactive in the dose of 400 μg (670 nanomole), i.e., in an amount 40 times higher than the effective dose of morphine.[5] Neither in other laboratories had Met-E any measurable analgesic effect giving it either i.c.v. to rats in very high doses[9,14] or to mice[15] or injecting 400 μg into the third ventricle of cats,[6] though the latter authors found in some cases a very short-lasting effect in a subsequent study.[7] As summarized in Table 1 also other groups found either a very weak effect or no activity at all.

In the cases of positive results, the observed antinociceptive action could be antagonized by naloxone. Unfortunately, pA_2 values have not been measured. Of course if the effect of the agonist is not stable and long-lasting enough, no pA_2 measurement can be performed, since in the case of an in vivo action lasting only for several minutes the tissue concentrations of agonist and antagonist will not be equilibrated. Thus, no data are available on the possible quantitative differences between morphine-like drugs and natural enkephalins in relation of antagonism by naloxone. Consequently, it is not possible to find out, whether or not the occasionally observed slight antinociceptive effect was elicited by those receptors also mediating the analgesic effect of classical opiates or not.

From this point of view the observations of Racagni et al.[20] are interesting, according to the relative sensitivities of different strains of mice to i.c.v. administered morphine and Met-E correlate with each other. Accordingly, a strain specifically sensitive to the analgesic action of morphine DABA(2J) exhibits lower analgesic ED_{50} to Met-E too.[20] This is the only work which may give information in relation to the question whether morphine and the natural enkephalins act on the same receptor population or not. This experiment indicates that at least in relation of antinociceptive action the answer is positive. In good agreement with our study, where the pA_2 value of morphine and that of a potent enkephalin analogue, D-Met, Pro-EA are very similar in the mouse hot plate test (see later).

Of course this inability of natural enkephalins to induce real analgesia might be due to their rapid catabolism during their transport from the site of application to the brain stem opiate receptors. Probably better activity might be expected if giving them directly into the periaqueductal gray matter (PAG). Actually Malick and Goldstein[16] found Met-E and Leu-E only four times weaker than morphine upon injecting them directly into the PAG, their ED_{50} being 15 μg. Also these authors found the action maximal already in the first minutes after treatment and the effect dissipated during the 7 min observation period.[16] However, Jacquet and Marks[10,13] found Met-E and Leu-E ineffective even upon direct administration into the PAG in the heroic doses of 54 to 150 μg, while Chang et al.[11] observed transient analgesia injecting 120 μg Met-E into this area but not by a higher dose (320 μg) or by Leu-E. Jacquet and Marks[13] supposed that even the seemingly positive effects were rather due to a transient impairment of motor functions than to a real antinociceptive action.

Probably the most unexpected findings (not confirmed so far by the subsequent studies) were presented by Leybin et al.[12]. These authors observed a slight but distinct hyperalgesic effect in the hot plate test if applying 100 μg Met-E i.c.v. to rats, while Leu-E elicited a dose-related inhibition of the hind paw licks.[12] They supposed that the hot plate testing itself resulted in mobilization of potent endogenous opioids, which were displaced by the exogenously given Met-E and Leu-E. Since Leu-E is a "purer" opiate receptor agonist than the former (Chapter 5, Volume III), it elicited a certain analgesic action of its own, while in the case of Met-E the competitive antagonism against the endogenous ligand prevailed.[12]

A careful examination of the data summarized in Table 1 reveals that a bit more consistent analgesic action could be detected if the enkephalins were administered intracisternally,[22] into the sensitive area of the medulla[19] or intrathecally.[18,23,24] Sufficiently steep dose-response curves have been obtained only upon lower medullary[19] or intrathecal administration.[18,24] Consequently, the existence of a real antinociceptive effect can be questioned only if administering them into the forebrain. In this relation it must be recalled that the most convincing electrographic demonstration of inhibition by enkephalins of the sensory pathways were made working on cells located in the spinal cord and the brain regions the most sensitive to local microinjection of opioids are also located in the caudal brain stem (Chapter 2, Volume II).

It is probably needless to emphasize that no analgesic effect has been observed upon peripheral application of natural enkephalins.[21] An "obvious" finding if considering that they are almost inactive also upon central application, but less evident if regarding the data on their alleged psychotropic effects observed upon systemic application (see later).

Thus, some of the data seem to confirm that the enkephalins have a certain analgesic potency. Nevertheless, the physiological relevance of the above-surveyed observations can be contested both that of the positive and negative results. It might be argued that the very weak and short-lasting analgesic activity, even if giving them in very high amounts, is the consequence of their rapid degradation, which can be circumvented by using of enzymatically stabilized analogues (Chapter 5, Volume III) or by administering them directly into the brain areas involved in pain transmission. However, the so-called "stabilized" enkephalin analogues are not only more resistant to metabolic inactivation, but they also have altered agonist properties in vitro, i.e., they have stronger agonist activity (Chapter 6, Volume III). Hence the efficacy of the potent enkephalin analogues does not necessarily prove that the natural enkephalins must have analgesic action under physiological conditions. Moreover, is it justified to draw any conclusion from data obtained after administering the substances in the dose-range of micrograms if their brain concentrations are not higher than several pg/g tissue? A considerable body of evidence points to their role in regulation of pain sensitivity, before all the close anatomical correlation of their occurrence with the afferent pathways of nociceptive reflexes and their presence in considerable quantities in areas the most sensitive to analgesia inducing actions of opiates or brain stimulation. Also the electrophysiological experiments (Chapter 2, Volume II) confirm their importance in regulation of pain sensitivity. Probably it is more correct to conclude that these in vivo studies are not suitable to decide the question. (Just as it would be rather strange to analyze the physiological functions of acetylcholine or noradrenaline by injecting them i.c.v. and measuring some behavioral parameters.)

B. β-Lipotropin and Its Fragments

As the amino acid sequence of Met-E is a part of β-lipotropin (LPH) and the latter was suspected to be a natural precursor of the former, other LPH fragments have also been tested for possible antinociceptive effect, since the relative in vivo ineffectiveness of natural enkephalins and the in vitro potencies of many other LPH fragments suggested, that more potent ones might be found among the other segments of the LPH chain.

As for the whole LPH molecule, it has been found completely inactive by Feldberg and Smyth,[7] while the presence of a very weak in vivo activity has not been excluded by Loh et al.[4] and Chretien (personal communication). In our laboratory[25] a rather weak but dose-dependent analgesic effect had been observed, the ED_{50} of LPH being 2.2 times higher than that of morphine on molar basis. This observation was in good agreement with the data reported by Loh et al.,[4] who found LPH 80 times weaker than β-EP and the latter 20 to 33 stronger than morphine. Thus, roughly similar morphine/LPH potency ratios have been calculated in the two different laboratories.[4,25] Since LPH is totally inactive in vitro in receptor binding assay and in opiate-sensitive isolated organ preparations (Chapter 5, Volume III), an in vivo enzymatic cleavage of LPH had to be supposed, which resulted in the production of active fragments. Since then an enzyme, splitting down β-EP from LPH has been found (Chapter 4, Volume III). Nevertheless, it must be taken into consideration that in all these above-mentioned laboratories natural LPH preparations were used. Working with LPH obtained from natural source, an autolytic process resulting in contamination with the highly active fragment LPH/61-91, i.e., β-EP cannot be ruled out with absolute certainty. Thus, until these observations will not be repeated with synthetic LPH, the possibility that these findings were simply experimental artefacts cannot be excluded. Moreover, as LPH is really broken down to β-EP, the whole question has little significance.

Returning to tbe LPH fragments, it must be stated in advance that only those fragments that were examined in vivo had a tyrosine residue corresponding to LPH/61 (in amino terminal position). As the presence of Tyr at the N-terminus is an absolute requirement of any opiate-like in vitro activity (Chapter 5, Volume III), other fragments have hardly been examined in vivo. In our unpublished experiments several LPH fragments were tested for analgesic activity in the rat tail flick assay in the i.c.v. dose of 60 to 80 nmoles. These totally inactive fragments were as follows LPH(1-79), LPH(1-60), LPH(1-58), LPH(1-51), LPH(48-91), LPH(52-79), and LPH(80-91).

LPH(61-69) is one of the fragments which may be obtained by tryptic digestion of LPH, As shown in Table 2, no analgesic effect has been found upon its i.c.v. application,[7,15,26] which is probably not surprising considering its decreased in vitro potency if comparing it to β-EP (Chapter 5, Volume III). Nevertheless, since the natural enkephalins have been unanimously reported to have certain in vivo analgesic activity upon intracisternal[22] or intrathecal[18,23,24] application, from theoretical point of view it would be interesting to know, what happens upon administering this fragment in similar way. Unfortunately no such data are available.

As for the α- and γ-endorphins, corresponding to LPH(71-76) and LPH(61-77), respectively, no analgesic effect has been observed upon applying them i.c.v. or directly into the PAG.[9,10,13] The fragment LPH (61-78) was found to have only weak activity by Feldberg and Smyth,[6,7] qualifying the much longer fragment LPH(61-87) totally inactive. We have found LPH(61-79) given i.c.v. to display a moderate activity in doses

Table 2
ANALGESIC ACTION OF β-ENDORPHIN FRAGMENTS

Fragment	Route of application	Dose	Species	Test	Effect	Reversal by opiate antagonists	Ref.
LPH(61-69)	i.c.v.	5.7-180.0 μg	Mouse	Tail flick	No effect	—	15
LPH(61-69)	i.c.v.	200 nmoles	Mouse	Tail flick, hot plate, writhing	No effect	—	4
LPH(61-91)	i.c.v.	ED_{50} = (a) 1.52 nmoles (b) 2.32 nmoles (c) 0.92 nmoles	Mouse	(a) Tail flick (b) Hot plate (c) Writhing	Analgesia	Yes	4
LPH(61-69)	i.c.v.	20 μg	Rat	Tail flick	No effect	—	26
LPH(61-79)	i.c.v.	40 nmoles	Rat	Tail flick	Very weak analgesia	Not examined	5
LPH(61-91)	i.c.v.	0.8 nmoles	Rat	Tail flick	Strong analgesia	Yes	5
LPH(61-76)	i.c.v.	1210 nmoles	Rat	Behavioral assessment	No effect	—	9
LPH(61-77)	i.c.v.	281 nmoles	Rat	Behavioral assessment	No effect	—	9
LPH(61-91)	i.c.v.	3.0-14.9 nmoles	Rat	Behavioral assessment	Strong analgesia	Yes	9
LPH(61-76)	PAG	5 μg	Rat	Behavioral assessment	No effect	—	10 13
LPH(61-91)	PAG	2-4 μg	Rat	Behavioral assessment	Strong analgesia	Yes	10 13
LPH(61-91)	i.v.	ED_{50} = 9.4 mg/kg	Mouse	Tail flick	Analgesia	Yes	27
LPH(61-91)	i.v.	ED_{50} = 12.0 mg/kg	Moue	Hot plate	Analgesia	Yes	27
LPH(61-69)	i.c.v.	30-100 μg	Cat	Tail flick	No effect	—	7
LPH(61-78)	i.c.v.	50-100 μg	Cat	Tail flick	Weak analgesia	Not examined	7
LPH(61-87)	i.c.v.	30-100 μg	Cat	Tail flick	No effect	—	7
LPH(61-91)	i.c.v.	10-20 μg	Cat	Tail flick	Strong analgesia	Yes	7

Table 2 (continued)
ANALGESIC ACTION OF β-ENDORPHIN FRAGMENTS

Fragment	Route of application	Dose	Species	Test	Effect	Reversal by opiate antagonists	Ref.
LPH(61-91)	i.c.v.	$ED_{50} = 7.7$ nmoles	Rat	Tail flick	Strong analgesia	Yes	8
LPH(61-91)	Intrathecally	$ED_{50} = 0.05$ μg	Rat	Tail flick	Strong analgesia	Yes	28
LPH(61-91)	Intrathecally	$ED_{50} = 0.07$ μg	Rat	Hot plate	Strong analgesia	Yes	28
LPH(61-91)	i.v.	0.01-1.00 mg/kg	Cat	Subjective assessment	No effect	—	29
LPH(61-76)	Intracisternally	$ED_{50} = 28.6$ nmoles	Mouse	Tail clip	Analgesia	Yes	22
LPH(61-77)	Intracisternally	$ED_{50} = 46.8$ nmoles	Mouse	Tail clip	Analgesia	Yes	22
LPH(61-79)	Intracisternally	$ED_{50} = 50.3$ nmoles	Mouse	Tail clip	Analgesia	Yes	22
LPH(61-91)	Intracisternally	$ED_{50} = 0.09$ nmoles	Mouse	Tail clip	Analgesia	Yes	22
LPH(61-91)	s.c.	136 mg/kg	Rat	Tail flick	No effect	—	30
LPH(61-91)	i.c.v.	25-50 μg	Cat	Jaw opening reflex	Analgesia	Yes	31 17
LPH(61-87)	i.c.v.	10-100 nmoles	Rat	Tail flick	Analgesia	Not examined	32
LPH(61-89)	i.c.v.	1-20 nmoles	Rat	Tail flick	Analgesia	Not examined	32
LPH(61-91)	i.c.v.	0.5-2.0 nmoles	Rat	Tail flick	Analgesia	Not examined	32
LPH(61-81)	i.c.v.	$E_{50} = 38.0$ μg	Mouse	Tail flick	Analgesia	Not examined	33
LPH(61-86)	i.c.v.	$ED_{50} = 1.6$ μg	Mouse	Tail flick	Analgesia	Not examined	33
LPH(61-88)	i.c.v.	$ED_{50} = 2.1$ μg	Mouse	Tail flick	Analgesia	Not examined	33

LPH(61-89)	i.c.v.	$ED_{50} = .7\ \mu g$	Mouse	Tail flick	Analgesia	Not examined	33
LPH(61-90)	i.c.v.	$ED_{50} = 0.2\ \mu g$	Mouse	Tail flick	Analgesia	Not examined	33
LPH(61-91)	i.c.v.	$ED_{50} = 0.15\ \mu g$	Mouse	Tail flick	Analgesia	Not examined	33

Note: PAG = periaquaductal gray.

about 100 times higher than the ED_{50} of β-EP.[5] Thus the data on the alleged antinociceptive potencies of these medium-sized LPH β-EP fragments seem to be contradictory. However, completely negative findings were obtained in experiments,[9,10,13] where the analgesic action was assessed more or less subjectively and not quantitatively measured by one of the standard analgesic assays. Ueda et al.[22] found the fragments LPH(61-76), LPH(61-77), and LPH(61-79) to act at dose levels about 300 to 500 times higher than β-EP. This finding confirms our previous finding[5] obtained with LPH(61-79). Therefore, it is probably justified to conclude that these medium-sized fragments of β-EP do have a certain analgesic action, but only in doses two to three orders of magnitudes greater than the ED_{50} value of β-EP. These observations are rather of theoretical significance, since these fragments are weaker upon in vitro testing than the natural enkephalins (Chapter 5, Volume III) showing that the in vivo activity might be correlated with the chain length but surely not with the in vitro potencies.

The active LPH fragments containing more than 20 residues have been systematically studied by Li et al.[33] and Geisow et al.[32] (Table 2). The latter group[32] reported that splitting down the last two or four carboxy terminal amino acids resulted in a roughly ten- and one hundredfold reduction in the analgesic activity, respectively. However, Li et al.[33] found the LPH(61-86) fragment to have about 8% of the activity of the parent compound β-EP and in their experiments even LPH(61-81) showed an activity of 0.3%. The contradiction is probably not so great considering that one of the groups[33] worked in mice, while the other in rats.[32] However, it must be added that Feldberg and Smyth[7] found even the fragment LPH(61-87) inactive in cats. It is worth mentioning that in one study[32] the decreased analgesic potencies of LPH(61-87) and LPH(61-89) in comparison to that of the LPH(61-91) (=β-EP) was correlated with diminished receptor binding capacities. All of these studies, summarized in Table 2 suggest that the presence of the last C-terminal amino acids is essential for the analgesic effect. As discussed in Chapters 2, 4, 5, and 6, Volume III, the last residues may fulfil multiple functions, probably they secure stability against certain proteolytic enzymes and also increase the affinity to the opiate receptors as shown above.[32] Probably both effects can be attributed to the spatial conformation of β-EP for which just the last residues are indispensable.

As for β-EP, as shown by the data summarized in Table 2, it is an extremely potent analgesic agent upon central administration in all tests and species so far examined. Actually β-EP is the most potent structurally identified endogenous analgesic ever extracted from the living organism but only upon direct cerebral application. Its analgesic potency is 90 to 180 times higher than that of morphine in cats on a molar basis.[7] In mice the potency ratio lies between 22 and 33 depending on the analgesic assay applied[4] and in rats we have found β-EP about 20 times more potent than morphine i.c.v., also on molar basis.[8] In mice upon intracisternal application the ratio was only about seven,[22] while giving them intrathecally to rats β-EP was found 24 to 27 times stronger than morphine.[28]

As shown in these studies the analgesic effect of β-EP is dose-dependent, its time course is similar to that of morphine, and the effect can be reversed by naloxone.

Thus, the extreme efficacy of β-EP as a central analgesic agent has been confirmed in many laboratories. However, the data are much less equivocal in relation of its analgesic activity upon peripheral application. First Tseng et al.[27] reported that β-EP was a potent analgesic by i.v. application in mice. Its activity was found about 3 to 4 times stronger than that of morphine on a molar basis.[27] Feldberg and Smyth[7] found it active in cats in the i.v. dose of 0.2 mg/kg. These experiments, however, have not been repeated either in other laboratories or by the authors themselves. β-EP has no analgesic activity in rats upon i.v. administration (Li, personal communication). In addition, in the numerous studies on the effect of β-EP in humans nobody observed analgesia upon i.v. application (see later). It has definitely no antinociceptive action upon s.c. injection.[30] In a later study Catlin et al.[29] found no antinociception in the i.v. dose range of 0.01 to 1.00 mg/kg in cats. The inactivity of β-EP as a peripheral analgesic agent has been reluctantly recognized in the scientific circles. This may be attributed among other factors to the high price of commercially available β-EP, which prompted the research workers not to administer it in high dosage and to the fact that in the cases of classical morphine congeners the pharmacological effects observed upon central and peripheral application do not differ qualitatively. This lack of effect upon peripheral application is obviously due to the low capacity of opioid peptides to pass the blood-brain barrier, as it has recently been demonstrated (Chapter 3, Volume III). However, β-EP in contrast to the natural enkephalins is more or less continuously secreted from the anterior pituitary together with ACTH (Chapter 3, Volume II and Chapter 10, Volume III). Thus, the possibility arises that β-EP has nothing to do with the regulation of the pain threshold and all the above-mentioned data on the analgesic effect of centrally administered β-EP are physiologically irrelevant. However, β-EP can be found not only in the anterior pituitary, but also beyond the blood-brain barrier in the midbrain and diencephalon (Chapter 3, Volume II). The majority of cell bodies of β-EP containing neurones beyond the blood-brain barrier are located in and around the n. arcuatus of the hypothalamus (Chapter 3, Volume III), and as discussed in Chapter 2, Volume I. hypothalamus is hardly involved in regulation of the pain sensitivity.

Consequently, the in vivo pharmacological data are rather puzzling; enkephalins which are considered to regulate the pain transmission have little antinociceptive activity, while the highly active β-EP could not be unanimously related to the natural analgesic mechanisms.

Based on the behavioral observations it could be argued that the endorphin-induced analgesia is a secondary phenomenon, a nonspecific manifestation of the general unresponsiveness of animals treated with endorphins (see later). β-EP does have analgesic action also upon intracisternal and intrathecal application,[22,28] when the antinociceptive action is not accompanied by the much-discussed behavioral effects. Thus, the research work on the enkephalins and endorphins is not beyond an early stage when a great part of the main pharmacological effects has already been described, but almost nothing is known on their physiological significance. Otherwise, in relation to their analgesic effect several questions deserve special interest.

C. Do the Opiates Act Via Mobilization of the Endogenous Opioids?

According to some data also the morphine congeners exert their effects by mobilizing the endogenous opioids. Bergmann et al.[34] reported that upon treatment with opiates the Leu-E content of the liquor increased. This elevation of the opioid content could be a consequence of its mobilization (displacement) from the nerve terminals. Nevertheless, this displacement might be rather the consequence of the opiate treatment than the cause of their analgesic effect due to the "stressor" action of opiate treatment (Chapter 5, Volume II). Fu and Dewey[35] reported that the analgesic effect of s.c. given morphine in mice was unaffected by spinalization (at T_{12}) if the dura mater was left intact, but its antinociceptive action was markedly reduced when the dura mater was removed (without lesioning of the neural elements) resulting in leakage of the liquor.[35] These observations[34,35] suggest that opiates might elicit the release of certain substances, which are transported in the liquor into the caudal regions of the spinal cord. The same group[36] found, that liquor taken from rabbits treated with morphine had antinociceptive effect in the phenylquinone writhing test and it also inhibited the electrically evoked contractions of the guinea pig ileum. Of course the unknown substance is not obviously an opioid. As discussed previously (Chapter 2, Volume I) opiates liberate serotonin in the brain and the pharmacological manipulations which increase or decrease this process also potentiate or attenuate the analgesic action, respectively. This might be valid also for the opioid peptides. The centrally injected serotonin or serotonin uptake inhibition by clomipramine was found to potentiate, while depletion of brain serotonin by reserpine or raphe lesions inhibited the analgesic effect of a potent enkephalin analogue (D-Ala^2, Leu^5)-E.[35]

Lin et al.[37] reported that the analgesic effect of β-EP was antagonized upon depletion of serotonin by *p*-chlorophenylalanine, 5.6-or 5.7-dihydroxytryptamine just as proven in the cases of classical opiates (Chapter 2, Volume I).

In this connection it is worth mentioning that Leu-E has been found to potentiate the analgesic effect of morphine, administering it either supraspinally or intrathecally.[24,38,39] Interestingly Met-E had no such effect.[24,39] Otherwise Leu-E also potentiated the antinociceptive action of Met-E. Morphine analgesia was potentiated by (D-Ala^2, D-Leu^5)-E, but not by (D-Ala^2, Met^5)-E, moreover Leu-E also enhanced the analgesic action of the former analogue.[39] All these potentiating actions were reversed by naloxone, thus also the potentiation was mediated by opiate receptors.[39]

According to the in vitro obtained data there are certain qualitative differences between Met-E and Leu-E. Thus Leu-E is a purer agonist of opiate receptors than Met-E according to the effect of sodium on their potencies in receptor binding assay (Chapter 5, Volume III). Moreover, (D-Ala^2, D-Leu^5)-E is a derivative known to have very strong potency in the mouse vas deferens preparation while being relatively weak in the guinea pig ileum (Chapter 5, Volume III). Consequently, Leu-E and (D-Ala^2, D-Leu^5)-E and Leu-E are "purer" opiate agonists and more specific ligands of the δ-receptors (mostly found in the mouse vas deferens) than Met-E and (D-Ala^2, Met^5)-E, which have a relatively stronger affinity to the μ-receptors (mostly found in the guinea pig ileum). Therefore Met-E and Leu-E might interact with at least partially different receptor populations, or more exactly the relative affinities of natural enkephalins to the different subclasses of opiate receptors are not the same. Thus, simultaneous stimulation by the two natural enkephalins of more than one subclasses of opiate receptors might result in enhanced biological actions, explaining the outcome of interaction experiments. Moreover, as morphine, similarly to many classical opiates is rather selective agonists of the μ-receptors, also the potentiation of its effect by Leu-E and (D-Ala^2, D-Leu^5)-E, i.e., by agents acting selectively on δ-receptors is easy to understand.

If the combined effect of Met-E and Leu-E is stronger than that of the Met-E alone and Leu-E potentiates the morphine-induced analgesia, then morphine should release either Met-E alone or it acts directly on the opiate receptors and not indirectly via enkephalin release. But it would be difficult to explain why morphine should liberate Met-E without concomitant release of Leu-E. Thus, the data on the interactions between morphine and enkephalins indirectly disprove the possibility that opiates would act via enkephalin liberation.

D. Do the Endogenous Opioids Also Act Peripherally?

Interestingly, also a peripheral site of action has been suggested for the opioid peptides. Ferreira and Nakamura[40,41] demonstrated, that the hyperalgesic effect of PGE_2 injected into the paws of rats could be reversed by injecting morphine, Met-E or (D-Ala^2, Met^5)-E also locally. Met-E reversed the hyperalgesic effect of locally administered carrageenan, prostacycline, and isoprenaline but not the effect of dibutyryl-cAMP.[40] As discussed in Chapter 2, Volume I and Chapter 1, Volume II opiates and opioid peptides inhibit the activation by prostaglandins of adenylate cyclase. As the local algogenic effect of inflammation-inducing agents is mediated to a large extent by prostaglandins, the above-mentioned effects are easy to interpret. Also the inability of opioids to inhibit the effect of dibutyryl-cAMP proves that opioids suppress an event ensuing after prostaglandin administration but before formation of cAMP. If so, these effects are obviously independent of opiate receptors, they may reflect a metabolic effect of opiates which take place after the receptor activation.

II. EFFECTS OF OPIATE ANTAGONISTS ON THE PAIN THRESHOLD

From a physiological point of view the main question is whether enkephalins and endorphins are really involved in regulation of pain sensitivity or the previously mentioned data reflect only pharmacological actions. As in the cases of other endogenous agonists, one of the most reliable ways to answer this question is to administer an antagonist and to follow up the consequences. If the endogenous ligands really decrease the pain sensitivity tonically, their displacement should result in enhanced pain sensitivity and in facilitation of the nociceptive reactions. Needless to emphasize that such a pain-enhancing effect was *á priori* expected upon treatment with naloxone-like drugs. The data summarized in Table 3 show that very heterogeneous findings have been obtained.

In this relation probably the most frequently quoted report is the one written by Jacob et al.,[42] who observed roughly dose-dependent facilitation by naloxone of the jumping reaction if mice were dropped on a hot plate. The authors[42] also demonstrated that the other characteristic nociceptive reaction observable in the hot plate test, licking of the limbs, was practically unaffected and even the jumping reaction was facilitated only at relatively low temperatures. Since almost everybody shared the conclusion of the authors, the hyperalgesic action was due to antagonism of endogenous morphine-like agents, only the final outcome of the study was remembered, i.e., the hyperalgesic action of naloxone and some important methodological details were simply overlooked. Thus instead of asking that under what experimental conditions can this hyperalgesic action be detected, the very existence of such an effect was debated.

First of all it must be taken into consideration that in spite of being the "purest" known opiate antagonist,[76] naloxone might have a certain agonist activity of its own. It would actually be surprising if a morphine derivative with a tremendous affinity to the opiate receptors had absolutely no intrinsic activity. Accordingly, Sewell and Spencer[47] found a dose-dependent analgesic effect in the tail immersion test in rats and mice applying naloxone whether s.c. or i.c.v. The dose-response curves were, however, flat, confirming that naloxone, like other mixed agonist-antagonists, showed a "ceiling effect" where the analgesic action could not be enhanced beyond a certain limit by increasing the dose.[47] Slight antinociceptive effect has been found also by Woolf[73] in the tail immersion test, giving naloxone into the third ventricle. The same group observed weak analgesia if naloxone was administered in relatively small doses intrathecally.[74] Separating these latter observations[74] on the intrathecal effect of naloxone from the other very conflicting reports concerning the action of systemically applied opiate antagonists, it may be concluded that in the spinal cord endogenous opioids really exert a tonic inhibitory action on nociceptive sensory transmission. In humans it was found to suppress the postoperative pain in doses of 0.4 to 2.0 mg, while to enhance it at higher dose level.[68]

Obviously, displacement of the endogenous ligands can result in a hyperalgesic effect, which confounds the inherent analgesic action. Moreover, if the dose-response curve of displacement of endogenous opiate receptor ligand is steeper than that of the intrinsic agonist action of antagonist examined, analgesic action may be expected at low dose level, while hyperalgesia in high dosage. Accordingly, the possible analgesic effect can be concealed by applying too high doses. Otherwise, the mixed agonist-antagonists are well known to induce analgesia only up to a certain dose-level. Hence

Table 3
EFFECT OF OPIATE ANTAGONISTS ON PAIN SENSITIVITY

Treatment	Dose	Species	Biological assay	Effect	Ref.
Naloxone	0.1-30.0 mg/kg s.c.	Rat and mouse	Hot plate test: jumping response	Hyperalgesia	42
Naloxone	0.1-30.0 mg/kg s.c.	Rat and mouse	Hot plate test: licking response	No effect	42
Naloxone	1.2 mg s.c. 1.0 mg i.v.	Human	Ischemic pain	No effect on hypnotic analgesia	43
Naloxone	0.4-0.8 mg i.v.	Human	Electrical stimulation of the forearm	No effect on the threshold of pain sensation and on maximal pain tolerance	44
Naloxone	0.5-3.0 mg/kg s.c.	Rat	Tail compression	No effect	45
Naloxone	2.0-25.0 mg/kg s.c.	Rat	Shock escape threshold	No effect	46
Naloxone	3.0-30.0 mg/kg s.c.	Rat and mouse	Tail immersion test	Weak analgesia	47
Naloxone	10.0-30.0 μg i.c.v.	Mouse	Tail immersion test	Weak analgesia	47
Naloxone	0.01-3.0 mg/kg s.c.	Dog	Skin twitch test	No effect	48
Naloxone	0.03-3.05 μmoles/kg s.c.	Mouse	Writhing test	No effect	49
Naloxone	2.0-10.0 mg i.v.	Human	Ischemic pain	No effect on pain rating but increased anxiety	50
Naloxone	0.5-4.0 mg/kg s.c.	Mouse	Hot plate: jumping and licking responses	Hyperalgesia	51
Naloxone	1.0-10.0 mg/kg i.p.	Rat	Analgesia induced by vaginal stimulation	No effect	52
Naloxone	0.4-2.0 mg i.v.	Human	Electrical stimulation of the forearm	No effect in pain sensitive subjects; hyperalgesia in pain insensitive subjects	53
Naloxone	0.03-10.0 mg/kg s.c.	Mouse	Hot plate: licking response	No effect	54
Naloxone	0.03-10.0 mg/kg s.c.	Mouse	Hot plate: jumping response	Hyperalgesia	54
Naloxone	10.0 mg/kg s.c.	Rat	Footshock induced startle reaction	Hyperalgesia	55
Naloxone	10.0 mg/kg s.c.	Rat	Footshock induced vocalization	Hyperalgesia	55
Naloxone	1.0-10.0 mg i.v.	Human	Cold water immersion induced pain	No effect	56
Naloxone	2.0-10.0 mg i.v.	Human	Ischemic pain	No effect	56
Naloxone	3.0 mg/kg s.c.	Mouse	Hot plate:jumping reaction	Inhibition of the diurnal fluctuation	57
Naloxone	0.5-16.0 mg/kg i.p., s.c.	Rat	Hot plate (44.5°C)	Analgesia	58
Naloxone	10.0 mg/kg i.p.	Rat	Hypersensitivity induced by local injection of formaline	No effect	59
Levallorphan	3.0-10.0 mg/kg s.c.	Mouse	Hot plate: jumping response	Hyperalgesia	60
Naloxone	3.0-10.0 mg/kg s.c.	Mouse	Hot plate: jumping response	Hyperalgesia	60
Naloxone	9.0 mg i.v.	Human	Pain intensity after tooth extraction	Hyperalgesia	61

Table 3 (continued)
EFFECT OF OPIATE ANTAGONISTS ON PAIN SENSITIVITY

Treatment	Dose	Species	Biological assay	Effect	Ref.
Naloxone	0.1-2.0 mg/kg i.v.	Rat	Intraarterially injected bradykinin induced flexor reflex	Reversal of adaptation	62
Naloxone	4.0 mg/kg i.p.	Rat	Tail flick test	Reversal of the food deprivation induced analgesia	63
Naloxone	50.0 mg i.v.	Human	Pin prick	No effect on hypnotic analgesia	64
Naloxone	1.0 mg i.v.	Human	Ischemic pain	No effect on hypnotic analgesia	65
Naloxone	0.05-0.1 mg/kg i.p.	Rat and mouse	Hot plate: licking of the hind limbs	Hyperalgesia	66
Naloxone	5.0 mg/kg i.p.	"Naive" rat	Response to intermittent footshocks	No effect	67
Naloxone	5.0 mg/kg i.p.	Rat[a]	Response to intermittent footshocks	Increase in the number of squeaks but not in that of the jumps	67
Naloxone	0.4-2.0 mg i.v.	Human	Pain intensity after tooth extraction	Analgesia	68
Naloxone	7.5-10.0 mg i.v.	Human	Pain intensity after tooth extraction	Hyperalgesia	68
Naloxone	1.0-100.0 mg/kg s.c.	Mouse	Acetic acid induced writhing	Hyperalgesia	69
Naltrexone	1.0-30.0 mg/kg s.c.	Mouse	Acetic acid induced writhing	Slight analgesia	69
Levallorphan	1.0-30.0 mg/kg s.c.	Mouse	Acetic acid induced writhing	Slight analgesia	69
Diprenorphine	1.0-10.0 mg/kg s.c.	Mouse	Acetic acid induced writhing	Slight analgesia	69
Naloxone	2.0 mg i.v.	Human	Ischemic pain	Reduction in pain tolerance only in the high stress group	70
Naloxone	5.0 mg/kg i.p.	Rat[b]	Flinch jump test	Hyperalgesia	71
Naloxone	1.0 μg into the PAG	Rat[b]	Flinch jump test	Hyperalgesia	71
Naloxone	0.8 mg i.v.	Human	Clinical pain due to neurological disorders	No effect	72
Naloxone	1.0 mg/kg i.v., s.c.	Rat	Tail immersion test	No effect	73
Naloxone	1.0 mg/kg i.v. or s.c.	Rat	Tail immersion test	No effect	73
Naloxone	2.0-50.0 μg into the ventricle III.	Rat	Tail immersion test	No effect	73
Naloxone	150.0 μg into the ventricle III.	Rat	Tail immersion test	Slight analgesia	73
Naloxone	0.25-5.0 μg into raphe nuclei	Rat	Tail immersion test	No effect	73
Naloxone	0.25-5.0 μg into PAG	Rat	Tail immersion test	No consistent effect	73
Naloxone	7.5-25.0 μg intrathecally	Rat	Tail immersion and vocalization threshold test	Slight analgesia	74
Naloxone	50.0-150.0 μg intrathecally	Rat	Tail immersion and vocalization threshold test	Hyperalgesia	74

Table 3 (continued)
EFFECT OF OPIATE ANTAGONISTS ON PAIN SENSITIVITY

Treatment	Dose	Species	Biological assay	Effect	Ref.
Naloxone	3.0-10.0 mg/kg s.c.	Rat	Effect of electric footshock on lever pressing for food reinforcement	Enhancement of behavioral disruption	75

Note: μM = micromole. PAG = periaqueductal gray matter.

[a] Previously shocked rat.
[b] Morphine dependent rat.

the possibility emerges that the dysphoric effect of pentazocine-, cyclazocine-, nalorphine-like drugs in high dosage is due to interference with endogenous opioids.

A further decisive factor seems to be the analgesic assay used. Numerous workers remarked that naloxone, if given alone, did not modify the pain threshold in the rat tail flick test. These reports are not incorporated into Table 3, since generally the authors did not go further than stating, that naloxone in the dose applied by them to antagonize the effects of opioids, did not exert any effect if it was given alone. Naloxone failed to induce hyperalgesia also in our laboratory in the rat tail flick test, whether giving it s.c. or i.c.v. as shown in Figures 4 and 5. Moreover, upon i.c.v. application even a slight but distinct prolongation of the tail flick latency could be observed just as by others.[15] Consequently, our data (Figure 5) confirm that even naloxone has a very weak but demonstrable agonist (analgesic) activity if applying it i.c.v. It might be assumed that upon peripheral application the consequences of displacement of endogenous opioids obscure the "hidden" and weak primary action.

As shown in Table 3, the hyperalgesic effect could be detected mostly in the hot plate test.[42,51,54,57,60,66] A further important detail is that mostly the latency of jumping reaction was shortened in the hot plate test, while the reaction time of paw licking was less uniformly affected.[42,54,60] It must, however, be taken into consideration, that according to the original description of the method[77] and also to the general usage, the paw licking movement is the nociceptive reaction measured and not the subsequent jumping.

Consequently, the hyperalgesic action of naloxone has been demonstrated by measuring the latency time of a reaction (jumping response) generally not used for measurement of the pain threshold. The latency time of the jumping reaction is longer than that of the licking reaction, thus it is reasonable to suppose that the first reaction is elicited by stimulation of the superficial receptors, while those mediating the jumping response are located deeper. Jacob and Ramabadran[60] suggested first that the superficial and the deep nociceptions were differentially influenced by naloxone. Such a differentiation is supported also by the electrophysiological experiments (Chapter 2, Volume II), where opiates inhibited only those units responding to nociceptive stimulation but not the ones activated by non-noxious stimuli. The distinction is also supported by the experimental data summarized in Table 3, especially if the conventional tail flick assay is supposed to reflect the activation of superficially located nociceptors. In the tail immersion test, where the analgesic action has been detected,[47,73,74] even weaker stimulus is applied. However, as for the writhing test, the data available are contradictory.[49,69] Even more difficult is the interpretation of experiments, where electric shocks served as painful stimuli,[44,46,53,55,67] as this kind of nociceptive stimulation is completely unnatural. In addition the human data are also difficult to compare with these animal experimental findings or to relate them to superficial or to deep pain, since in the

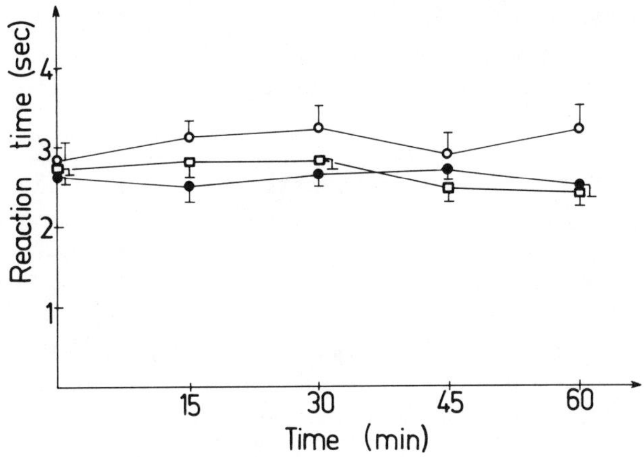

FIGURE 4. Effect of 2 (●) or 10 (O) mg/kg naloxone as compared with saline (□) on the latency time in the rat tail flick test. On the abscissa the time elapsed since treatment, on the ordinate the reaction time (Mean ± S.E.).

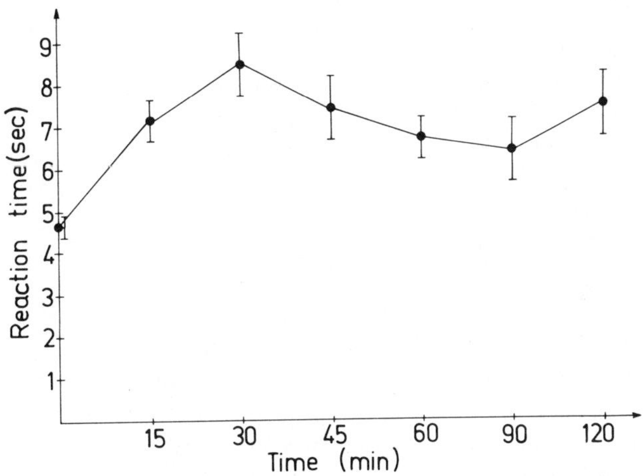

FIGURE 5. Effect of 25 μg naloxone (given i.c.v.) on the latency time in the rat tail flick test. See Figure 4, for further details.

clinical studies the rather intensive pain sensation has been unaffected by naloxone.[43,46,50,56,65,70] The most frequently used ischemic pain is really a deep sensation and independent of the superficial receptors of the skin. Thus, if naloxone had selectively potentiated the deep pain, just the ischemic pain should have been enhanced by naloxone. Naloxone also failed to modify unambiguously the pain experienced after tooth-extraction.[61,68] Accordingly, the "effectiveness" of naloxone does not seem to be correlated with either the pain modality or the localization of the nociceptors affected. Nevertheless, the data summarized in Table 3 do suggest that the outcome of these examinations is somehow related to the experimental conditions. Of the other possible variables probably three further factors deserve attention.

The direction of naloxone-induced alteration in the pain sensation might be correlated with the stimulus intensity. In experiments where analgesic effect was detected, relatively weak stimuli were applied, e.g., hot plate of relatively low temperature

(44.5°C),[58] or the tail was immersed into a not too hot bath.[47,73,74] In the much-debated work of Jacob et al.[42] naloxone induced hyperalgesia at relatively low temperature, and on increasing the temperature of the hot plate it gradually lost its hyperalgesic effect.

According to Bass et al.[55] the naloxone-induced facilitation of a startle response was enhanced if relatively more intense foot shocks were applied. Thus, its effectiveness seemed to be positively correlated with the stimulus strength. These seemingly contradictory data might be understood by supposing that the more intense stimuli maximally activate the pain pathways, which cannot be facilitated further by suppression of the endogenous pain-killing mechanisms.

The outcome of these experiments also might be dependent on the site of naloxone application. As seen in Table 3 hypo- and hyperalgesic effects have been observed upon central and peripheral drug administration.

A further important factor might be the "aversive character" of pain stimulus applied. In the tail flick or tail immersion assays the nociceptive reaction, i.e., withdrawal of the tail results in interruption of the pain stimulus. Theoretically the motor response ensues just at the moment when the thermal stimulus becomes aversive. In the hot plate test, however, licking of the paws gives only a temporary relief. The altered, nociceptive character of the pain sensation is reflected by the modified motor reaction: instead of licking a jumping response appears. Similarly, if electric shocks were applied and also an escape contingency was given as in experiments of Goldstein et al.,[46] where the rats found shelter jumping on a pole, no effect has been observed. In other experiments, however, the electrical stimulation could not be avoided[67,71] and certain behavioral manifestations of pain were enhanced, especially if the experimental conditions were already familiar to the animals.[67] These findings are in good agreement with the old clinical experience; opiates do not suppress the primary pain sensation (e.g., that induced by a pinprick) but only its unpleasant character, i.e., the unpleasant torturing feeling accompanying the pain sensation.

Such a relationship between aversive character and facilitation by naloxone of pain sensation is partially confirmed partially refuted by the competent human studies. Grevert and Goldstein[50,56] repeatedly reported that naloxone did not enhance the pain sensation in humans induced by ischemization of the upper arm or by immersing it into cold water. Their subjects did not experience stronger pain according to their rating.[50,56] Maybe in these carefully designed experiments the healthy subjects, otherwise free of any pain of pathological origin, rated the pain intensity itself and not its unpleasant character. However, in the same study[50] naloxone did increase the tension-anxiety score. Increased psychic tension-anxiety is surely an important manifestation of the aversive stimulation.

Interestingly Lindblom and Tegner[72] found the chronic clinical pain of neurological origin uninfluenced upon naloxone treatment, but it is possible that this uneffectiveness was due to the low dose (0.8 mg i.v.) applied. On the other hand, Levin et al.,[61,68] studying the effect of naloxone on pain rating in subjects recovering from anesthesia applied during surgical removal of impacted mandibular third molars, did observe an aggrevation of suffering. The obvious "advantage" of these observations[61,68] was that the suffering was of natural origin and the expectation of the subjects were probably worth more than in the laboratory experiments, as the pain sensation could not be stopped according to their wish. However, it cannot be excluded that the pain enhancing effect was due to the antagonism of the residual effects of general anesthetics used during the operation. Naloxone has already been suggested to antagonize the effects of certain general anesthetics.[78]

A further important variable is the pain "expectation", which can surely not be separated from its aversive character. Thus, Kaplan and Glick[67] found the incidence of squeaking and jumping responses elicited by intermittent footshocks uninfluenced by naloxone in "naive" rats. If the animals were repeatedly exposed to noxious stimulation, naloxone treatment enhanced the frequency of squeak reaction. On the other hand, in another experimental situation where the animals could avoid the footshock by jumping on a pole, naloxone had no effect.[46] In human studies naloxone did enhance the pain sensation in clinical situations,[61,68] but it did not suppress significantly the hypnotic analgesia.[64,65] It may be supposed that pain was anticipated after the surgical operation but this anticipation had been eliminated by the hypnosis. Expectation of otherwise unavoidable painful stimuli means anxiety in humans and the corresponding behavioral manifestations in animals.

Consequently, naloxone exerts its much-debated hyperalgesic effect if the nociceptive stimuli applied are intensive enough, aversive and inescapable, i.e., they induce anxiety. Probably the common denominator of these properties is their stressful character. As discussed in Chapter 3, Volume III stressful events result in instantaneous release of endogenous opioids, which may relieve suffering and tension. The expression "stressful event" may be interpreted in very different ways, but in the clinical practice it is used for such stimuli which arouse anxiety due to their excessive strength and/or aversive character. It is easy to suppose that the excessive thermal stimulation used in the hot plate test or the unavoidable electric footshocks or the lasting clinical pain are stressful, but the experimentally induced ischemic pain in humans, which may be interrupted at any time at the wish of the experimental subject or the mild stimuli used in the tail immersion assay or the tail flick test during which animals actively interrupt the noxious stimulation are not very stressful. Frid and Singer[65] reported that... "the decrement in hypnotic analgesia responses as a function of naloxone was much greater for the "stress" condition than for the "nonstress" condition". This connection between the hypnosis inhibitory effect of naloxone and stess was later extended to other experimental conditions as well,[70] showing that those subjects responded to naloxone for whom the experimental situation was stressful and vice versa.

Let us mention here that the repetition of analgesia testing alone resulted in a gradual lengthening of the latency time in the tail flick or hot plate assays.[79,80] Obviously the stress caused by handling of the animals resulted in release of endogenous opioids. In another very informative experiment Colpaert[81] induced chronic pain in rats by intradermal inoculation of heat inactivated Mycobacterium culture. This exposure to chronic pain resulted in a significant elevation of pain threshold which could be prevented by naloxone administration.[81] Thus, there is little doubt that under pathological conditions accompanied by lasting suffering, endogenous opioids are released.

In a recent study naloxone was found to enhance the effect of punishment, i.e., in rats trained to lever press for food reinforcement the disruptive action of electric footshocks was enhanced.[75] Of course this effect might be either due to decrease in pain threshold or to an enhancement of the aversive character of electric shocks. This observation proves that endogenous opioids are mobilized upon painful and/or stressful stimulation.

In humans extreme psychic stress may totally abolish the pain sensation. It is a well-known phenomenon that soldiers wounded on the battlefield frequently do not experience pain, but they do suffer after having been transported away to a safe place, where they are not exposed to danger of death.

Supposing that naloxone acts via antagonism of the stress-induced endorphin release, it is easy to understand why the analgesia induced by food deprivation was inhibited by it.[63] Similarly, if the adaptation to noxious stimuli is a consequence of the release of endogenous opioids, it should be reversed by naloxone. Thus Satoh et al.[62] observed that the gradual attenuation of flexor response elicited by repeated bradykinin injection could be inhibited by naloxone. Naloxone did not modify the flexor responses in those animals which did show this adaptation to noxious stimuli.[62] Consequently, the same stimuli might activate the endogenous opioids only in a subgroup of subjects. Individual differences in reactivity may exist also in humans. Levine et al.[68] reported that a subgroup of subjects showed a certain analgesic effect to placebo injection during dissipation of surgical anesthesia, while other individuals did not react to placebo. When naloxone was administered, it caused dose-dependent analgesia or hyperalgesia but only in the group of placebo responders.[68] In a related study Buchsbaum et al.[53] have shown that naloxone decreased the pain threshold in the relatively pain-insensitive subjects, while the pain-sensitive individuals did not show any reaction.

Summarizing these data it seems that naloxone does not modify the pain threshold in functionally intact individuals. However, if under stressful conditions endogenous opioids are mobilized, their effects will be antagonized by opiate antagonists. Such a conclusion is more or less in agreement with the above-cited data. Of course this conclusion implies that endogenous opioids do not act tonically, they are not involved in the continuous physiological regulation of the pain threshold, they are mobilized only in the cases of stressful external stimulation.

There are some data which suggest a continuous presence of endogenous opiate receptor ligands. After prolonged administration of naloxone the analgesic effect of morphine was enhanced.[82] In addition, after chronic exposure to another opiate antagonist (naltrexone), the foot-shock-induced analgesia was enhanced.[83] Supposing that the foot-shock-induced analgesia was due to the mobilization of endogenous opioid peptides,[83] this observation suggests that opiate antagonist treatment increases the sensitivity of opiate receptors not only to exogenous but also to endogenous opioids. The increased potency of exogenous opiates might also be a consequence of inhibited drug metabolism. The hepatic mixed function oxydase activity was reduced in mice treated chronically with naltrexone.[84] Thus, the above-observed changes in the sensitivity to opiates could be due to certain nonspecific metabolic effects. However, the increased sensitivity to endogenous peptides could hardly be explained in this way. On the other hand Lahti and Collins[85] demonstrated that chronic naloxone administration resulted in an increase in the number of opiate receptors in the brain, while the affinity of the receptors to their ligands did not change. Any apparent increase in the number of receptors might be a consequence of decreased occupancy by its natural ligands, especially if such changes are observed after prolonged administration of an antagonist. Nevertheless, this development reflects that the receptors are more or less continuously exposed to their natural agonist ligands, otherwise no such reaction characteristic of events ensuing upon displacement of physiological ligands would have developed, unless supposing that the living organism is more or less constantly exposed to stressful stimuli.

Moreover, there are some data on diurnal fluctuation of opioid content in the brain which is connected with a concomitant oscillation of pain threshold. Measuring the latency of jumping response in hot plate test, significantly longer latency times were recorded in the afternoon than in the morning.[86] The decrease in the pain sensitivity was accompanied by an increase in the whole brain total opioid content.[86] This regular diurnal increase and decrease in the enkephalin/endorphin contents suggests that these substances are occasionally liberated under particular environmental conditions. Nevertheless, it is difficult to suppose that endogenous substances synthesized and released regularly would not exert any tonic influence on their receptors.

Thus, though a considerable amount of data are available on the antinociceptive effects of endogenous opioids, very little is known on their physiological significance in the unstressed, normally functioning organism.

III. BEHAVIORAL EFFECTS OF OPIOID PEPTIDES

From a historical point of view, it is interesting that the behavioral effects of endorphins were described later than their antinociceptive actions. When Bloom et al.[9] and Jacquet and Marks[13] reported on the motor and autonomic effects of endorphins, which reminded them of those of the neuroleptic drugs, already several papers had been published on the analgesic effects of enkephalins[2,3] and endorphins.[4-7,25,27,87] Why had these latter authors failed to describe the striking behavioral symptoms (catalepsy, deep sedation, muscular rigidity, etc.) which could hardly be unnoticed by them? We do think that the failure of the first groups (among them ourselves too) to report on the conspicuous behavioral effects reflects conceptual differences. For pharmacologists who had been working with opiates previously, it was well known that morphine-like drugs elicited not only analgesia, but also a peculiar complex of symptoms where antinociception was only an easily measurable indicator of altered reactivity of the organism to external stimulation. Moreover, the first publications on the antinociceptive action of the newly discovered endogenous substances were probably composed in a certain haste (it was surely so in the cases of our reports),[5,8] where attention had been focused on the main effect. Moreover, the appearance of the well-known syndrome also accompanying the analgesic effect of opiates might be too evident to be worthy of special discussions. Nevertheless, it was the distinct merit of the latter authors[9,13] that by focusing the attention to the behavioral symptoms and describing the action of endorphins as a peculiar syndrome and not simply as antinociception, marked a new direction of endorphin research. In addition, in these pioneering works[9,13] even certain conclusions have been drawn on the possible physiological functions of endorphins in the psychic homeostasis of the organism and also on their role in the pathogenesis of some mental illnesses. Before going into details let us mention that in the first studies the similarities between the effects induced by opiates and those elicited by the opioid peptides were emphasized, while in the subsequent works the subtle but undeniable differences were stressed.

A. Behavioral Effects Upon Central Administration

As in some species (including humans) very different behavioral effects appear upon central and peripheral administration of opioids, it is probably better to discuss them separately. It must be mentioned in advance that upon central application of opioids, opiate-like effects can be observed in all species examined so far, while peripherally striking differences have been found.

Catalepsy, locomotor depression, and muscular rigidity are the most striking behavioral effects of β-EP[9,10,13,88-92] injecting it into the lateral ventricle of rats or directly into the PAG. In these animals no spontaneous movements can be seen, though they are not paralyzed, they may be startled by strong auditory stimuli. The rats do not react to noxious stimuli, thus analgesia is recorded in all tests. Also the postural reflexes (righting, contact placing) and many other reflexes (biting, grasping) are lost.[9,10] Lack of the spontaneous movements, loss of postural reflexes, and dramatically increased muscular tone result in a characteristic state called catalepsy. This catalepsy is extremely strong, the rats may be molded in different awkward positions, as they do not correct any unnatural posture, i.e., they may be laid on their back or sides or stretched on sharp edged objects.[9,10] Such a high degree of catalepsy cannot be evoked by neuroleptic drugs, thus Jacquet and Marks[13] introduced a new term "waxy flexibility". The same state was called "catatonia" by Tseng et al.[88] and "akinesia" by Izumi et al.[93] Later Segal et al.[92] and Browne et al.[94] introduced some new tests by the aid of which this peculiar state could be dissolved into two components such as "immobility" and "rigidity". In the classical tests to measure the catatonia inducing effects of neuroleptics the animals have to cling to a horizontal[95] bar or to a vertically placed column.[96] Otherwise, the animals treated with neuroleptic drugs are not particularly rigid, they cannot be placed on their side or back, i.e., they do not lose the righting reaction.[95,96] However, the rats treated with opioids or opiates slide off from the grid and also the postural reactions are lost due to the extreme rigidity and total areflexia, i.e., they may be set in any position just as piece of wood.[9,10,13,88-94,97]

The close correlation between the antinociceptive and cataleptogenic actions have been observed by subjective observation of the rats in the above-cited studies. The interdependence of these actions can also be demonstrated by comparing the corresponding ED_{50} values. Determining the analgesic ED_{50} values in the rat tail flick test, β-EP was found about 6 to 10 times stronger than morphine and D-Met, Pro-EA about 8 times more potent than β-EP upon i.c.v. treatment. Comparing the analgesic ED_{50} values of these substances to their cataleptogenic ED_{50} values, tested according to Murpurgo,[96] the ratios were around unity (0.5 to 3.0)[87,93] (Figure 6). Thus, according to the studies made in our laboratory (Figure 6) and those mentioned above roughly the same amounts are needed to elicit analgesia and catalepsy with morphine or opioid peptides.

Probably there is a good reason why rats have been used in these studies. Opiates (and also opioid peptides) are known to elicit primarily CNS depressant actions in this species (Chapter 2, Volume I) and the majority of behavioral experiments to measure the cataleptogenic actions of neuroleptic drugs were performed also in rats. Nevertheless, these symptoms, i.e., cessation of spontaneous locomotion, increase in muscular tone, and loss of certain postural reflexes have also been observed in golden hamsters,[89] in cats[6,7,17,98] and in mice,[97,99,100] but in these species the depressant actions differ from those observed in rats. In golden hamsters the loss of the righting reflex is the leading symptom.[89] The cats do not seem to be particularly sedated, they remain apparently alert, but "gazing into the emptiness" they do not react to the external stimuli and no spontaneous movements can be seen.[6,7,17,98] Mice show a characteristic hunched back posture and the spontaneous motility is inhibited for a short period.[97,100] A certain loss of the postural reflexes and an increase in the muscular tone can be observed in these species too,[7,89,97] though these effects are not so conspicuous as in rats. As mice and cats are known to react with excitation to opiates, they have not been specifically examined for the depressant symptoms upon treatment with opioid peptides either. In these species, if examined as mentioned above, rather the excitatory symptoms have been emphasized just as in the case of classical opiates (Chapter 2, Volume I). However, as mentioned previously, even in mice and in cats first the symptoms of sedative type appear.

It is worth mentioning that also in humans treated with β-EP i.c.v. to suppress the pain due to meningeal carcinomatosis, the analgesic action was accompanied by drowsiness and mild confusion just as upon treatment with morphine congeners.[101]

Otherwise, some autonomic symptoms have also been observed in the above-quoted studies such as piloerection, exophthalmos, and mydriasis in rodents and cats as well. Probably these symptoms (mydriasis and piloerection) reflecting increased sympathetic activity led some pharmacologists to overlook the initial sedation in mice and in cats treated whether with opiates or opioid peptides. Also the Straub phenomenon, a conspicuous manifestation of increased muscular tone, has been detected in mice, rats, and cats as well in the above-mentioned studies.

Upon repeated treatment, tolerance develops not only to the antinociceptive but also to the other behavioral effects of depressant type induced by opioids.[88,98] Actually upon repeated application of opioid peptides the inhibitory effects are replaced step by step by excitatory symptoms just as in the cases of morphine derivatives (Chapter 2, Volume I).

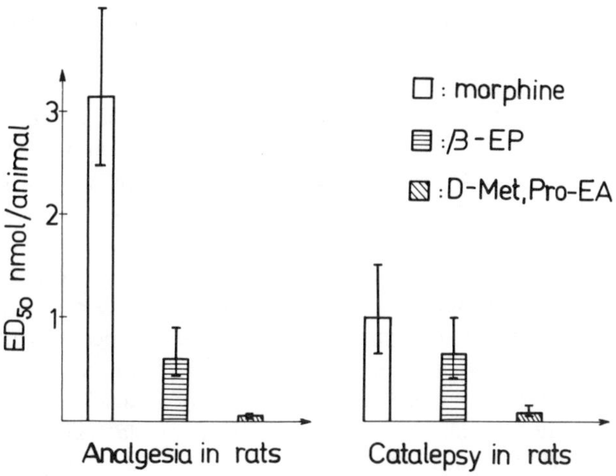

FIGURE 6. Comparison of the analgesic and cataleptogenic actions of morphine, β-EP and D-Met, Pro-EA upon i.c.v. treatment.

All these effects could be reversed by naloxone or prevented by it. Thus not only the analgesic but also the general depressant and cataleptogenic effects are induced primarily by the opiate receptors. As it will be discussed later, these effects are mediated mostly by the opiate receptor-induced alterations in other neurotransmitter functions and the different symptoms within this syndrome are brought about by modulation of different neurotransmitter activities just as in the case of classical opiates (Chapter 2, Volume I).

As for the structure activity relationships, at the beginning all the natural enkephalins and endorphins were reported to elicit symptoms similar to those induced by β-EP only in much higher dosage.[9,10,13] In addition, the potent enkephalin analogues were found to elicit these CNS depressant actions.[90,94,97,102,103] Thus, the analgesic and general behavioral depressant actions seem to be correlated also in the cases of the opioid peptides. The initial studies on these correlations were written in the first years of the endorphin research, when all the opioid peptides (including enkephalins) were thought to be generated from a common precursor (LPH) and also the opiate receptors mediating all the naloxone-sensitive in vivo effects were considered homogeneous.

That the situation is not so simple, was first suggested upon describing the "irregular" CNS action of γ-endorphin.[104] Then many other reports were published on the differential behavioral actions of opiates and opioid peptides. Thus, Tseng et al.[89] reported that in golden hamsters β-EP and (D-Ala^2, D-Leu^5)-E induced loss of the righting reflex, but morphine had no such effect. In rats the β-EP and morphine-induced analgesia and catalepsy were accompanied by muscular rigidity, while the animals treated with (D-Ala^2, D-Leu^5)-E were flaccid.[89] Also in rats, injecting the substances directly into the PAG, DALA and (D-Ala^2, D-Leu^5)-E, and β-EP induced "wet dog shake" before the development of analgesia and catalepsy, while morphine and etorphine did not elicit such a response.[90] Administering them into the lateral ventricles, β-EP induced "wet dog shake" but morphine did not, while administering them directly into the amygdala both elicited this movement.[94]

Otherwise *wet dog shake* is a phenomenon, which can hardly be related to the opiate receptors mediating analgesia and behavioral depression. It can also be elicited by serotonin and its effect reversed by mianserine and methysergide and by morphine too.[105] Moreover, wet dog shake can be elicited by many other neuropeptides as well.[94] Thus, besides accepting that this behavioral response is specifically mediated by the amygdala, the pharmacological analysis of this phenomenon does not provide any help in elucidation of the opiate receptor mechanisms.

Returning to the catalepsy, injecting the substances into the striatum, etonitazene elicited rigidity and loss of the righting reflex, while β-EP and D-Met, Pro-EA did not.[94] Giving them i.c.v. or into the PAG both opiates and opioid peptides elicited these responses.[94,92] Thus, contrary to the neuroleptic drugs, the striatal mechanisms are not essential for the cataleptic effect. Such a conclusion is substantiated further by the experiments of Dill and Costa,[106] who observed analgesia and catatonia if injecting morphine into the n. accumbens, while only moderate catalepsy upon striatal treatment. On the other hand, the cataleptogenic action of peripherally administered morphine could be reversed by injecting naltrexone into the n. accumbens, while striatal treatment had no effect.[106] According to Pert[107] and others[108] intranigral injection of morphine, β-EP and (D-Ala^2, Met^5)-E results in contralateral rotation, which can be blocked by haloperidol administration into the n. caudatus. Consequently, opioid peptides might actually facilitate the nigro-striatal system, while as discussed above they exert inhibition via the n. accumbens.

Also several drug interaction studies show that the opioid-induced catalepsy differs from that elicited by the classical neuroleptics.

It is well known that anticholinergic drugs inhibit the cataleptogenic action of neuroleptic drugs while leaving the similar action of opiates unaffected (Chapter 2, Volume I). In our laboratory the cataleptogenic action of several potent enkephalin analogues among them that of D-Met, Pro-EA was unaffected by atropine treatment (Miglécz and Székely unpublished observations). Stimulation by apomorphine[93] and 1-DOPA[91] of the dopaminergic receptors resulted in attenuation of β-EP-induced catalepsy, while its analgesic effect was not affected. Accordingly, in spite of the noninvolvement of striatal mechanisms, opioid-induced catalepsy is somehow connected with a decrease of the central dopaminergic tone just as in the case of antipsychotic drugs. Similarly to the neuroleptics, pretreatment with α-methyl-*p*-tyrosine, i.e., the blockade of the catecholamine synthesis resulted in potentiation of β-EP-induced rigidity.[93] In experiments of van Loon and Kim[109] the cataleptogenic effect of i.c.v. administered β-EP was potentiated by haloperidol and inhibited by apomorphine. Moreover, β-EP increased the striatal levels of DOPAC and HVA similarly to haloperidol, and apomorphine inhibited this metabolic effect of β-EP too.[109] Thus, though striatum is not the primary site responsible for the β-EP-induced catalepsy, it might be secondarily involved probably via pathways connecting the PAG with the striatum. If such a connection really exists, opioids exert a direct stimulant action on the dopamine releasing neurones in the substantia nigra and simultaneously an indirectly mediated inhibitory effect via the PAG.

Stimulation of the serotoninergic receptors by 5-hydroxytryptophan enhanced the cataleptogenic[91] and analgesic[37,98] effects of β-EP. Just as in the case of morphine-induced catalepsy, which is enhanced by drugs facilitating the serotoninergic transmission (clomipramine, quipazine, 1-tryptophan) and inhibited by serotonin receptor blocking drugs (methysergide).[110] Pobably the serotoninergic cells located in the PAG and activated by the opioids send inhibitory collaterals into the nigro-striatal circuitry as suggested by Izumi et al.[93] As for n. accumbens, Dill and Costa[106] suggested that the decrease in acetylcholine turnover in this area resulted in disinhibition of the GABAergic neurones impinging upon neurones in the PAG. Moreover, if the inhibition of dopaminergic functions is only a secondary consequence of activation by opioids of the cells in the PAG, it is easy to understand that 1-DOPA, apomorphine, and *p*-chlorophenylalanine suppress the opioid induced catalepsy without modifying the analgesic action. (Interestingly, naloxone was reported to potentiate the amphetamine-induced gnawing.[111] Therefore, the endogenous opioids might tonically suppress the dopaminergic neurones responsible for the amphetamine-induced gnawing.)

B. Hypermotility and Stereotypy

Opioid peptides similarly to the classical morphine-derivatives induce not only CNS depression but also excitatory symptoms. As in the case of the opiates, the opioid peptides have also been reported to elicit exclusively depressant actions in certain species, such as in rats (see above) and mostly excitation in other species (see below). However, as discussed in Chapter 2, Volume I a careful analysis of data revealed a biphasic pattern of action in almost all commonly used laboratory animals.

In rats generally only the sedative effects are emphasized, but as demonstrated by Tortella et al.[112] both morphine and DALA produced biphasic effects in rats upon i.c.v. treatment. The initial phase of behavioral stupor and synchronized EEG activity was followed by behavioral and EEG arousal.[112]

Segal et al.[92] observed that β-EP-induced amphetamine-like hypermotility after the dissipation of initial phase of rigidity in rats. Our laboratory also reported[97,99] that in rats (and in mice too) the behavioral depression induced by morphine, β-EP, and D-Met, Pro-EA was invariably followed by hypermotility. This phase of hyperactivity could be observed only after dissipation of the initial behavioral stupor the duration of which was positively correlated with the doses of opioids applied. Thus, it is not easy to find out why this late phase of hyperactivity has been more or less overlooked in many laboratories where the first phase of the behavioral action was so carefully analyzed. Probably the hyperactivity was actually observed but not considered a really specific effect, not without foundation, since the similar excitatory effects of opiates cannot be reversed by opiate antagonists if they have already been sufficiently evolved and no tolerance develops to the hypermotility (Chapter 2, Volume I). This biphasic character of action might, however, be a consequence of systemic, i.e., peripheral or i.c.v. application, when the substances may simultaneously reach very different brain regions. As discussed in Chapter 2, Volume I local application of opiates into the PAG or striatum resulted in monophasic CNS depression and excitation, respectively. For this reason it is conceivable that direct infusion of opioids into the PAG[10,13,16] really did not result in any "late hyperactivity". On the other hand, implanting enkephalins into the ventral thalamus, rubbing and licking have been observed reminding one of the similar action of morphine congeners, but the full picture of stereotypy did not evolve.[113] In addition also in rats intranigral microinjection of β-EP or DALA (and that of morphine) induced contralateral turning, continuous biting, and gnawing.[107,108] In the same studies injecting these opioids into the PAG, profound catalepsy and antinociception could be elicited.[107,108] If DALA was injected bilaterally into the substantia nigra, a clear-cut facilitation of the spontaneous locomotion was observed.[114] All these excitatory effects could be reversed by naloxone,[107,108,114] thus these effects are mediated by opiate receptors. Consequently, opioid peptides induce hypermotility via striatal and ventral thalamic mechanisms, while catalepsy and antinociception via the PAG. As discussed in Chapter 2, Volume I morphine induced an explosive hyperactivity syndrome in several cases if implanting it into the PAG, but nobody reported such an activity upon application of opioid peptides into this area.

Interestingly the data concerning the n. accumbens are contradictory. Dill and Costa observed catalepsy (and analgesia) upon injecting morphine into the n. accumbens, while Pert and Sivit[115] reported on hypermotility upon administering (D-Ala^2, Met^5)-E and morphine into the same area. A careful inspection of the figures published in the latter[115] report reveals that the excitatory effect appeared only 1 to 2 hr after treatment. Therefore, it cannot be excluded that in one study[106] only the initial phase was recorded, while in the other one[115] following up the events for 6 hr just the relatively short-lasting initial depression has been neglected. Being a part of the mesencephalo-diencephalic dopaminergic system, the stimulatory action via the n. accumbens is easy to interpret, while yet unidentified connections between this nucleus and the PAG might explain the opposite effects. It may be concluded that opioids can also induce excitatory behavioral effects in rats which are due to a direct stimulation of nigro-striatal and mesolimbic structures. An opposite effect exerted directly on the PAG would conceal this action in the phase of initial behavioral depression.

In mice the opioid peptides have been reported to cause mostly locomotor facilitation as also morphine-like opiates are believed to induce only the much discussed "running fit" as motor action (Chapter 2, Volume I).

First the natural enkephalins were reported to elicit dose-related hypermotility and this effect has probably been demonstrated more reliably than their alleged antinociceptive action.[116,117] Also the synthetic enkaphalin analogues elicit this hypermotility.[90,99] According to Wei et al.[90] the potency order of enkephalin derivatives and opiate alkaloids for stimulation of locomotor activity is correlated with their analgesic activity. Frigeni et al.,[117] however, obtained the opposite effects upon comparing the morphine and Met-E-induced analgesic and locomotor excitatory actions in different strains of mice. According to their experience the stronger was the locomotor stimulation, the weaker analgesic response could be recorded upon morphine treatment, while Met-E (given in high i.c.v. doses) did not alter the spontaneous activity at all.[117] Of course in this case again the locomotor facilitation might have an aspecifical influence, distorting the outcome of analgesia testing. In our experiments,[97,99] (see also Table 1 in Chapter 2, Volume I) both morphine and D-Met, Pro-EA elicited a biphasic effect, i.e., an initial hypoactivity was followed by hypermotility.[97,99] The initial locomotor inhibition was of roughly similar degree if comparing the effects of morphine and D-Met, Pro-EA in similar doses (upon s.c. application), but the locomotor facilitatory action was stronger with morphine.[99] Recently morphine has been reported to induce biphasic effect on locomotion also in another laboratory.[118] Unfortunately no data are available whether tolerance develops to the motility increasing effects of opioid peptides in mice (as to their cataleptogenic action in rats[88]) or not. It may probably be concluded *per analogiam* that opioid peptides induce biphasic alteration of the spontaneous motility in mice too and the frequently discussed excitatory action is not related to their analgesic effect, since the former might be rather a manifestation of the dopaminergic neurones but not of those mediating the analgesic effect.

In cats, treating them with opioid peptides, the excitatory symptoms prevail[6,7,17,98] but immediately after treatment a catalepsy-like action can be observed.[7]

Grooming — an interesting action of opiates and opioid peptides is the induction of grooming behavior. Grooming can be elicited by low i.c.v. doses of β-EP, morphine but also by β-MSH and ACTH(1-24).[119] Since this effect of β-EP can be reversed by naloxone,[120] the action might be mediated by a special subclass of opiate receptors, sensitive not only to opioids but also to ACTH fragments.

Seizure activity (wet dog shake) — as discussed in Chapter 2, Volume II, opioid peptides induce seizure activity if given in sufficiently high doses, similarly to the morphine derivatives. Consequently, the above-discussed increase in muscular tone, jerking movements of different parts of the body, and chronic convulsions are actually motor manifestations of this activity.[121,122] It has been suggested[121] that the much-discussed "wet dog shake" behavior, which can be more easily elicited by opioid peptides than by opiates,[90] was also a motor manifestation of convulsive behavior. Finally let us mention here, that for the time being almost nothing is known about the contingent behavioral actions of the newly discovered dynorphin. In the only related study Katz[123] reported on grooming and eating behavior and on barrel rotation observed in mice upon its i.c.v. administration.

C. Behavioral Effects Upon Peripheral Administration

1. Analgesia

Contrary to the classical morphine congeners, opioid peptides elicit significant analgesic action only upon central application. The antinociceptive action of natural enkephalins is rather uncertain also upon central administration (Table 1 in Chapter 3, Volume II) and upon peripheral application nobody observed such an activity. As for β-EP, its ineffectiveness in rodents upon i.v. and s.c. application has already been mentioned (Chapter 3, Volume II). In cats giving β-EP i.v. up to the dose of 1.0 mg/kg some peculiar autonomic reactions were elicited such as licking, vomiting, and relaxation of the nictitating membrane but neither analgesia nor the other well-known manifestations of feline mania.[29,97] Moreover, no antinociceptive effect has been observed in humans upon its i.v. administration[101] (see also Chapter 10, Volume III).

Thus, the natural opioid peptides have no significant analgesic action upon peripheral administration. This lack of effect is due either to their rapid catabolism in the plasma and in other peripheral tissues and/or to their incapability to pass through the blood-brain barrier (Chapter 3, Volume III). However, of the synthetic enkephalin analogues the most potent ones do have significant analgesic potency also upon i.v., or s.c. application. These derivatives having much stronger affinity to the opiate receptors and being more resistant to the metabolic degradation have more chance to activate the centrally located receptors responsible for the analgesic and behavioral effects than the natural opioids (Chapters 5 and 7, Volume III).

2. Other Behavioral Effects

In spite of their incapability to elicit analgesia and the related behavioral actions, natural enkephalins have been reported to induce some behavioral effects upon peripheral application. (Their effects on conditioned behavior will be discussed separately.)

First Plotnikoff et al.[124] observed dose-dependent potentiation of 1-DOPA induced motor responses in mice upon i.p. application of Met-E. In the same study[124] also the serotonin-induced motor effects were weakly potentiated, while the audiogenic seizures and footshock elicited fighting were slightly inhibited.

In the open-field test minute amounts of s.c. injected α-endorphin were found to increase the ambulation in one study,[125] while no effect was found in another laboratory.[126] According to the latter group (D-Ala^2)-γ-endorphin increased the spontaneous locomotion and defecation, which were interpreted as indices of heightened emotionality.[126] In another study 5 μg LPH(61-69) given s.c. resulted in fastening of the hippocampal EEG activity, also a sign of increased alertness.[125] DALA and β-EP were found to reduce the passive immobility in the swimming test, where otherwise the tricyclic antidepressants display a remarkable potency.[127,128] In this test, however, morphine was found inactive, while some enkephalin analogues void of any analgesic potency proved to be active.[128]

At present it is extremely difficult to interpret these motley effects of peripherally administered opioid peptides. Unfortunately in these experiments neither the dose-dependency of these actions nor their antagonizability by naloxone have been studied. For this reason it is questionable whether these actions are related to the opiate receptors or not.

IV. EFFECTS OF OPIOID PEPTIDES ON CONDITIONED REACTIONS

Contrary to naloxone and similar to the morphine derivatives, opioid peptides have been found to inhibit the operant behavior and other types of conditional reactions.[127,129-139] Two enzyme-resistant enkephalin analogues, similar to morphine caused a dose-related inhibition of lever pressing under a fixed ratio schedule of water reinforcement in water-deprived rats.[136] This response-rate decreasing effects could be reversed by naloxone. Consequently, the action was mediated by opiate receptors.[136] Similarly, β-EP and morphine elicited a dose-dependent depression of operant behavior of rats trained on a fixed ratio schedule of food reinforcement.[137,138] On a molar basis β-EP proved to be 21 times more potent than morphine.[137] It is interesting that the relative operant behavior depressant potency of β-EP is about 20 times higher than that of the morphine,[137,138] i.e., similar to its relative analgesic potency (Chapter 3, Volume II). Also this observation suggests that the receptors mediating this depressant action might be rather similar to those responsible for the analgesic effect. Otherwise, the inhibition by opioids of operant responding can be explained in more than one way. All substances having neuroleptic-cataleptogenic potency are potent inhibitors of the conditioned reactions as well. But this inhibition may also be a consequence of the drive-reduction induced by opioids. All substances having a euphorigenic effect decrease the sensation of thirst and hunger or their behavioral motivating actions and opioid peptides do have certain euphorigenic potency (see later). Furthermore, it is conceivable that the antinociceptive action is related to all kinds of unpleasant feelings also to those induced by food and water deprivation. Accordingly, opioids, like the opiates, may modulate the conditioned reactions in many ways all independent of any selective memory modulating action.

Naloxone, if given alone, had no effect on operant responding,[139] which might indicate that endogenous opioids do not play a very important role in the maintenance of this type of conditioned reactions.

Met-E, Leu-E, and (D-Ala^2, D-Leu^5)-E given i.p. before training inhibited the *acquisition of an active avoidance* response, the effect being naloxone insensitive.[129-131] It is difficult to stress enough that in these experiments[129-131] the natural enkephalins were administered peripherally in very small doses. Thus, though being almost inactive upon i.c.v. application in relation of analgesia, they do induce marked behavioral actions upon peripheral application. (Needless to emphasize that in these cases the actions should be mediated by primarily peripheral mechanisms.) The incapability of naloxone to reverse these actions suggests that the corresponding peripheral mechanisms are independent of the opiate receptors.

Training rats in the shuttle box, β-EP has been found to inhibit the formation of conditional reactions.[132] In this study[132] β-EP was administered i.p. in the minute amounts of 0.4 to 10 μg. Unfortunately the effect of naloxone on β-EP induced attenuation of performance has not been examined. As discussed in Chapter 2, Volume I morphine was reported to inhibit the acquisition of active avoidance behavior. Giving naloxone, in one work[129] no effect was found while in others[133,134] facilitation of acquisition was reported. Accordingly, the endogenous opioids might be supposed to inhibit the acquisition just as the exogenous opiates do. As discussed in Chapter 2, Volume I, among others the opiate receptors (and their ligands) located in the amygdala might be involved in these effects. Any interference with the acquisition of a new behavior does not necessarily mean that the memory formation itself is directly affected. A decrease in motivation might be suggested as a likely explanation of this inhibition.

There are, however, experimental situations, where the memory formation was rather facilitated by the opioids. Thus, both DALA and an enkephalin analogue with negligible opiate agonist potency; (D-Phe^4, Met^5)-E facilitated the learning in a 12 choice Warden maze upon i.p. application, just as heroin improved the performance in a 5-choice apparatus for pattern discrimination.[127] It must be emphasized that both the inhibition of acquisition of active avoidance[129] and the facilitation of performance in Warden maze[127] upon peripheral application of enkephalins were naloxone-resistant effects. Therefore, these actions should be independent of opiate receptors. This conclusion is probably valid for the experiments where naloxone-insensitive facilitation of *passive avoidance behavior* has been observed. Rigter[140,141] reported that in rats using a one-trial passive avoidance paradigm (step through apparatus), minute amounts of Met-E and Leu-E (0.0003 to 0.3 μ s.c.!!) attenuated the amnesic effect of exposure of the animals to 100% carbon dioxide atmosphere. Met-E displayed this antiamnesic action whether giving it before the acquisition or the retrieval, while Leu-E was effective only in the latter case and naloxone failed to reverse this action.[140] Consequently, opiate receptors are not involved in this action, since similar memory facilitation was observed in previous studies upon s.c. treatment with ACTH(4-10), α-MSH, and LPH (10 μg/rat).[125,142] The facilitation of passive avoidance behavior was observed also on i.c.v. treatment with natural enkephalins or morphine otherwise the action of Met-E being stronger than that of morphine.[143] Others reported on facilitation and attenuation of passive avoidance upon treatment with morphine[144] and naloxone,[145] respectively (see also Chapter 2, Volume I). Consequently, opioid peptides, similarly to the opiate facilitate the passive avoidance by peripheral or central application, as well, but the effect is not naloxone sensitive.[146] How could the natural enkephalins exert any action upon peripheral application in doses which are too low to elicit any significant action via central administration? Moreover, if the effects are not naloxone sensitive, they are not mediated by opiate receptors. In addition, why would be the actions of enkephalins and those of the opiates similar if these actions are not mediated by opiate receptors? Probably a yet unknown peripheral system should mediate this effect where enkephalins are stronger agonists than morphine. This system should be activated also by ACTH, α-MSH, and LPH, which have a common sequence (ACTH/4-10). However, the sequence of Met-E is included only in the chain of LPH but not in the α-MSH and ACTH. Moreover, the sequence of Leu-E which would have a similar action cannot be found in any of the parent molecules.

In partly related studies the extinction of conditional pole jumping reaction was found delayed by s.c. given Met-E, LPH(61-69), α-endorphin, β-EP, and ACTH(4-10).[125,147] Morphine also inhibited the extinction, while naltrexone accelerated the process, the effect of the former one not being dose-dependent.[125] ACTH(4-10) and opioid peptides decelerated the extinction upon i.c.v. treatment too and of the LPH fragments α-endorphin was found the most active.[125] This action could not be reversed by opiate antagonists contrary to the grooming, which is also a naloxone-sensitive action of ACTH and LPH fragments.[119]

De Wied and his co-workers[125] regard the extinction of unreinforced jumping response as a manifestation of memory loss and the facilitation of this process as a proper behavioral test of neuroleptic activity. However, thinking in pure Pavlovian terms this conclusion may be contested, since the extinction of an unreinforced conditioned response is also a kind of learning. Accordingly, attenuation of this process means decelerated acquisition of a new conditional reaction. Nevertheless, as haloperidol also facilitated the extinction under these experimental conditions,[148] it seems to be justified to regard the opposite effect (i.e., attenuation of the extinction) a symptom of improved memory function in this experimental paradigm. Moreover, as this behavioral effect of LPH fragments (having opioid activity) appeared at lower dose levels than analgesia the authors supposed that antinociception was not their primary physiological action,[125] a conclusion fully justified considering that their antiamnesic action was apparent upon peripheral application but these interesting findings of the Dutch group still require further confirmation.

Probably the most exciting finding of these authors is that α- and γ-endorphins, two LPH fragments differing from each other only by the presence (γ-endorphin) or absence (α-endorphin) of the residue LPH(77) have opposite behavioral effects.[147,148] γ-

endorphin given either s.c. or i.c.v. facilitated the extinction of pole jumping response and inhibited the acquisition of the above-mentioned step-through avoidance reflex, while α-endorphin had the opposite effect.[147,148] However, in open field[148] none of them had any significant effect, contrary to haloperidol. Similarity with haloperidol was found only in one behavioral assay, in the so-called "grip test", where γ-endorphin was "positive" similarly to haloperidol but α-endorphin was not.[148] Removing the amino terminal tyrosine residue of γ-endorphin (corresponding to LPH/61), the resulting des-tyrosine-γ-endorphin (LPH/62-771) proved to be an even more potent inhibitor of the conditional reflexes.[148,149] Subsequently the same group demonstrated that enzymes producing (des-Tyr¹)-γ-endorphin are present in the brain tissue (Chapter 3, Volume III). (Des-Try¹)-γ-endorphin similarly to γ-endorphin has no effect in the classical behavioral tests used for assaying the neuroleptic activity apart from its distinct action on conditioned reactions, i.e., it does not inhibit the amphetamine-or apomorphine-induced stereotypy, it does not elicit catalepsy and it inhibits only the vertical component of spontaneous locomotion.[148,150] Therefore, the authors supposed that LPH(62-77) was a natural neuroleptic peptide of the brain, maintaining the normal homeostasis of the brain functions and its reduced bioavailability (due to reduced synthesis or increased catabolism) were an etiological factor in the genesis of certain mental illnesses.[147,149] They also suppose that the other behavioral effects of neuroleptic drugs (e.g., catalepsy, strong sedation) are unspecific side effects not related to their antipsychotic actions[147,148] and the lack of effect of (des-Tyr¹)-γ-endorphin in receptor binding assay (against dopamine agonists or antagonists in vitro) is due to methodological difficulties.[151] Otherwise in a recent study (des-Tyr¹)-γ-endorphin was found to inhibit the spiperone binding to striatal and mesolimbic tissues in vivo.[152]

These data have been only partially confirmed so far. Moal et al.[153] found that s.c. administered α-endorphin inhibited while γ-endorphin facilitated the extinction of a conditional avoidance behavior but both inhibited the extinction of a thirst motivated conditional reaction. Thus, the outcome of these conditional reflex experiments depends to a large degree on the kind of reinforcement applied, but this group[153] confirmed that minute amounts (1 μg/rat) of s.c. administered opioid peptides may markedly modify the conditional behavior and the effects of different LPH fragments are diverse. According to a recent study β-EP, α-, and γ-endorphins and (des-Tyr¹)-γ-endorphin all elicited locomotor activation upon infusing them into the ventral tegmental area.[154] This locomotor activation could be antagonized by naloxone or by destruction of the mesocorticolimbic dopamine system.[154] Consequently, these data confirm that the different active fragments of LPH affect the central dopaminergic neurones, but not the above-discussed hypothesis on the differential actions of α- and γ-endorphins.

Thus, based on its general behavioral effects (upon i.c.v. application) β-EP seems to be an endogenous neuroleptic. Regarding its action on conditional reflex activity γ-endorphin might be a natural antipsychotic agent. Nevertheless, it cannot be excluded that all these behavioral effects reflect their opiate-like euphorigenic actions, thus they do not play specific role in the maintenance of normal psychic equilibrium.

V. EUPHORIGENIC EFFECTS AND STIMULUS PROPERTIES OF OPIOID PEPTIDES

A basic difference between the neuroleptics and opiates is that the formers are not euphorigenic. They have no rewarding properties, animals cannot be trained to self-administer neuroleptic drugs (see textbooks), and they cannot substitute for the opiates in state dependent learning paradigms (Chapter 2, Volume I). These observations are in good agreement with the clinical experience according to which haloperidol-like drugs contrary to opiates, psychostimulants, minor tranquilizers, and certain hypnotics have no addiction liability; an important point which should be considered upon discussing the supposed "endogenous antipsychotic action" of opioid peptides. Now it is difficult to guess whether the endogenous opioid peptides sharing the pharmacological properties of opiates and neuroleptics as well are euphorigenic or not.

First Belluzzi and Stein[143,155] reported that rats could be trained in a Skinner apparatus to press the lever for enkephalin infusion delivered directly into the lateral ventricle. Rewarding each lever pressing with 1 μg of peptide, higher rate of self-administration was observed than with 0.5 μg/lever press of morphine.[143] Rewarding with Leu-E resulted in a higher rate of responding than with Met-E and morphine, but the tetrapeptide Tyr-Gly-Gly-Phe had no effect.[143,155] The lack of reinforcing property in the case of the tetrapeptide corresponding to LPH(61-64) proves that this action is really related to the opiate receptors, as this fragment is practically inactive in all opiate-specific in vitro and in vivo assays (Chapter 5, Volume III). The stronger rewarding action of Leu-E if comparing it to Met-E, can be attributed to its "purer" agonist character (Chapter 5, Volume III). However, van Ree et al.[156] were unable to elicit self-administration of Met-E also by infusing it into the lateral ventricle. For the time being it is difficult to explain these discrepancies. Differences in the dosage cannot explain them, since in one laboratory 1 to 10 μg Met-E was offered,[143,155] while in the other 1 μg.[156] It must be stressed that no unambiguous dose-dependence was found in the laboratory reporting on the positive results.[143,155] Moreover, it would be difficult to explain, how enkephalins induce a rewarding effect in such a small dosage which was much lower than needed to elicit analgesia. Just the same authors[2] applied 100 to 200 μg for detection of the antinociceptive actions of Met-E and Leu-E.

However, van Ree et al.[156] did observe self-administration if β-EP was offered (0.05 μg/lever press). As the analgesic effect of β-EP upon i.c.v. administration is indisputable (Chapter 3, Volume II), this positive finding is easier to interpret. Let us mention here that according to the illustrations attached to these reports, a relatively low rate of self-administration has been found in both laboratories with all the opioid peptides examined.[143,155,156] Otherwise, β-EP has not been found to induce "craving" in human subjects upon i.v. injection (Chapter 10, Volume III), but this might be attributed to its incapability to pass the blood-brain barrier. However, no significant addictive properties were observed with the potent enkephalin analogue FK 33-824 either (Chapter 10, Volume III), i.e., with an opioid peptide surely able to get into the brain. Examining another peripherally active enkephalin analogue, D-Met, Pro-EA, no stable self-administering behavior could be established in monkeys upon its i.v. infusion.[157] Thus, in spite of the expectations based on the similarities between opiates and opioid peptides, even the very existence of addictive potency of opioid peptides requires further careful reexamination.

A related question is whether opioid peptides have stimulus properties at all. In one study β-EP acted as discriminative stimulus in rats trained to differentiate fentanyl and saline, while Met-E had no effect, injecting them into the n. raphe magnus.[156] Consequently, Met-E contrary to β-EP might have no cuing action at all in the dose applied or its subjective effects are different from those induced by β-EP. Considering the ineffectiveness of Met-E as analgesic, the first explanation seems to be acceptable. Browne and Fondren[158] trained rats to discriminate between morphine and saline, and stimulus generalization was observed if the former was substituted with methadone or etonitazene but not if D-Met, Pro-EA was given. In this experiment the substances were administered s.c., but if they were injected i.c.v., stimulus generalization could be observed between the opiates and the potent enkephalin analogue.[158] This observation is in good agreement with the lack of addictive potential of this peptide upon i.v. administration[157] but also with its capacity to induce tolerance development upon i.c.v.

treatment.[159] Hence the possibility arises that opioid peptides and opiates act on similar or identical receptors upon i.c.v. treatment, resulting not only in analgesia but also in addiction, while upon peripheral administration the two classes of opiate receptor ligands bind to at least partially different populations of receptors. However, as reported by Miglécz et al.,[159] in the case of this potent enkephalin analogue (D-Met, Pro-EA) dependence may develop to it upon peripheral administration (Chapter 9, Volume III). Accordingly, it is better to conclude that in the cases of opioid peptides the analgesic potency and their capacity to induce tolerance and dependence do not correlate with the addictive property especially if they are administered peripherally.

Yet the endogenous ligands of opiate receptors might be natural euphorigenic neurotransmitters (neuromodulators) of the brain. Electrical self-stimulation of brain areas otherwise rich in opiate receptors may be antagonized by naloxone.[143,155] Consequently, if such ligands really work in the brain, they may be either enkephalins, endorphins, or related yet unidentified substances. The possible euphorigenic potency of enkephalins and endorphins is also proven by the vast body of experimental data on their tolerance and physical dependence inducing capacity (Chapter 9, Volume III).

The inhibition by opioid peptides of thirst, hunger, and sexually motivated behavior (see later) is further indirect evidence of their euphorigenic potency. Obviously euphorigenic substances are expected to have a general drive reduction action. Of course if the opioid peptides are natural endogenous factors reducing all kinds of motivation induced by unpleasant external or internal stimuli, naloxone should exert an opposite effect. However, as discussed in the corresponding chapters, naloxone has no universal motivation increasing action, which means that endorphins do not have such an euphorigenic effect or they are not released tonically.

Thus, though there is a considerable amount of indirect evidence indicating that the endogenous opioids should be of addictive character, the direct experimental findings obtained in animal experiments are not unequivocal, while the corresponding human studies failed to detect such effects. It is worth mentioning that according to the human tolerability studies, giving FK 33-824 i.v., in spite of the presumed euphorigenic action rather unpleasant autonomic "side effects" can be observed.[160,161] However, giving β-EP intracerebrally to humans,[101] the observed psychic effects were more or less similar to those induced by peripheral application of the classical morphine derivatives. The unexpected autonomic reactions and the corresponding unpleasant subjective sensations upon peripheral application might be due to certain differences in the relative distribution in the organism of opiates and opioid peptides. The latter, including the peripherally active potent enkephalin analogues, cross the blood-brain barrier with relative difficulty (Chapter 3, Volume III). Consequently, their relative peripheral/brain concentration is always higher, i.e., to attain a certain brain concentration they must have a higher blood concentration than morphine. The higher peripheral concentration may give rise to such autonomic reactions, which being dysphoric suppress the centrally elicited euphoric actions.

VI. ANTAGONISM BY NALOXONE OF THE ANALGESIC AND BEHAVIORAL EFFECTS OF OPIOID PEPTIDES

As discussed previously, in spite of the basic similarity, there are certain differences between opiates and opioid peptides in relation to their central actions. Moreover, different natural and synthetic opioid peptides induce in part qualitatively different behavioral effects. These differences might be attributed to interaction of opioid peptides with nonopiate receptors (e.g., in relation to their effects on the conditional reactions). A further explanation of these differences is surely connected with their differential distribution in the organism, especially upon peripheral application. Even a third possibility should be counted with according to opiates and opioid peptides, or different opioid peptides would interact with different subclasses of the opiate receptors.

Such differences might be examined more precisely by receptor binding assay or by the specific isolated organ preparations. Actually different receptor populations might be proposed for the different actions if the corresponding pA_2 values also are different. In vitro just the careful determination of pA_2 (or Ke) values revealed that morphine derivatives and opioid peptides interacted with at least three different classes of opiate receptors (μ, δ, ε) (Chapter 1, Volume III) and the careful studies in spinal dogs revealed the existence of receptor types κ and σ which selectively bind certain morphine congeners (*N*-allyl-norphenazocine) and exert their actions in part by dopaminergic mechanisms respectively.[162] Therefore, the possibility arises that the observed differences are due to differential activation by opioids of these receptor subpopulations not only in vitro but also in vivo, since it has already been demonstrated that the pA_2 value of naloxone against opiates and opioid peptides are different in relation of several autonomic reactions (Chapter 4, Volume II). Thus, different receptor populations might mediate the behavioral and several autonomic actions of opioid peptides.

As for the antagonizability by naloxone of the analgesic actions of opiates and opioid peptides, not many data are available in the literature.

In our laboratory the pA_2 values of naloxone against morphine, β-EP and D-Met, Pro-EA were compared in the rat tail flick and mouse hot plate assays.[163] As shown in Figures 1 to 3 in Chapter 2, Volume I naloxone pretreatment resulted in parallel shifts of the dose-response curves of all the three opioids to the right i.e., the antagonism was of competitive type. All three agonists have been applied centrally and not peripherally since their capabilities to pass the blood-brain barrier are different (Chapter 3, Volume III). The pA_2 value of s.c. administered naloxone against the i.c.v. injected morphine, β-EP and D-Met, Pro-EA were 6.86, 6.79, and 6.73 in mice and 7.17, 7.55, and 7.51 in rats, respectively. The statistical comparison of these values led to the conclusions as follows.

- The pA_2 values obtained in mice did not differ significantly of each other, i.e., the same receptor population might mediate the antinociceptive effects of morphine and opioid peptides examined in mice (Figure 7).
- In rats the pA_2 values of naloxone against β-EP and D-Met, Pro-EA were almost identical but both proved to be higher than that of morphine (Figure 7). Consequently, in rats different receptors might be involved in the analgesic action of the opiate and opioid peptides under study.

In further experiments the relative cataleptogenic, conditional reflex inhibitory and analgesic actions displayed by the same substances were compared upon their central and peripheral application. The cataleptogenic effect has been assayed according to Morpurgo[96] in rats. For measuring the conditional reflex inhibitory action in rats, previously trained in single lever Skinner boxes, a discrete-trial avoidance escape schedule was used as described previously.[164,165] As mentioned in Chapter 3, Volume II and summarized in Figure 1 in Chapter 3, Volume II, the ED_{50} values for analgesia in the rat tail-flick test and catalepsy were roughly similar if giving morphine β-EP, or D-Met, Pro-EA i.c.v., though their relative potencies in both assays were very different. (The enkephalin analogue was about 80 times stronger than morphine on molar basis.) However, upon s.c. application D-Met, Pro-EA was twice weaker than morphine on weight basis (Figure 8) proving that also this potent enkephalin analogue passes the blood-brain barrier with difficulty. A further surprising finding was that in the case of morphine the cataleptogenic and conditional reflex inhibitory ED_{50} values were

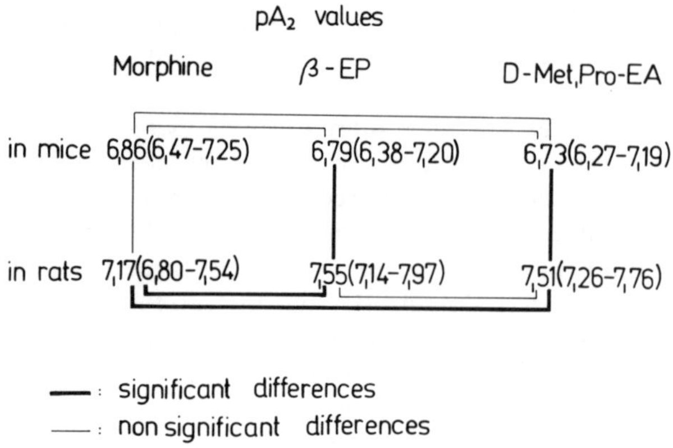

FIGURE 7.

about two and four times higher than its analgesic ED_{50}, while of D-Met, Pro-EA about six and ten times more was needed to elicit these "neuroleptic" effects than for analgesia (Figure 8). Consequently, different relative potencies were measured in the same behavioral test upon peripheral and central application. These findings are difficult to interpret otherwise than to assume that different receptor subpopulations are responsible for these different behavioral actions (analgesia, catalepsy, and inhibition of the previously established conditional reactions), and the relative affinities of morphine and opioid peptides to these receptors are also dissimilar.

For this reason in accordance with the in vitro studies, the opiate receptors should be regarded as a collective term, where the different subpopulations mediate different pharmacological effects and the relative affinities of different opioids to these subclasses are also different.

Finally, just as in the cases of many other effects in relation to the behavioral actions, the question arises, what happens if the endogenous opioids are displaced by naloxone of their receptors. Naloxone has been found to decrease the spontaneous locomotion, exploratory behavior, and rearing reactions under different experimental conditions.[166-169] As in subsedative doses anxiolytic agents are known to increase the locomotion and exploratory behavior in novel environment due to their disinhibitory action, these effects of naloxone were attributed to the loss of anxiolytic effect of endogenous opioids, i.e., to increased emotionality.[167-169] The validity of such interpretation was confirmed by Amir et al.,[166] who examined the effect of naloxone administration on motor behavior when rats were repeatedly exposed to a novel environment. The saline-treated controls displayed increased exploratory behavior with a concurrent decrease in several measures of emotional stress, such as defecation and freezing, while the animals treated with naloxone before each session, did not show any modification of behavior.[166] Therefore, endorphins might be involved in the adaptation to unusual external stimuli, i.e., without them a novel environment does not lose its anxiety-inducing effect.

VII. MODULATION BY OPIOID PEPTIDES (AND OPIATE ANTAGONISTS) OF FOOD AND WATER INTAKE

First Holtzman[170] reported that narcotic antagonists (naloxone, naltrexone, nalorphine) induced a dose-related *decrease in water consumption.* This action is stereospecific and the relative potencies of different antagonists were found proportional to their antagonist capacities observed in other opiate specfic tests.[171] Accordingly, the

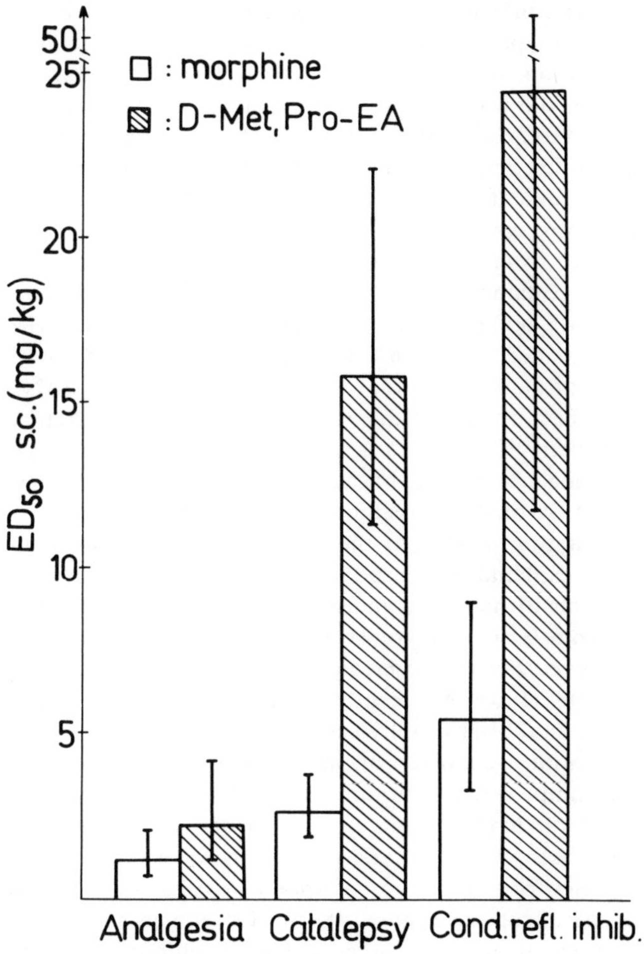

FIGURE 8. Comparison of the analgesic, cataleptogenic, and conditional reflex inhibitory actions of morphine and D-Met, Pro-EA upon s.c. treatment.

effect seems to be elicited by inhibition of opiate receptor-mediated mechanisms. However, also some narcotic agonists inhibit the water intake, though some of them have biphasic effect, i.e., increase in fluid intake in small dosage and a decrease at higher dose level.[170] In a subsequent study the partial agonists showed this biphasic action, while pure agonists such as morphine elicited a dose-related decrement.[172] Actually inhibition by naloxone of fluid intake is one of the few in vivo assays where naloxone, given it alone to drug naive animals, elicits a clear-cut dose-dependent effect. This antidipsogenic action was found unaltered in morphine-tolerant animals, which raises the question whether this effect was related to "regular" opiate receptors or not.[173] Unfortunately due to certain technical difficulties these observations were made 7 days after removal of the s.c. implanted morphine pellet, but at this time the animals still showed decreased analgesic response to morphine.[173] It would be easy to conclude that thirst is an enkephalin-mediated drive and naloxone acted in these experiments by displacement of the endogenous ligands. Supposing that this endogenous ligand has stronger affinity to the opiate receptors than the opiates, also the inhibitory action of "agonists" could be interpreted. However, the situation is not so simple. Injecting morphine into the hypothalamus, the dipsogenic effects of numerous procedures were inhibited such as intrahypothalamic application of carbachol, loading the organism with 10% saline solution, or angiotensin administration.[174] The natural enkephalins were found to inhibit the water intake.[175] De Caro et al.[176] reported that Leu-E and (D-Ala^2, D-Leu^5)-E given i.c.v. inhibited the spontaneous or angiotensin induced water intake in very low dosage and this effect was only partially antagonized by naloxone. (It is tempting to speculate that this action is mediated by δ-receptors, where morphine and naloxone are relatively weak agonist and antagonist ligands, respectively, and the strong action of the above-mentioned enkephalin analogue is due to its selective affinity to the δ-receptors (Chapter 5, Volume III). Of course the antidipsogenic effects of enkephalins and opiate agonists might be attributed to their antinociceptive action if this effect can be extended to all kinds of unpleasant stimuli, among them to thirst. However, the effect of naloxone cannot be explained in this way. Also the possibility of ADH release has been suggested due to a concomitant antidiuretic effect upon enkephalin administration.[175] Nevertheless, as discussed in Chapter 4, Volume II naloxone has no antidiuretic effect, and this action of opioid agonists is not accompanied with ADH liberation. A further possibility is to relate the inhibition of drinking to a general motor debilitating action. In the above-cited studies such small doses were applied that such an explanation can be refused. The main difficulty is obviously the similarity in the effects of naloxone and opioid agonists. Probably the only common effect of narcotic agonists and antagonists is that they are stressful. As stress is accompanied by ACTH and corticosteroid-mediated antidiuretic action, such an explanation seems to be acceptable. However, recently Brown et al.[177] demonstrated the inhibition by naloxone of water intake in hypophysectomized rats.

As for the modulation by opioids of *spontaneous food intake*, the data available are probably less incomplete and contradictory. Examining the interaction of naloxone with amphetamine, Holtzman[178] observed first that the former also inhibited the spontaneous food intake but only in rats and not in mice. Since the rat is one of the most commonly used species for examination of appetitive behavior, while the mouse is not, this finding obtained in rats determined the further course of experiments. Injecting naloxone in a slow release vehicle resulted in lasting inhibition of food intake, which was followed by increased caloric intake upon dissipation of the drug effect.[179]

Naloxone also inhibits the stress-related feeding.[180] A reduction in spontaneous food intake was observed in mice too, the effect being more pronounced in genetically obese than in normal "lean" mice.[181] The data of Sewell[182] probably deserves special attention as he observed an antagonism of suckling behavior in newborn rats upon naloxone treatment. Consequently, endogenous opioids might be involved in the stimulation of feeding in the early periods of postnatal life. Upon repeated administration, tolerance developed to the anorexigenic effect.[179] In good agreement with these findings[179] Yanaura and Suzuki[183] reported that morphine modified the eating pattern of rats, animals originally eating only during darkness ingested food also in daytime after treatment, i.e., its effect is opposite to that of the naloxone. As for the opioid peptides, Grandison and Guidotti[184] injecting β-EP into the ventromedial hypothalamus of satiated rats observed stimulated food intake, and this effect could be antagonized by naltrexone. As also muscimol a GABAergic agonist increased the food intake, and bicuculline inhibited both the muscimol and β-EP induced eating behavior, opiate receptor agonists were supposed to exert their effect via facilitation of the GABAergic mechanisms.[184] Moreover, considering that the satiety center is located in the ventromedial hypothalamus, and also the glucosensitive neurones can be found in this region, these data seem to reflect basic physiological mechanisms, since as discussed in Chapter 2, Volume II, the glucosensitive neurones of this area are sensitive to opiate agonists and antagonists. As opioids were reported to antagonize the glucose-induced facilitation of the cells in the satiety center (Chapter 2, Volume II), endogenous opioids might increase the eating by disinhibition of the reciprocally innervated eating center.

In a frequently quoted study Margules et al.[185] reported that in genetically obese mice and rats naloxone selectively abolished the overeating behavior. In these obese animals the pituitary β-EP content and its plasma level were found elevated, while its brain concentration and that of Leu-E were normal.[185] Thus, we are again faced with a contradiction, since the hypothalamic β-EP content was not found elevated,[185] but β-EP induced overeating if injected into this area.[184] Moreover, a subsequent study[186] revealed that the abnormally rapid weight gain, overeating, and the increase in pituitary β-EP content did not correlate with each other during the ontogenetic development. In this study[186] just the Leu-E content of the posterior pituitary correlated with the body weight. Of course the induction of overeating by β-EP[184] does not exclude that actually enkephalins are the physiological regulators of the food intake,[186] since the posterior hypophyseal nerve terminals belong to cell bodies located in the ventromedial hypothalamus (Chapter 3, Volume III). In a recent study[187] naloxone inhibited the 2-deoxy-D-glucose induced hyperphagia, indicating again the role of hypothalamic opioids.

There are, however, several other mechanisms, which also may contribute to the effects of opioids on eating behavior. As mentioned previously, Belluzzi and Stein[155,143] reported that i.c.v. administered enkephalins inhibited the lever pressing in Skinner apparatus for food reward. This effect was attributed to their euphorigenic action since in euphoric state all kinds of motivation also the one induced by food deprivation are reduced. Consequently, this effect should counteract the direct hypothalamic facilitation of food intake.

A further site of action is the endocrine pancreas. Its activity has been found altered by opioids (Chapter 5, Volume III), and any change in insulin release results in modified pattern of food intake.

In addition, the direct effects of opioids exerted on the secretory and motor activities of the gastrointestinal tract should be taken into consideration (Chapter 4, Volume II). Thus, McCloy and McCloy[188] proposed the enkephalinergic neurones in the gut initiated impulses resulting in feeling of the postprandial well being. This is obviously connected with the inhibition by enkephalins of secretory and motor activities of the bowels. If the intestinal opioids are released in increased amounts and tolerance develops to their inhibitory actions,[185,188] an overeating behavior might be the consequence. Of course this is only a hypothesis, but there is more than one way in the physiological regulation through which endogenous opioids can influence the eating behavior.

REFERENCES

1. **Hughes, J., Smith, T. W., Kosterlitz, H. W., Fothergill, L. A., Morgan, B. A., and Morris, H. R.,** Identification of two related pentapeptides from the brain with potent opiate agonist activity, *Nature (London)*, 258, 577, 1975.
2. **Belluzzi, J. D., Grant, N., Garsky, V., Sarantakis, D., Wise, C. D., and Stein, L.,** Analgesia induced *in vivo* by central administration of enkephalin in rat, *Nature (London)*, 260, 625, 1976.
3. **Büscher, H. H., Hill, R. C., Römer, D., Cardinaux, F., Closse, A., Hauser, D., and Pless, J.,** Evidence for analgesic activity of enkephalin in the mouse, *Nature (London)*, 261, 423, 1976.
4. **Loh, H. H., Tseng, L. F., Wei, E., and Li, C. H.,** β-endorphin is a potent analgesic agent, *Proc. Natl. Acad. Sci. USA*, 73, 2895, 1976.
5. **Gráf. L., Székely, J. I., Rónai, A. Z., Dunai-Kovacs, Z., and Bajusz. S.,** Comparative study on analgesic effect of Met[5]-enkephalin and related lipotropin fragments, *Nature (London)*, 263, 240, 1976.

6. **Feldberg, W. and Smyth, D. G.**, The C-fragment of lipotropin — a potent analgesic, *J. Physiol. (London)*, 260, 31P, 1976.
7. **Feldberg, W. and Smyth, D. G.**, C-fragment of lipotropin-an endogenous potent analgesic peptide, *Br. J. Pharmacol.*, 60, 445, 1977.
8. **Székely, J. I., Rônai, A. Z., Dunai-Kovács, Z., Grâf, L., and Bajusz, S.**, C-terminal fragment (residues 61-91) of β-lipotropin: is it the natural opiate-like neurohormone of the brain? *Experientia*, 33, 54, 1977.
9. **Bloom, F., Segal, D., Ling, N., and Guillemin, R.**, Endorphins: profound behavioral effects in rats suggest new etiological factors in mental illness, *Science*, 194, 630, 1976.
10. **Jacquet, Y., Marks, N., and Li, C. H.**, Behavioral and biochemical properties of "opioid" peptides, in *Proc. Int. Narcotic Res. Club Meeting*, Kosterlitz, H. W., Ed., North-Holland, Amsterdam, 1976, 411.
11. **Chang, J.-K., Fong, B. T. W., Pert, A., and Pert, C. B.**, Opiate receptor affinities and behavioral effects of enkephalin: structure-activity relationship of ten synthetic peptide analogues, *Life Sci.*, 18, 1473, 1976.
12. **Leybin, L., Pinsky, C., LaBella, F. S., Havlicek, V., and Rezek, M.**, Intraventricular Met^5-enkephalin causes unexpected lowering of pain threshold and narcotic withdrawal signs in rats, *Nature (London)*, 264, 258, 1976.
13. **Jacquet, Y. F. and Marks, N.**, The C-fragment of β-lipotropin: an endogenous neuroleptic or antipsychotogen? *Science*, 194, 632, 1976.
14. **Cowan, A., Doxey, J. C., and Metcalf, G.**, A comparison of pharmacological effects produced by leucine-enkephalin, methionine-enkephalin, morphine and ketocyclazocine, in *Proc. Int. Narcotic. Res. Club Meeting*, Kosterlitz, H. W., Ed., North-Holland, Amsterdam, 1976, 95.
15. **Dewey, W. L., Chau-Pham, T. T., Day, A., Lujan, M., Harris, L. S., and Freer, R. J.**, The effects of enkephalins on the isolated guinea pig ileum stereospecific binding of dihydromorphine and antinociception in mice, in *Proc. Int. Narcotic Res. Club Meeting*, Kosterlitz, H. W., Ed., North-Holland, Amsterdam, 1976, 103.
16. **Malick, J. B. and Goldstein, J. M.**, Analgesic activity of enkephalins following intracerebral administration in the rat, *Life Sci.*, 20, 827, 1977.
17. **Meglio, M., Hosobuchi, Y., Loh, H. H., Adams, J. E., and Li, C. H.**, β-endorphin: Behavioral and analgesic activity in cats, *Proc. Natl. Acad. Sci. USA*, 74, 774, 1977.
18. **Yaksh, T. L., Huang, S. P., Rudy, T. A., and Frederickson, R. C. A.**, The direct and specific opiate-like effect of Met^5-enkephalin and analogues on the spinal cord, *Neuroscience*, 2, 593, 1977.
19. **Takagi, H., Satoh, M., Akaike, A., Shibata, T., Yajima, H., and Ogawa, H.**, Analgesia by enkephalins injected into the nucleus reticularis gigantocellularis of rat medulla oblongata, *Eur. J. Pharmacol.*, 49, 113, 1978.
20. **Racagni, G., Bruno, F., Iuliano, E., Longiave, D., Mandelli, V., and Berti, F.**, Comparative *in vitro* and *in vivo* studies between morphine and methionine-enkephalin: genotype dependent response in two different strains of mice, in *Advances in Biochemical Psychopharmacology*, Vol. 18, Costa, E. and Trabucchi, M., Eds., Raven Press, New York, 1978, 289.
21. **Shaw, J. S., Turnbull, M. J., Dutta, A. S., Gormley, J. J., Hayward, C. F., and Stacey, G. J.**, A structure-activity study with enkephalin analogs: further evidence for multiple opiate receptor types, in *Proc. Int. Narcotic Res. Conf.*, van Ree, J. M. and Terenius, L., Eds., Elsevier/North-Holland, Amsterdam, 1978, 185.
22. **Ueda, H., Amano, H., Shiomi, H., and Takagi, H.**, Comparison of the analgesic effects of various opioid peptides by a newly devised intracisternal injection technique in conscious mice, *Eur. J. Pharmacol.*, 56, 265, 1979.
23. **Woolf, C. J. and Mitchell, D.**, Analgesic action of Met-enkephalin and 5-HT microinjected on to the spinal cord of the rat, *S. African J. Sci.*, 75, 37, 1979.
24. **Larson, A. A., Vaught, J. L., and Takemori, A. E.**, The potentiation of spinal analgesia by leucine-enkephalin, *Eur. J. Pharmacol.*, 61, 381, 1980.
25. **Rônai, A. Z., Székely, J. I., Grâf, L., Dunai-Kovács, Z., and Bajusz, S.**, Morphine-like analgesic effect of a pituitary hormone, β-lipotropin, *Life Sci.*, 19, 733, 1976.
26. **Grâf, L., Rônai, A. Z., Bajusz, S., Cseh, G., and Székely, J. I.**, Opioid agonist activity of β-lipotropin fragments: a possible biological source of morphine-like substances in the pituitary, *FEBS Lett.*, 64, 181, 1976.
27. **Tseng, L. F., Loh, H. H., and Li, C. H.**, β-endorphin as a potent analgesic by intravenous injection, *Nature (London)*, 263, 239, 1976.
28. **Yaksh, T. L. and Henry, J. L.**, Antinociceptive effects of intrathecally administered human β-endorphin in the rat and cat, *Can. J. Physiol. Pharmacol.*, 56, 754, 1978.
29. **Catlin, D. H., George, R., and Li, C. H.**, β-endorphin: pharmacologic and behavioral activity in cats after low intravenous doses, *Life Sci.*, 23, 2147, 1978.

30. **Tseng, L. F., Loh, H. H., and Li, C. H.,** (D-Thr[2], Thz[5])-enkephalinamide: a potent analgesic by subcutaneous and oral administration, *Life Sci.*, 23, 2053, 1978.
31. **Li, C. H.,** β-endorphin: a pituitary peptide with potent morphine-like activity, *Arch. Biochem. Biophys.*, 183, 592, 1977.
32. **Geisow, M. J., Deakin, J. F. W., Dostrovsky, J. O. and Smyth, D. G.,** Analgesic activity of lipotropin C-fragment depends on carboxyl terminal tetrapeptide, *Nature (London)*, 269, 167, 1977.
33. **Li, C. H., Tseng, L. F., and Yamashiro, D.,** β-endorphin: complete primary structure is required for full analgesic activity, *Biochem. Biophys. Res. Commun.*, 85, 795, 1978.
34. **Bergmann, F., Altstetter, R., and Weissman, B. A.,** *In vivo* interaction of morphine and endogenous opiate-like peptides, *Life Sci.*, 23, 2601, 1978.
35. **Fu, T. C. and Dewey, W. L.,** Morphine antinociception: evidence for the release of endogenous substance/s/, *Life Sci.*, 23, 53, 1979.
36. **Dewey, W. L., Fu, T. C., Izazola-Conde, C., and Halenda, S.,** Characterization of cerebrospinal fluid from morphine treated rabbits on rodents, in *Proc. Int. Narcotic Res. Conf.*, van Ree, J. M. and Terenius, L., Eds., Elsevier/North-Holland, Amsterdam, 1978, 271.
37. **Lin, M. T., Chi, M. L., Chandra, A., and Tsay, B. L.,** Serotoninergic mechanisms of beta-endorphin- and clonidine-induced analgesia in rats, *Pharmacology*, 20, 323, 1980.
38. **Larson, A. A., Vaught, J. L., and Takemori, A. E.,** Potentiation of the analgesic effects of morphine (M) and Met-enkephalin (ME) at the spinal level by pretreatment with Leu-enkephalin, *Pharmacologist*, 21, 198, 1979.
39. **Vaught, J. L. and Takemori, A. E.,** A further characterization of the differential effects of leucine enkephalin, methionine enkephalin and their analogs on morphine-induced analgesia, *J. Pharmacol. Exp. Ther.*, 221, 280, 1979.
40. **Ferreira, S. H. and Nakamura, M.,** Prostaglandin hyperalgesia: the peripheral analgesic activity of morphine, enkephalins and opioid antagonists, *Prostaglandins*, 18, 191, 1979.
41. **Ferreira, S. H. and Nakamura, M.,** Prostaglandin hyperalgesia; relevance of the peripheral effect for the analgesic action of opioid-antagonists, *Prostaglandins*, 18, 201, 1979.
42. **Jacob, J. J., Tremblay, E. C., and Colombel, M. C.,** Facilitation de réactions nociceptives par la naloxone chez la souris et chez le rat, *Psychopharmacologia (Berl.)*, 37, 217, 1974.
43. **Goldstein, A. and Hilgard, E. R.,** Failure of opiate antagonist naloxone to modify hypnotic analgesia, *Proc. Natl. Acad. Sci. USA*, 72, 2041, 1975.
44. **El-Sobky, A., Dostrovsky, J. O., and Wall, P. D.,** Lack of effect of naloxone on pain perception in humans, *Nature (London)*, 263, 873, 1976.
45. **Harakal, C., Tallarida, R. J., Geller, E. B., Biunno, I., Adler, M. W., and Maslow, J.,** Morphine analgesia in rats as a function of time. I. Modification of morphine dose-response curves by naloxone. II. Kinetic analysis for correcting apparent pA_2 values *in vivo*, *Pharmacologist*, 18, 174, 1976.
46. **Goldstein, A., Pryor, G. T., Otis, L. S., and Larsen, F.,** On the role of endogenous opioid peptides: failure of naloxone to influence shock escape threshold in the rat, *Life Sci.*, 18, 599, 1976.
47. **Sewell, R. D. E. and Spencer, P. S. J.,** Antinociceptive activity of narcotic agonist and partial agonist analgesics and other agents in the tail-immersion test in mice and rats, *Neuropharmacology*, 15, 683, 1976.
48. **Jacob, J. J. and Michaud, G. M.,** Production par la naloxone d'effects inverses de ceux de la morphine chez le chien éveillé, *Arch. Int. Pharmacodyn.*, 222, 332, 1976.
49. **Elliott, H. W., Spiehler, V., and Navarro, G.,** Effect of naloxone on antinociceptive activity of phenoxybenzamine, *Life Sci.*, 19, 1637, 1976.
50. **Grevert, P. and Goldstein, A.,** Effects of naloxone on experimentally induced ischemic pain and on mood in human subjects, *Proc. Natl. Acad. Sci. USA*, 74, 1291, 1977.
51. **Frederickson, R. C. A., Burgis, V., and Edwards, J. D.,** Hyperalgesia induced by naloxone follows diurnal rhythm in responsitivity to painful stimuli, *Science*, 198, 756, 1977.
52. **Crowley, W. R., Rodriguez-Sierra, J. F., and Komisaruk, B. R.,** Analgesia induced by vaginal stimulation in rats is apparently independent of a morphine-sensitive process, *Psychopharmacology*, 54, 223, 1977.
53. **Buchsbaum, M. S., Davis, G. C., and Bunney, W. E., Jr.,** Naloxone alters pain perception and somatosensory evoked potentials in normal subjects, *Nature (London)*, 270, 620, 1977.
54. **Grevert, P. and Goldstein, A.,** Some effects of naloxone on behavior in the mouse, *Psychopharmacology*, 53, 111, 1977.
55. **Bass, M. B., Friedman, H. J., and Lester, D.,** Antagonism of naloxone hyperalgesia by ethanol, *Life Sci.*, 22, 1939, 1978.
56. **Grevert, P. and Goldstein, A.,** Endorphins: naloxone fails to alter experimental pain or mood in humans, *Science*, 199, 1093, 1978.
57. **Frederickson, R. C. A., Wesche, D. L., and Richter, J. A.,** Mouse brain enkephalins: study of diurnal changes correlated with changes in nociceptive sensitivity, in *Proc. Int. Narcotic Res. Conf.*, van Ree, J. M. and Terenius, L., Eds., Elsevier/North-Holland, Amsterdam, 1978, 169.

58. **Pinsky, C., LaBella, F. S., Havlicek, V., and Dua, A. K.,** Apparent central agonist actions of naloxone in unrestrained rat, *Proc. Int. Narcotic Res. Conf.*, van Ree, J. M. and Terenius, L., Eds., Elsevier/North-Holland, Amsterdam, 1978, 439.
59. **North, M. A.,** Naloxone reversal of morphine analgesia but failure to alter reactivity to pain in the formaline test, *Life Sci.*, 22, 295, 1978.
60. **Jacob, J. J. C. and Ramabadran, K.,** Enhancement of a nociceptive reaction by opioid antagonists in mice, *Br. J. Pharmacol.*, 64, 91, 1978.
61. **Levine, J. D., Gordon, N. C., Jones, R. T., and Fields, H. L.,** The narcotic antagonist naloxone enhances clinical pain, *Nature (London)*, 272, 826, 1978.
62. **Satoh, M., Kawajiri, S., Yamamoto, M., Makino, H., and Takagi, H.,** Reversal by naloxone of adaptation of rats to noxious stimuli, *Life Sci.*, 24, 685, 1979.
63. **McGivern, R., Berka, C., Berntson, G. G., Walker, J. M., and Sandman, C. A.,** Effect of naloxone on analgesia induced by food deprivation, *Life Sci.*, 25, 885, 1979.
64. **Nasrallah, H. A., Holley, T., and Janowsky, D. S.,** Opiate antagonism fails to reverse hypnotic-induced analgesia, *Lancet*, 1, 1355, 1979.
65. **Frid, M. and Singer, G.,** Hypnotic analgesia in conditions of stress is partially reversed by naloxone, *Psychopharmacology*, 63, 211, 1979.
66. **Carmody, J. J., Carrol, P. R., and Morgans, D.,** Naloxone increases pain perception in rats and mice, *Life Sci.*, 24, 1149, 1979.
67. **Kaplan, R. and Glick, S. D.,** Prior exposure to foot-shock induced naloxone hyperalgesia, *Life Sci.*, 24, 2309, 1979.
68. **Levine, J. D., Gordon, N. C., and Fields, H. L.,** Naloxone dose dependently produces analgesia and hyperalgesia in postoperative pain, *Nature (London)*, 278, 740, 1979.
69. **Ramabadran, K. and Jacob, J. J. C.,** Stereospecific effects of opiate antagonists on superficial and deep nociception and on motor activity suggest involvement of endorphins on different opioid receptors, *Life Sci.*, 24, 1959, 1979.
70. **Frid, M., Singer, G., and Rana, C.,** Interactions between personal expectations and naloxone: effects on tolerance to ischemic pain, *Psychopharmacology*, 65, 225, 1979.
71. **Wilcox, R. E., Mikula, J. A., and Levitt, R. A.,** Periaqueductal gray naloxone microinjections in morphine-dependent rats: hyperalgesia without "classical" withdrawal, *Neuropharmacology*, 18, 639, 1979.
72. **Lindblom, U. and Tegner, R.,** Are the endorphins active in clinical pain states? Narcotic antagonism in chronic pain patients, *Pain*, 7, 65, 1979.
73. **Woolf, C. J.,** Intracerebral naloxone and the reaction to thermal noxious stimulation in the rat, *Brain Res.*, 190, 578, 1980.
74. **Woolf, C. J.,** Analgesia and hyperalgesia produced in the rat by intrathecal naloxone, *Brain Res.*, 189, 593, 1980.
75. **Young, G. A.,** Naloxone enhancement of punishment in the rat, *Life Sci.*, 26, 1787, 1980.
76. **Foldes, F. F., Lunn, J. N., Moore, J., and Brown, J. M.,** *N*-allyl-noroxymorphone: A new potent narcotic antagonists, *Am. J. Med. Sci.*, 245, 23, 1963.
77. **Eddy, N. B., Fuhrmeister Touchberry, C., and Lieberman, J. E.,** Synthetic analgesics. I. Methadone isomers and derivatives, *J. Pharmacol.*, 98, 121, 1950.
78. **Furst, Z., Foldes, F. F., and Knoll, J.,** The influence of naloxone on barbiturate anesthesia and toxicity in the rat, *Life Sci.*, 20, 921, 1977.
79. **Havdala, H. S., Comaty, J. E., Borison, R. L., and Diamond, B. I.,** Chronic naltrexone enhancement of morphine action, *Anesthesiology*, 51, 235, 1979.
80. **Havdala, H. S., Comaty, J. E., Borison, R. L., and Diamond, B. I.,** Chronic naltrexone sensitization to morphine analgesia, *Pharmacologist*, 21, 159, 1979.
81. **Colpaert, F. C.,** Can chronic pain be suppressed despite purported tolerance to narcotic analgesia? *Life Sci.*, 24, 1201, 1979.
82. **Thang, A. H. and Collins, R. J.,** Enhanced analgesic effect of morphine after chronic administration of naloxone in the rat, *Eur. J. Pharmacol.*, 47, 473, 1978.
83. **Amir, S. and Amit, Z.,** Enhanced analgesic effects of stress following chronic administration of naltrexone in rats, *Eur. J. Pharmacol.*, 59, 137, 1979.
84. **Lehman, T. M., Pyati, P., and Peterson, G. R.,** Inhibition of drug metabolism by chronically administered naltrexone, *Life Sci.*, 25, 1591, 1979.
85. **Lahti, R. A. and Collins, R. J.,** Chronic naloxone results in prolonged increases in opiate binding sites in brain, *Eur. J. Pharmacol.*, 51, 185, 1978.
86. **Wesche, D. L. and Frederickson, R. C. A.,** Diurnal differences in opioid peptide levels correlated with nociceptive sensitivity, *Life Sci.*, 24, 1861, 1979.
87. **Teschemacher, Hj., Bläsig, J., and Kromer, W.,** Porcine pituitary peptides with opiate-like activity: partial purification and effects in the rat after intraventricular injection, *Naunyn-Schmiedeb. Arch. Pharmacol.*, 294, 293, 1976.

88. **Tseng, L.-F., Loh, H. H., and Li, C. H.**, Human β-endorphin: development of tolerance and behavioral activity in rats, *Biochem. Biophys. Res. Commun.*, 74, 390, 1977.
89. **Tseng, L.-F., Ostwald, T. J., Loh, H. H., and Li, C. H.**, Behavioral activities of opioid peptides and morphine sulfate in golden hamsters and rats, *Psychopharmacology*, 64, 215, 1979.
90. **Wei, E. T., Tseng, L. F., Loh, H. H., and Li, C. H.**, Comparison of the behavioral effects of β-endorphin and enkephalin analogs, *Life Sci.*, 21, 321, 1977.
91. **Motomatsu, T., Lis, M., Seidah, N., and Chretien, M.**, Cataleptic effect of 61-91 beta-lipotropin in rat, *Clin. Res.*, 24, 660A, 1976.
92. **Segal, D. S., Browne, R. G., Bloom, F., Ling, N., and Guillemin, R.**, β-Endorphin: endogenous opiate or neuroleptic? *Science*, 198, 411, 1977.
93. **Izumi, K., Motomatsu, T., Chretien, M., Butterworth, R. F., Lis, M., Seidah, N., and Barbeau, A.**, β-endorphin induced akinesia in rats: effect of apomorphine and α-methyl-*p*-tyrosine and related modifications of dopamine turnover in the basal ganglia, *Life Sci.*, 20, 1149, 1977.
94. **Browne R. G., Derrington, D. C., and Segal, D. S.**, Comparison of opiate- and opioid-peptide-induced immobility, *Life Sci.*, 24, 933, 1979.
95. **Costall, B. and Naylor, R. J.**, Neuroleptic and non-neuroleptic catalepsy, *Arzneim.-Forsch. (Drug Res.)*, 23, 674, 1973.
96. **Morpurgo, C.**, Effects of antiparkinson drugs on a phenotiazine-induced catatonic reaction, *Arch. Int. Pharmacodyn.*, 137, 84, 1962.
97. **Székely, J. I., Dunai-Kovács, Z., Miglécz, E., and Tarnawa, I.**, The *in vivo* pharmacology of (D-Met², Pro⁵)-enkephalinamide, in *Endorphins'78*, Gráf, L., Palkovits, M., Rónai, A. Z., Eds., Publishing House of the Hungarian Academy of Science, Hungary, 1978, 319.
98. **Hosobuchi, Y., Meglio, M., Adams, J. E., and Li, C. H.**, β-endorphin: development of tolerance and its reversal by 5-hydroxytryptophan in cats, *Proc. Natl. Acad. Sci. USA*, 74, 4017, 1977.
99. **Székely, J. I., Miglécz, E., and Rónai, A. Z.**, Biphasic effects of a potent enkephalin analog (D-Met², Pro⁵)-enkephalinamide and morphine on locomotor activity in mice, *Psychopharmacology*, 71, 299, 1980.
100. **Bläsig, J.**, On the role of brain catecholamines in acute and chronic opiate action, in *Developments in Opiate Research*, Herz, A., Ed., Marcel Dekker, New York, 1978, chap. 6.
101. **Foley, K. M., Inturrisi, C. E., Kourides, I. A., Kaiko, R. F., Posner, J. B., Houde, R. W., and Li, C. H.**, Intravenous (IV) and intraventricular (IVT) administration of beta endorphin (β_h-EP) in man: safety and disposition, in *Proc. Int. Narcotic Res. Club Meeting*, van Ree, J. M. and Terenius, L., Eds., Elsevier/North-Holland, Amsterdam, 1978, 421.
102. **Roemer, D., Buescher, H. H., Hill, R. C., Pless, J., Bauer, W., Cardinaux, F., Closse, A., Hauser, D., and Huguenin, R.**, A synthetic enkephalin analogue with prolonged parenteral and oral analgesic activity, *Nature (London)*, 268, 547, 1977.
103. **Székely, J. I., Rónai, A. Z., Dunai-Kovács, Z., Miglécz, E., Berzétei, I., Bajusz, S., and Gráf, L.**, (D-Met², Pro⁵)-enkephalinamide: a potent morphine-like analgesic, *Eur. J. Pharmacol.*, 43, 293, 1977.
104. **Guillemin, R., Ling, N., Burgus, R., Bloom, F., and Segal, D.**, Characterization of the endorphins, novel hypothalamic and neurohypophyseal peptides with opiate-like activity: evidence that they induce profound behavioral changes, *Psychoneuroendocrinology*, 2, 59, 1977.
105. **Drust, E. G., Sloviter, R. S., and Connor, J. D.**, Effect of morphine on "wet-dog" shakes caused by cerebroventricular injection of serotonin, *Pharmacology*, 18, 299, 1979.
106. **Dill, R. E. and Costa, E.**, Behavioral dissociation of the enkephalinergic systems of nucleus accumbens and nucleus caudatus, *Neuropharmacology*, 16, 323, 1977.
107. **Pert, A.**, The effects of opiates on nigrostriatal dopaminergic activity, in *Proc. Int. Narcotic Res. Club Meeting*, van Ree, J. M. and Terenius, L., Eds., Elsevier/North-Holland, Amsterdam, 1976, 389.
108. **Iwamoto, E. T. and Way, E.**, Circling behavior and stereotypy induced by intranigral opiate microinjections, *J. Pharmacol. Exp. Ther.*, 203, 347, 1977.
109. **Van Loon, G. R. and Kim, C.**, Dopaminergic mediation of β-endorphin-induced catalepsy, *Res. Comm. Chem. Pathol. Pharm.*, 21, 37, 1978.
110. **Sewell, R. D. E. and Lee, R. L.**, Opiate receptors, endorphins and drug therapy, *Postgraduate Med. J.*, 56, 25, 1980.
111. **Weiner, W., Moon, B. H., Feigenbaum, J. J., Carson, P. E., and Klawans, H. L.**, Naloxone potentiation of amphetamine stereotypy and its possible implications, *Pharmacologist*, 20, 166, 1978.
112. **Tortella, F. C., Moreton, J. E., and Khazan, N.**, EEG and behavioral effects of D-Ala²-Meth-enkephalinamide in the rat, *Pharmacologist*, 19, 189, 1977.
113. **Bergmann, F., Altstetter, R., Pasternak, V., Chaimovitz, M., Oreg, M., Roth, D., Hexter, C., and Wilchek, M.**, Cerebral application of enkephalins, *Experientia*, 33, 217, 1977.

114. **Broekkamp, C. L. E., Phillips, A. G., and Cools, A. R.,** Stimulant effects of enkephalin microinjection into the dopaminergic A 10 area, *Nature (London),* 278, 560, 1979.
115. **Pert, A. and Sivit, C.,** Neuroanatomical focus for morphine and enkephalin-induced hypermotility, *Nature (London),* 265, 645, 1977.
116. **Radouco-Thomas, C., Garcin, F., Mack, G., and Radouco-Thomas, S.,** Comparative *in vitro* and *in vivo* studies between morphine and Met-enkephalin. Effects on naive, acute pretreated and dependent animals, in *Proc. Int. Narcotic Res. Club Meeting,* Kosterlitz, H. W., Ed., North-Holland, Amsterdam, 1976, 111.
117. **Frigeni, V., Bruno, F., Carenzi, A., Racagni, G., and Santini, V.,** Analgesia and motor activity elicited by morphine and enkephalins in two inbred strains of mice, *J. Pharm. Pharmacol.,* 30, 310, 1978.
118. **Schiørring, E. and Hecht, A.,** Behavioral effects of low, acute doses of morphine in nontolerant groups of rats in an open-field test, *Psychopharmacology,* 64, 67, 1979.
119. **Gispen, W. H., Wiegant, V. M., Bradbury, A. F., Hulme, E. C., Smyth, D. G., Snell, C. R., and de Wied, D.,** Induction of excessive grooming in the rat by fragments of lipotropin, *Nature (London),* 264, 794, 1976.
120. **Gispen, W. H., Ormond, D., Ten Haaf, J., and de Wied, D.,** Modulation of ACTH-induced grooming by (des-Tyr^1)-γ-endorphin and haloperidol, *Eur. J. Pharmacol.,* 63, 203, 1980.
121. **Frenk, H., Urca, G., and Liebeskind, J. C.,** Epileptic properties of leucine- and methionine-enkephalin: comparison with morphine and reversibility by naloxone, *Brain Res.,* 147, 327, 1978.
122. **Henriksen, S. J., Bloom, F. E., McCoy, F., Ling, N., and Guillemin, R.,** β-endorphin induces nonconvulsive limbic seizures, *Proc. Natl. Acad. Sci. USA,* 75, 5221, 1978.
123. **Katz, R. J.,** Behavioral effects of dynorphin-a novel opioid neuropeptide, *Neuropharmacology,* 19, 801, 1980.
124. **Plotnikoff, N. P., Kastin, A. J., and Coy, D. H.,** Neuropharmacological actions of enkephalin after systemic administration, *Life Sci.,* 19, 1283, 1976.
125. **De Wied, D., Bohus, B., van Ree, J. M., and Urban, I.,** Behavioral and electrophysiological effects of peptides related to lipotropin (β-LPH), *J. Pharmacol. Exp. Ther.,* 204, 570, 1978.
126. **Veith, J. L., Sandman, C. A., Walker, J. M., Coy, D. H., and Kastin, A. J.,** Systemic administration of endorphins selectively alters open field behavior of rats, *Physiol. Behav.,* 20, 539, 1978.
127. **Olson, R. D., Kastin, A. J., Olson, G. A., and Coy, D. H.,** Behavioral effects after systemic injection of opiate peptides, *Psychoneuroendocrinology,* 5, 47, 1979.
128. **Kastin, A. J., Olson, R. D., Schally, A. V., and Coy, D. H.,** CNS effects of peripherally administered brain peptides, *Life Sci.,* 25, 401, 1979.
129. **Messing, R. B.,** Enkephalins interfere with active avoidance acquisition, opiates do not, *Pharmacologist,* 21, 199, 1979.
130. **Rigter, H.,** Enkephalins interfere with acquisition of an active avoidance response, *Life Sci.,* 26, 337, 1980.
131. **Izquierdo, I., Paiva, A. C. M., and Elisabetsky, E.,** Post-training intraperitoneal administration of leu-enkephalin and beta-endorphin cause retrograde amnesia for two different tasks in rats, *Behav. Biol.,* 28, 246, 1980.
132. **Izquierdo, I., Souza, D. O., Carrasco, M. A., Dias, R. D., Perry, M. L., Eisinger, S., Elisabetsky, E., and Vendite, D. A.,** Beta-endorphin causes retrograde amnesia and is released from the rat brain by various forms of training and stimulation, *Psychoparmacology,* 70, 173, 1980.
133. **Izquierdo, I.,** Effect of naloxone and morphine on various forms of memory in the rat: possible role of endogenous opiate mechanisms in memory consolidation, *Psychopharmacology,* 66, 199, 1979.
134. **Kostowski, W. and Plaznik, A.,** Naloxone-induced facilitation of conditioned avoidance behavior in rats, *Pol. J. Pharmacol. Pharm.,* 31, 293, 1979.
135. **Castellano, C.,** Dose-dependent effects of heroin on memory in two inbred strains of mice, *Psychopharmacology,* 67, 235, 1980.
136. **Carney, J. M. and Rosecrans, J. A.,** A comparison of the effects of intraventricular morphine and two enzyme resistant enkephalins on operant behavior in the rat, *Pharmacologist,* 19, 189, 1977.
137. **Lichtblau, L., Fossom, L. H., and Sparber, S. B.,** Suppression of fixed-ratio operant behavior by β-endorphin, *Pharmacologist,* 19, 190, 1977.
138. **Lichtblau, L., Fossom, L. H., and Sparber, S. B.,** β-Endorphin: dose-dependent suppression of fixed-ratio operant behavior, *Life Sci.,* 21, 927, 1977.
139. **Young, A. M. and Thompson, T.,** Naloxone effects on schedule-controlled behavior in morphine-pelleted rats, *Psychopharmacology,* 62, 307, 1979.
140. **Rigter, H.,** Attenuation of amnesia in rats by systemically administered enkephalins, *Science,* 200, 83, 1978.
141. **Rigter, H., Jensen, R. A., Martinez, J. L., Jr., Messing, R. B., Vasquez, B. J., Liang, K. C., and McGaugh, J. L.,** Enkephalin and fear-motivated behavior, *Proc. Natl. Acad. Sci. USA,* 77, 3729, 1980.

142. **Rigter, H., Shuster, S., and Thody, A. J.,** ACTH, α-MSH, and β-LPH: pituitary hormones with similar activity in an amnesia test in rats, *J. Pharm. Pharmacol.*, 29, 110, 1977.

143. **Stein, L. and Belluzzi, J. D.,** Brain endorphins: possible role in reward and memory formation, *Fed. Proc. Fed. Am. Soc. Exp. Biol.*, 38, 2468, 1979.

144. **Mondadori, C. and Waser, P. G.,** Facilitation of memory processing by posttrial morphine: possible involvement of reinforcement mechanisms, *Psychopharmacology*, 63, 297, 1979.

145. **Vasquez, B. J., Jensen, R. A., Messing, R. B., Martinez, J. L., Jr., Rigter, H., and McGaugh, J. L.,** Naloxone impairs memory in aged rats, *Pharmacologist*, 21, 269, 1979.

146. **Rigter, H., Greven, H., and van Riezen, H.,** Failure of naloxone to prevent reduction of amnesia by enkephalins, *Neuropharmacology*, 16, 545, 1977.

147. **De Wied, D., Bohus, B., van Ree, J. M., Kovács, G. L., and Greven, H. M.,** Neuroleptic-like activity of (des-Tyr^1)-γ-endorphin in rats, *Lancet*, 1, 1046, 1978.

148. **De Wied, D., Kovács, G. L., Bohus, B., van Ree, J. M., and Greven, H. M.,** Neuroleptic activity of the neuropeptide β-LPH(62-77)[(des-Tyr^1)-γ-endorphin; DT γ E], *Eur. J. Pharmacol.*, 49, 427, 1978.

149. **De Wied, D., van Ree, J. M., and Greven, H. M.,** Neuroleptic-like activity of peptides related to (des-Tyr^1-)γ-endorphin: structure activity studies, *Life Sci.*, 26, 1575, 1980.

150. **Weinberger, S. B., Arnsten, A., and Segal, D. S.,** Des-tyrosine-γ-endorphin and haloperidol: behavioral and biochemical differentiation, *Life Sci.*, 24, 1637, 1979.

151. **Van Ree, J. M., Witter, A., and Leysen, J. E.,** Interaction of des-tyrosine-γ-endorphin DT γ E, β-LPH(62-77) with neuroleptic binding sites in various areas of rat brain, *Eur. J. Pharmacol.*, 52, 411, 1978.

152. **Pedigo, N. W.,** Inhibition of in vivo ^{3}H-spiperone binding by the proposed antipsychotic (des-Tyr^1)-γ-endorphin, *Eur. J. Pharmacol.*, 60, 359, 1979.

153. **Moal, M., Koob, G. F., and Bloom, F. E.,** Endorphins and extinction: differential actions on appetitive and aversive tasks, *Life Sci.*, 24, 1631, 1979.

154. **Stinus, L., Koob, G. F., Ling, N., Bloom, F. E., and Le Moal, M.,** Locomotor activation induced by infusion of endorphins into the ventral tegmental area: evidence for opiate-dopamine interactions, *Proc. Natl. Acad. Sci. USA*, 77, 2323, 1980.

155. **Belluzzi, J. D. and Stein, L.,** Enkephalin may mediate euphoria and drive-reduction reward, *Nature (London)*, 266, 556, 1977.

156. **Van Ree, J. M., Smyth, D. G., and Colpaert, F. C.,** Dependence creating properties of lipotropin C-fragment (β-endorphin): evidence for its internal control of behavior, *Life Sci.*, 24, 495, 1979.

157. **Harrigan, S., Labay, R., and Gluckman, M. I.,** (D-Met^2, Pro^5-enkephalinamide)-Self-administration by monkeys. Report from the Warner-Lambert/Parke-Davis Pharmaceutical Research Division, 1978.

158. **Browne, R. G. and Fondren, B.,** β-endorphin and the narcotic cue, *Psychopharmacology*, 58, 3, 1978.

159. **Miglécz, E., Székely, J. I., and Dunai-Kovács, Z.,** Comparison of tolerance development and dependence capacities of morphine, β-endorphin and (D-Met^2, Pro^5)-enkephalinamide, *Psychopharmacology*, 62, 29, 1979.

160. **Graffenfried von, B., del Pozo, E., Roubicek, J., Krebs, E., Pöldinger, W., Burmeister, P., and Kerp, L.,** Effects of the synthetic enkephalin analogue FK 33-824 in man, *Nature (London)*, 272, 729, 1978.

161. **Stubbs, W. A., Delitala, G., Jones, A., Jeffcoate, W. J., Edwards, C. R. W., Ratter, S. J., Besser, G. M., Bloom, S. R., and Alberti, K. G. M. M.,** Hormonal and metabolic responses to an enkephalin analogue in normal man, *Lancet*, 2, 1225, 1978.

162. **Martin, W. R., Eades, C. G., Thompson, J. A., Huppler, R. E., and Gilbert, P. E.,** The effects of morphine- and nalorphine-like drugs in the nondependent and morphine-dependent chronic spinal dog, *J. Pharmacol. Exp. Ther.*, 197, 517, 1976.

163. **Székely, J. I., Dunai-Kovács, A., Miglécz, E., Ronái, A. Z., and Bajusz, S.,** *In vivo* antagonism by naloxone of morphine, β-endorphin and a synthetic enkephalin analog, *J. Pharmacol. Exp. Ther.*, 207, 878, 1978.

164. **Székely, J. I., Dunai-Kovács, Z., and Borsy, J.,** The effect of anti-parkinsonian drugs on chlorpromazine-induced depression of operant behavior, *Pharmacology*, 14, 240, 1976.

165. **Székely, J. I., Borsy, J., and Király, I.,** Potentiation of tetrabenazine-induced behavioral depression by imipramine on a discrete-trial avoidance escape schedule, *Psychopharmacologia (Berl.)*, 38, 339, 1974.

166. **Amir, S., Solomon, M., and Amit, Z.,** The effect of acute and chronic naloxone administration on motor activation in the rat, *Neuropharmacology*, 18, 171, 1979.

167. **Rodgers, R. J. and Deacon, R. M. J.,** Effect of naloxone on the behavior of rats exposed to a novel environment, *Psychopharmacology*, 65, 103, 1979.

168. **Arnsten, A. T. and Segal, D. S.,** Naloxone alters locomotion and interaction with environmental stimuli, *Life Sci.*, 25, 1035, 1979.

169. **Haber, S., Hatsukami, T., Berger, P. A., Barchas, J. D., and Akil, H.,** Naloxone blocks amphetamine-induced rearing: potential interaction between catecholamines and endorphins, *Prog. Neuro-Pharm.*, 2, 425, 1978.
170. **Holtzman, S. G.,** Effects of narcotic antagonists on fluid intake in the rat, *Life Sci.*, 16, 1465, 1975.
171. **Brown, D. R. and Holtzman, S. G.,** Evidence that opiate receptors mediate suppression of hypertonic-saline-induced drinking in the mouse by narcotic antagonists, *Life Sci.*, 26, 1543, 1980.
172. **Maickel, R. P., Braude, M. C., and Zabik, J. E.,** The effects of various narcotic agonists and antagonists on deprivation-induced fluid consumption, *Neuropharmacology*, 16, 863, 1977.
173. **Holtzman, S. G.,** Suppression of appetitive behavior in the rat by naloxone: lack of effect of prior morphine dependence, *Life Sci.*, 24, 219, 1979.
174. **Chance, W. T. and Rosecrans, J. A.,** Inhibition of drinking by intrahypothalamic administration of morphine, *Nature (London)*, 270, 167, 1977.
175. **Cantalamessa, G., De Caro, G., and Perfumi, M.,** Water intake inhibition and vasopressin release induced by eledoisin and leu-enkephalin in rats of the Brattleboro and Wistar strain, *Pharm. Res. Comm.*, 12, 375, 1980.
176. **De Caro, G., Micossi, L. G., and Venturi, F.,** Drinking behavior induced by intracerebroventricular administration of enkephalins to rats, *Nature (London)*, 277, 51, 1979.
177. **Brown, D. R., Blank, M. S., and Holtzman, S. G.,** Suppression by naloxone of water intake induced by deprivation and hypertonic saline in intact and hypophysectomized rats, *Life Sci.*, 26, 1535, 1980.
178. **Holtzman, S. G.,** Behavioral effects of separate and combined administration of naloxone and *d*-amphetamine, *J. Pharmacol. Exp. Ther.*, 189, 51, 1974.
179. **Brands, B., Thornhill, J. A., Hirst, M., and Gowdey, C. W.,** Suppression of food intake and body weight gain by naloxone in rats, *Life Sci.*, 24, 1773, 1979.
180. **Lowy, M. T., Maickel, R. P., and Yim, G. K. W.,** Naloxone reduction of stress-related feeding, *Life Sci.*, 26, 2113, 1980.
181. **Shimomura, Y., Oku, J., Glick, Z., and Bray, G. A.,** Opiate receptors, food intake and obesity, *Clin. Res.*, 28, 82A, 1980.
182. **Sewell, R. D. E.,** Naloxone antagonism of suckling behavior in newborn rats suggests a possible endorphin modulation, *IRCS Pharmacol.*, 8, 224, 1980.
183. **Yanaura, S. and Suzuki, T.,** Eating pattern of morphine dependent rats, *Jpn. J. Pharmacol.*, 29, 753, 1979.
184. **Grandison, L. and Guidotti, A.,** Stimulation of food intake by muscimol and beta endorphin, *Neuropharmacology*, 16, 533, 1977.
185. **Margules, D. L., Moisset, B., Lewis, M. J., Shibuya, H., and Pert, C. B.,** β-endorphin is associated with overeating in genetically obese mice (ob/ob) and rats (fa/fa), *Science*, 202, 988, 1978.
186. **Rossier, J., Rogers, J., Shibasaki, T., Guillemin, R., and Bloom, F. E.,** Opioid peptides and α-melanocyte-stimulating hormone in genetically obese (ob/ob) mice during development, *Proc. Natl. Acad. Sci. USA*, 76, 2077, 1979.
187. **Sewell, R. D. E. and Jawaharlal, K.,** Antagonism of 2-deoxy-D-glucose-induced hyperphagia by naloxone: possible involvement of endorphins, *J. Pharm. Pharmacol.*, 32, 148, 1980.
188. **McCloy, J. and McCloy, R. F.,** Enkephalins, hunger, and obesity, *Lancet*, 2, 156, 1979.

Chapter 4

THE ROLE OF ENDOGENOUS OPIOIDS IN THE VEGETATIVE REGULATION

J. I. Székely

TABLE OF CONTENTS

I. Cardiovascular Effects 156
 A. Centrally Mediated Effects of Opioid Peptides 156
 B. Antidiuretic Action 157
 C. Peripherally Mediated Vascular Effects of Opioid Peptides 158
 D. On the Role of Endogenous Opioid Peptides in Physiological Regulation and the Pathology of Cardiovascular System 159
 E. Summary 164

II. Respiratory Effects 164

III. Gastrointestinal Effects 171

IV. Modulation of the Thermoregulation 181
 A. Effects of Opiates and Opioid Peptides on the Core Temperature: Structure-Activity and Dose-Response Relationships 181
 B. Possible Mechanism of Alterations in Thermoregulation Induced by Opioids 187
 C. Are the Endogenous Opioids Involved in the Physiological Thermoregulation? 191
 D. Summary 193

V. Effects on the Eye 193
 A. Effect on Pupillary Size 193
 B. Relaxation of Nictitating Membrane 196
 C. Eyelid Retraction and Induction of Lenticular Opacity 196

References 199

I. CARDIOVASCULAR EFFECTS

A. Centrally Mediated Effects of Opioid Peptides

Since their discovery endogenous opioid peptides have been suspected to take part in the physiological regulation of the cardiovascular functions. This supposition was based mainly on the well-known and intensively investigated vasodepressant effects of opiates and on the relatively high concentrations of opiate receptors and their ligands in the lower brain stem and other brain areas primarily involved in cardiovascular regulation (Chapters 1 and 3, Volume III). As morphine congeners are qualified as vasodepressants in the standard pharmacological textbooks and the enkephalin containing neurones are especially abundant around certain medullary and pontin parasympathetic centers (Chapter 3, Volume III), obviously centrally mediated vasodilatation and bradycardia were expected. Though the pharmacological effects of exogenous opioids do not necessarily reflect the physiological functions of the endogenous ones, such experiments may be informative. The data available in this aspect are scanty but more or less unanimous.

Met-E has been found practically inactive even in the heroic dose of 0.4 to 0.5 mg, if giving it intracisternally to dogs[1] or i.c.v. to cats.[2] In rats intracisternally administered Met-E, Leu-E, and α-endorphin produced dose-related moderate pressor effect.[3] As reported by Bolme et al.[3] this pressor effect of Met-E and β-endorphin was followed by bradycardia and could be antagonized by naloxone if giving the latter in relatively high doses. Thus, the natural enkephalins did not elicit the "expected" hypotensive action. This might be ascribed to their relative metabolic instability and/or their inability to reach the opiate receptors responsible for the depressant effects. Nevertheless, the fact that in the same experiments[3] the initial hypertensive action of morphine could not be antagonized by naloxone and the latter induces similar action if given alone,[4] might indicate that this pressor effect was not related to the classical opiate receptors. An alternative explanation might be that the well-known hypotensive and heart-rate decelerating actions of opioids are mediated by μ-receptors, while the above-discussed pressor action of the natural enkephalins and β-endorphin is mediated by δ-receptors. Since the natural enkephalins are rather selective agonists of the δ-receptors, their irregular behavior if comparing it to that of the morphine is easy to interpret.

The experimental data obtained with the potent "stabilized" enkephalin analogues and β-EP do not give definite answer to these questions though they challenge the latter assumption. β-EP which equally stimulates the δ- and μ-receptors (Chapters 1 and 5, Volume III) gave biphasic effect (see below), while DALA and D-Met, Pro-EA displayed only hypotensive action.[3] These findings fit nicely into the hypothesis which relates the hypertensive action to δ- and the hypotension to μ-receptors, since the in vitro character of D-Met, Pro-EA lies closer to the profile of morphine than to that of β-EP (Chapter 7, Volume III). However, Laubie et al.[1] observed an initial pressor effect in dogs upon intracisternal application of (D-Ala^2, Met^5)-E and DALA. Moreover, Feldberg and Wei[2] found (D-Ala^2, D-Leu^5)-E to be 30 times more potent vasodepressant agent than morphine in cats if giving it intracisternally. As (D-Ala^2, D-Leu^5)-E is an enzymatically stabilized potent and selective agonist of the δ-receptors (Chapters 1 and 5, Volume III), the receptor specificity seems to be of little importance, as also morphine, the relatively pure agonist of μ-receptors, produced an initial pressor effect.[3] Thus, the cardiodepressant action of opioid peptides rather parallels their in vivo antinociceptive potency than their receptor specificity.

The relative activities of opioid peptides may also depend on the site of application within the ventricular system. All peptides are more active upon intracisternal administration than via the third ventricle.[2] There are also qualitative differences; they elicit tachycardia if giving them into the third ventricle but intracisternally bradycardia.[2]

As for β-EP, it was repeatedly shown to induce a transitory pressor action followed by distinct hypotensive and cardiodepressant effects in rats,[3] in cats,[2] and in dogs[1] as well, the actions being naloxone sensitive. Interestingly β-EP lowers the blood pressure also upon i.v. injection even in relatively low doses in rats;[5] a very surprising finding considering that it has no detectable analgesic activity in rats upon i.v. application in this species (Chapter 3, Volume II). Consequently, a part of the opiate receptors involved might be located on the peripheral side of the blood-brain barrier.

In a recent study Wei et al.[6] observed bradycardia upon i.v. application of Met-E and Leu-E in the i.v. doses of 1.3 and 1.5 μmoles/kg, while β-EP and FK 33-824 exerted their similar effects in the doses of 0.07 and 0.001 μmol/kg, respectively. These data[6] suggest that bradycardia is a more specific action of opioid peptides than hypotension, since it appears at lower dose level and also upon i.v. application.

B. Antidiuretic Action

β-EP, Met-E, Leu-E, (D-Ala^2, D-Leu^5)-E, (D-Ala^2, Met^5)-E have been shown to possess strong antidiuretic action upon i.v. administration[7,8] and upon i.c.v. treatment as well.[7,9,10] The most surprising fact is that the natural enkephalins which are otherwise almost inactive as analgesics also upon i.c.v. application (Chapter 3, Volume II), allegedly due to their rapid catabolism, elicit antidiuretic action via i.v. administration.

Thus, the question arises, how can these i.c.v. inactive natural enkephalins significantly modify the renal activity upon peripheral treatment? It could be suggested that this action is a peripheral effect probably not related to the central opiate receptors. Nevertheless, all authors[7-10] agree that this action of opioid peptides is naloxone sensitive. Furthermore, it has been shown that both Leu-E and β-EP were more potent upon i.c.v. than via i.v. application.[7,8] For this reason opioid peptides should act centrally, probably via vasopressin eliberation as suggested by Bisset et al.[8] and by others.[7,11] The alleged increase in vasopressin release could be demonstrated directly only by one group.[11] However, this group[11] demonstrated this action only for β-EP and not for the enkephalins. Fujimoto et al.[12] injecting morphine into the n. supraopticus recorded decreased urine outflow, but similar injection of Met-E and Leu-E resulted in an increase in the urine production. Moreover, according to several recent studies (Chapter 5, Volume II) enkephalins rather decrease than enhance the vasopressin liberation. Consequently, the antidiuretic action of these opioids cannot be attributed to this hormone. Otherwise, morphine and the opioid peptides, β-EP, DALA decrease not only the total volume of urine excreted but also its sodium and potassium content.[9,13] Since the vasopressin-induced oliguria is associated with dose-dependent increase in Na^+, K^+, and Cl^- excretion, its release cannot explain the antidiuretic action of opioids, since simultaneous administration of morphine counteracts the changes in the urine composition caused by vasopressin and morphine-tolerant animals were not found tolerant to vasopressin.[13]

A further possiblity is that this antidiuretic effect evolves as a secondary consequence of decreased renal blood flow. However, in a recent study[14] morphine was found to increase and not to decrease the renal blood flow in dogs. A centrally mediated change in the renal function might easily explain the antidiuretic effect. Furthermore, the endogenous creatinin clearance decreases upon morphine treatment proving that the antidiuretic action is due to decreased glomerular filtration.[13] However, this hypothesis holds only for those opioid peptides which cause hypotension, but as discussed above also Leu-E has antidiuretic action though it does not modify the systemic blood pressure. Accordingly, central modulation of the vasomotor centers cannot explain the antidiuretic action of Leu-E. This finding proves that certain circulatory actions of the opioid peptides are mediated via peripheral mechanisms. Thus, probably opiate receptor located in the renal tissue might explain the action.

From this point of view it is important that high affinity saturable opiate binding has been detected in the renal tissue.[15] These renal opiate receptors are rather peculiar, they show reversed stereospecificity and do not bind the endogenous opioids.[15] However, the absence of classical opiate receptors in the homogenized renal tissue does not necessarily exclude the possibility of more or less direct local effects. Opiate receptors mediating the antidiuretic action might be located somewhere else in the peripheral vegetative ganglions. On the other hand, opioid peptides might act directly on the intrarenal vascular smooth muscle cells. Nevertheless, in this case opiate receptors should be present in the renal tissue. Since the neural elements constitute only a tiny fraction of the whole renal homogenate, the amount (concentration) of opiate receptors may be well below the detection threshold. It should be considered that the effective filtration depends not only on the plasma osmolality, which does not change significantly upon treatment with opioids, and on the systemic blood pressure, which also remains constant during the Leu-E induced antidiuresis, but above all on the difference in the diameters of arteriolae afferens and efferens. Consequently, even a slight dilation of the arteriolae efferens or constriction of the arteriolae afferens can result in a significant decrease in filtration and diuresis. Unfortunately no data are available in this relation but some studies indicate that opioids exert their circulatory effects at least in part via peripheral mechanisms.

C. Peripherally Mediated Vascular Effects of Opioid Peptides

Opiates are known to induce histamine release, the action being allegedly independent of opiate receptors and not reversible by opiate antagonists.[16] Accordingly, the morphine-induced fall of blood pressure in spinal cats could be reversed by mepyramine, a histamine receptor blocking agent.[17] Consequently, the opiate-induced vasodilatation is in part due to histamine liberation. Opioid peptides have not been examined yet in this direction in animal experiments. However, some clinical data indicate that opioid peptides do have a certain histamine liberating action stronger than that induced by morphine (Chapter 10, Volume III).

A further important aspect of their mode of action is the adrenal mobilization (Chapter 5, Volume II). It is well known that different kinds of stress result in decrease of urine production since the intrarenal blood flow can be diverted into vessels (called shunts) not taking part in the glomerular filtration. This process is generally initiated by activation of the sympathetic efferents or by increased catecholamine output from the adrenals. (In these cases most probably a constriction of the renal preglomerular (afferent) arteriolae is the central event elicited by enhanced (nor)adrenaline secretion from the adrenal medulla.)

Moreover, even a direct effect of opioids exerted on the vascular smooth muscles should be counted with. First Knoll[18] reported that Met-E inhibited the electrically induced contractions of isolated rabbit ear artery and this effect could be reversed by high doses of naloxone. Interestingly morphine and many of its congeners were ineffective.[18] Hence the author concluded on the existence of a separate receptor entity which was sensitive to opioid peptides but not or much less to the classical opiates.[18] Later Rónai and Berzétei[19] demonstrated that this inhibitory effect could be antagonized by sulpiride, i.e., by a dopamine receptor antagonist, while the similar action of β-EP and some potent enkephalin analogues could be reversed only by naloxone-like opiate antagonists[19] (see also Chapter 1, Volume I). As the rabbit ear artery preparation has adrenergic innervation, they[19] concluded on a partial overlapping of presynaptic dopamine and opiate mechanisms controlling the noradrenaline release. In this connection it must be recalled that opioids inhibit the (nor)adrenaline release in several in vitro preparations (Chapter 1 and 2, Volume I and Chapter 1, Volume II). Thus, opioids might exert their vasodilatatory action via presynaptic inhibition of noradrenaline liberation. Some years ago Dressler et al.[20] reported that morphine antagonized the dopamine-induced renal and mesenteric vasodilatation in dogs similarly to haloperidol. In cats (pretreated with the α-receptor blocker phenoxybenzamine) morphine also reversed the dopamine-induced depressor response.[20] Consequently, an inhibition of presynaptic dopamine receptors or dopamine receptor-mediated events might explain many actions of the opioids.

Let us mention here that enkephalins also inhibit the $PGF_{2\alpha}$-induced contractions of pial arteries of the cat in vitro, an effect which can be reversed by naloxone.[21] In the isolated aorta of the rabbit natural enkephalins antagonized the contractions induced by substance P, but did not modify the noradrenaline-induced contractions.[22] Since enkephalins alone did not affect the activities of these isolated organs,[21,22] the opioid peptides may change the vascular tone only by presynaptic modulation of transmitter release.

The inhibition of synaptic transmission may take place not only in the peripheral organs but also in the sympathetic ganglia. Konishi et al.[23] found that Met-E and Leu-E depressed the excitatory postsynaptic potentials evoked by preganglionic stimulation in the celiac superior and inferior mesenteric ganglia. The inhibitory effect of enkephalins was enhanced at low Ca^{++} and high Mg^{++} concentrations and was abolished by naloxone.[23] In these ganglions acetylcholine is the neurotransmitter acting on nicotinic-cholinergic receptors, the synaptic transmission being inhibited by noradrenaline but enhanced substance P.[23] As the sympathetic ganglia do contain enkephalin-like material,[24] also enkephalins may be involved in the physiological regulation of the synaptic transmission. Otherwise, di Giulio et al.[24] suggested that the enkephalins were stored in and released from noradrenergic cells. Kumakura et al.[25] reported very similar findings working on cultured adrenal chromaffin cells, which also contain nicotinic-cholinergic receptors. In their study the release of catecholamines elicited by stimulation of postsynaptic receptors by nicotine was inhibited both by hexamethonium and opioids, i.e., by morphine, β-EP, Met-E, FK 33-824.[25] The detailed analysis of the data, i.e., Scatchard analysis showed that the antagonism by opioids of nicotinic receptors was not of the competitive type.[25]

The data reviewed suggest that opioid peptides modulate, notably inhibit the transmitter release in the peripheral autonomic ganglions and at the postganglionic nerve terminals as well. As Met-E and Leu-E have been detected in the autonomic nervous system but no β-EP and LPH (Chapter 3, Volume III), only the formers might be involved in the peripheral opiatergic mediation of vascular activity.

A general inhibition of impulse transmission in the sympathetic ganglions and at the neuroeffector synapses would result in a decrease in the sympathetic tone explaining bradycardia, vasodilatation, and lowering of the blood pressure. However, the stress-mediated mobilization of adrenal catecholamines induces opposite effects. The importance of adrenal mobilization should not be neglected. Grundy[26] demonstrated that morphine congeners did not induce pressor response and their depressor effect was potentiated upon ligation of the adrenal vessels. Thus, it would be very interesting to learn what happens to the pressor action of the centrally administered enkephalins and to the antidiuretic effects of opioids if the adrenal functions are blocked.

D. On the Role of Endogenous Opioid Peptides in Physiological Regulation and the Pathology of Cardiovascular System

As in the cases of their endocrine or behavioral actions the above-discussed data only raise the possibility that the endogenous opioids might play a certain role in the cardiovascular regulation but do not prove unambiguously that they are really involved in this function. Also in this aspect experiments with pure opiate antagonists may be informative, as the consequences of displacement of endogenous ligands from their binding with the opiate receptors might mirror their physiologic actions. It must be stated in advance that the contingent absence of an effect does not necessarily prove that the endogenous opiates are not involved in the function under study. In the cases of the vital functions such as maintenance of normal blood pressure or respiration multiple neurogenic and hormonal mechanisms take part in the regulation. The different mechanisms resulting in similar effects may more or less substitute for each other. Obviously the above-discussed blood pressure lowering action of endorphins does not mean that there are no other depressor mechanisms in the organism. Conversely, the failure of naloxone to modify the blood pressure in intact animals does not prove that endorphins and/or enkephalins do not modify the vascular tone. Obviously other depressor mechanisms may be either more sensitive, i.e., they might be activated upon smaller deviation of the blood pressure from its normal level or they may be efficient in themselves. They might be made apparent if the other compensatory mechanisms are switched off due to a pathological process or under experimentally created conditions. Even a superficial survey of the data on the naloxone-induced alterations in vasomotor functions indicates that the endogenous opioid mechanisms represent a "second" line of regulation, i.e., their effect becomes visible if the more sensitive regulatory functions are for some reasons inefficient.

Naloxone (and naltrexone) have been reported to induce tachycardia in dogs[27] and slight diuretic effect in rats,[13,28] but nobody observed hypertension upon their administration to intact animals. Zilm[29] observed actually a slight hypotension in healthy males upon administering it in small doses. This lack of stimulant actions apparently excludes any vasodepressant role of the endogenous opioids. However, recently Dashwood and Feldberg[30] found that otherwise ineffective doses of naloxone-induced pressor response with tachycardia if performing certain surgical procedures such as tying the sinus nerves, removal of the stellate ganglia, cutting the vagal nerves, and evisceration. An especially strong pressor effect was observed upon combining all four operations, but even the strongest action could be abolished by removing the adrenals.[30] These findings can be explained by supposing a tonic inhibitory control by endogenous opioids of sympathetic activity. By displacing these substances from their receptor binding sites naloxone elicited an increase in the sympathetic activity and catecholamine release from the adrenals. However, naloxone had no effect at all if the baroreceptor mediated and other depressor mechanisms were intact. Thus, in this fine study[30] the physiological function of opioids could be revealed by previous abolition of the more sensitive compensatory mechanisms.

Experiments[30] demonstrating the enhanced ability of naloxone to induce pressor response upon visceral deafferentation, also indicate that the main point of attack may be located either centrally or along the afferent pathways of the autonomic nervous system. The extreme efficacy of opioid peptides to elicit depressor reactions upon administering them into the ventricular system points to a primarily central site of action. (This refers only to hypotension and bradycardia but not to the antidiuretic action, which is easy to elicit by i.v. injection of opioid peptides as discussed above.)[7,8,10] Furthermore, since β-EP and the enkephalins elicit their depressor effects in smaller doses if given intracisternally than via the third or lateral ventricles,[1-3] the main site of action should be located somewhere in the caudal brain stem; a not very surprising finding considering that the main vasoregulatory centers can be found in the medulla or the pons, and inducing analgesia by microinjection of opioids the smallest amounts are needed if injecting them into the rhombencephalic areas (Chapter 2, Volume I and Chapter 2, Volume II). The first synapse of baroreceptor pathway is located in the nucleus tractus solitarius (NTS).[31] This nucleus is rich in enkephalins (Chapter 3, Volume III), which suggests that this area might be involved in mediation of the cardiovascular effects of opioids. However, though bilateral destruction of this nucleus caused hypertension and tachycardia, similar to those observed upon severing the baroreceptor afferents,[31] and attenuated the cardiovascular actions of clonidine,[32] this lesion did not abolish the cardiodepressant and hypotensive effects of fentanyl.[31] The same group found no change in the blood pressure, heart rate, or splanchnic discharges if this opiate was microinjected into the NTS.[33] Though opioid peptides have not been examined in these studies,[31,33] this area might be discounted.

A further area rich in opiate receptors and enkephalins (Chapters 1 and 3, Volume III) and also involved in regulation of the blood pressure is the locus coeruleus (LC). The interaction between noradrenergic and enkephalinergic neurones within this nucleus has been demonstrated by electrophysiological methods (Chapter 2, Volume II). Clonidine, the selective agonist of α_2-receptors, is supposed to exert its hypotensive action via this area (see the corresponding textbooks). In this connection are the data of Farsang and Kunos[34] particularly interesting, who could reverse by naloxone the blood pressure lowering effects of clonidine and α-methyldopa in spontaneously hypertensive rats.

The same group[35] reported that naloxone did not influence the slight hypotensive actions of clonidine and α-methyldopa in normotensive rats. Thus, the clonidine-induced blood pressure lowering effect is mediated, at least in part, by opioidergic mechanisms in the genetically hypertensive animals but not in the normotensive ones. It is tempting to speculate that certain endorphin-mediated processes are activated only under pathological conditions. Considering that clonidine suppresses the opiate withdrawal symptoms both in animal experiments (Chapter 2, Volume II) and in humans (Chapter 9, Volume III), it is easy to suppose that opioids and clonidine elicit their hypotensive actions acting on the same or overlapping neuronal population.

Probably it is worthwhile to note that the opioids generally inhibit the synaptic transmission, the transmitter release, and such action in the afferent baroreceptor pathways may result only in hypertension but not in circulatory depression. Therefore, the site of action should lie somewhere in the efferent part of the autonomic nervous system. A further region suspected to mediate the depressant action of opiates is the n. ambiguus. This nucleus which contains the origin of preganglionic vagal neurones responsible for the cardiovascular action of the vagus is rich in opiate receptors and enkephalin neurones as well (Chapters 1 and 3, Volume III). Microinjection of fentanyl or DALA into this area induced bradycardia and hypotension but the splanchnic discharges were not modified.[33] Injection of the same substances into the n. reticularis gigantocellularis or n. raphe magnus elicited no cardiovascular action.[33] Thus, n. ambiguus seems to be a principal and specific site of centrally induced circulatory depressant action of the opiates and opioid peptides as well. As the splanchnic discharges were not influenced[33] and n. ambiguus belongs to the central efferent pathways of vagus nerve, this site might be responsible only for the modification of parasympathetic activity upon opiate treatment and other brain regions might mediate the concomitant inhibition of sympathetic activity. Otherwise, the relative participation of increased vagal and decreased sympathetic activities in mediation of the cardiovascular depression might be different in the different species. Cutting the vagi totally abolishes the morphine-induced hypotension and the bradycardia in dogs[36] and rats[17] but not in cats.[17] In rats vagal transection prevented both the bradycardia and hypotension induced by FK 33-824.[6] Nevertheless, due to the well-known mutual reciprocal antago nism between sympathetic and parasympathetic brain stem centers responsible for vasoconstriction and vasodilatation, respectively, opioids do not have to act on both systems simultaneously.

However, the well-established parasympatholytic effects of opioids and the data on inhibition by them of acetylcholine liberation (Chapters 1 and 2, Volume I and Chapter

1, Volume II) seem to contradict the existence of a centrally mediated activation of the vagal tone. Therefore, it must be supposed that the vagal tone is enhanced centrally but the opioids reaching the peripheral tissue exert a direct inhibitory action on the acetylcholine release from the nerve endings. In the cardiovascular system mostly the central while in other organs, i.e., in the gastrointestinal tract, rather the peripheral action should prevail. The picture is only further complicated by other peripheral actions such as local inhibition of synaptic transmission in the autonomic synapses, mobilization of adrenal catecholamines due to the stressor effect of opioids, not speaking on the histamine releasing effect.

Finally it must be taken into consideraton that bradycardia is a more consistent and dose-dependent effect of the opioids than hypotension and the latter appears only at relatively high dose levels.[33] Consequently, the lowering of the blood pressure might be only a secondary consequence of decreased cardiac output.

Very few data are available on the participation of different brain neurotransmitters in mediation of opioid-induced cardiovascular actions. Feldberg and Wei[4] observed that naloxone applied i.c.v. or intracisternally induced pressor response if the animals were pretreated with atropine. Pretreatment with atropine blocked the bradycardia elicited by FK 33-824, but only attenuated its blood pressure lowering action.[6] Thus, bradycardia might be exclusively due to the increase in the vagal tone, while the hypotensive effect is brought about by more than one mechanism. The endogenous opioids may act via muscarinergic neurones. According to Lamaire et al.[5] the hypotension elicited by i.v. β-EP was facilitated by fluoxetine, a serotonin uptake inhibitor and antagonized upon depletion of brain serotonin or blocking the serotonin receptors by cyproheptadine, methergoline, or mianserine. Hence the authors concluded on the participation of serotoninergic pathways in mediation of β-EP-induced cardiovascular effects.[5] Interestingly in a previous study methysergide did not inhibit the vasodepressant effect of fentanyl.[31]

The physiological role of endogenous opioids in regulation of cardiovascular functions is demonstrated also by the studies in which experimentally induced pathological processes were associated with alteration in the metabolism of opioids or modified by naloxone. First di Giulio et al.[37] reported that the Met-E and Leu-E-like immunoreactivities were decreased in the celiac and superior cervical ganglia and also in the salivary and adrenal glands of spontaneously hypertensive rats, while their hypothalamic concentrations were normal. As also the concentration of the yet unidentified high molecular weight precursor of Met-E was normal, the author supposed a hereditary deficit in the enzyme splitting down enkephalin from this precursor as the cause of hypertension.[37]

However, Zamir et al.[38] measured increased endogenous opioid activity in the hypothalamus, spinal cord, and pituitary of genetically hypertensive rats by receptor binding assay, which does not permit the exact identification of opioid peptide examined. In renal hypertensive animals the endogenous opioid content was similarly enhanced but only in the cervical spinal cord.[38] The same authors[39] reported that the different forms of experimental hypertension (spontaneous, renal, and that induced by DOCA + salt treatment) were accompanied by an increase in the pain threshold. Naloxone treatment normalized the pain sensitivity, but did not modify the blood pressure.[38] Excision of the stenotic kidney resulted in the normalization of the pain threshold and blood pressure but the two processes were not quite parallel.[39] As naloxone reversed the elevation of pain threshold but did not decrease the blood pressure,[38] the elevated concentration of opioid peptides could not be the immediate cause of hypertension. Their elevated level may reflect a compensatory mechanism or altered activity of interoceptive sensory pathways due to the hypertension.

It is not excluded that under certain pathological conditions the overproduction of endorphins and/or enkephalins is an important link in the chain of events. The precipitous fall in blood pressure induced by the endotoxin of *Escherichia coli* could be reversed by naloxone as shown by Holaday and Faden.[40] Stressful events mobilize ACTH and the ACTH release is always accompanied by a concomitant release of β-EP (Chapter 3, Volume III and Chapter 5, Volume II) as they are split down from the same precursor (proopiocortin). Considering the previously discussed hypotensive action of β-EP, its excessive liberation might be a cause of the endotoxin-induced shock. Unfortunately in this interesting study[40] the plasma β-EP level has not been measured. However, as it has been shown previously,[5] at least 0.03 mg/kg must be given i.v. to elicit hypotensive action, which means 0.6 /μg/mℓ concentration if the plasma constitutes 5% of the total body volume of a rat weighing 200 g. As discussed in Chapters 3 and 10, Volume III such high β-EP concentration has not been measured even under stress conditions. Moreover, due to the poor capacity of opioid peptides to pass the blood-brain barrier (Chapter 3, Volume III) β-EP secreted into the plasma has little chance to enter into the brain and to depress the vasopressor centers. Considering, however, that exposure of the organism to detrimental external stimuli results in mobilization of the endogenous opioids also in the brain (Chapter 8, Volume III) and giving it intracisternally it elicits hypotension in the minute amount of 0.9 unanomoles/rat,[3] this effect might be mediated by direct interaction with the vasomotor centers, since according to a recent study[41] ventriculocisternal administration of naloxone protects against severe hypotension during endotoxin shock.

Of course the opioid peptides mobilized during shock might act also at the opiate receptors located in the autonomic ganglia or on the terminals of postganglionic autonic nerves. A further possibility is that endogenous opioids mobilized during different kinds of shock exert direct vascular effects via the nerve endings in the wall of certain arteries. Such a mode of action is substantiated by their ability to induce antidiuresis in low i.v. doses.[8] Otherwise, the above-discussed antidiuretic action itself might be a manifestation of their shock inducing action. The efficacy of naloxone in prevention of endotoxin-induced shock proves that endogenous opioids are involved in the genesis of shock-states. All the more as corticosteroids, which are known to have positive therapeutic effect in shock, directly inhibit the ACTH and endorphin synthesis in the pituitary (Chapter 4, Volume III).

Faden and Holaday[42] demonstrated that naloxone was effective also in the treatment of hypovolemic shock. In these experiments[42] the mean arterial pressure of rats was kept at a very low value (40 mmHg) for a period by intermittent bleeding which resulted in rather high mortality in the control group but not in that pretreated with naloxone. This finding[42] was recently confirmed and extended; the reversal by naloxone of hemorrhagic shock was observed also in cats and the effect was especially conspicuous in anephric animals.[43]

Naloxone and naltrexone exerted beneficial action also in canine hemorrhagic shock in the experiments of Gurll et al.[44,45] Recently naloxone was found to prevent the precipitous fall of arterial blood pressure in certain cases of spinal shock.[46] Naloxone was reported to reverse also the halothane-induced hypotension, bradycardia, and depression of carotid sinus reflex activity.[47]

The diversity of experimental models of shock where naloxone proved to substantially improve the survival rate might have practical implications. The effects of opiate antagonists might be useful also in certain kinds of clinical shock states. Thus, the endorphin research might contribute to the perfection of human therapy not only by helping to develop new drugs but also by better understanding of certain pathological processes, i.e., by widening the scope of application for drugs already known in the everyday clinical practice.

E. Summary

1. Endogenous opioid peptides (EOP) are involved in the physiological regulation of cardiovascular functions at various levels. They increase the vagal tone, decrease the central sympathetic activity resulting in bradycardia and hypotension (Table 1). Moreover, EOPs inhibit the synaptic transmission in the autonomic ganglia and the transmitter release from certain postganglionic sympathetic nerve endings. These effects are modified by their stressor effects resulting in catecholamine release from the adrenals and by their histamine releasing action.
2. It is not clear whether EOPs play any role in the homeostasis in the normal intact organism. However, they are surely mobilized if the baroreceptor mediated and other hypotensive mechanisms are insufficient to keep the blood pressure under a certain limit.
3. Excessive mobilization of EOPs are involved in certain pathological conditions leading to hypovolemia, bradycardia, and shock.
4. Opiate antagonists might be useful in the treatment of the above-mentioned pathological conditions (Table 1).

II. RESPIRATORY EFFECTS

Respiratory depression is probably the most intimate autonomic "side effect" of opiates.[52-55] Discounting their addiction liability, suppression of normal breathing which can be observed sometimes already at analgesic dose level is probably the main contraindication of their clinical application. Also this effect is mediated by opiate receptors. The respiratory depressant action of opioids can be reversed by narcotic antagonists, which actually explains their efficacy in the treatment of acute opiate poisoning. Upon repeated application tolerance develops to their respiratory depressant effect.[55-58] Considering the considerable bulk of experimental and clinical findings on the respiratory depression elicited by opiates, the endogenous ligands of opiate receptors were supposed to play a crucial role in the physiological regulation of respiration.

However, for the time being very little is known on the physiological role of endogenous opioids. As a matter of fact, hardly more has been demonstrated yet that opioids really depress the respiration if they are administered in sufficiently high dosage.

The exact measurement of their respiratory suppressant action is confused by several factors. Working in unanesthetized animals, the diminution of respiratory movements can be only a manifestation of a general decrease in spontaneous movements ending in catalepsy and muscular rigidity if opioids are administered in high amounts. Alternatively as commented by Bloom et al.[59] "respiratory movements are shifted to the abdomen". Cessation of spontaneous movements and a decrease in body temperature obviously reduce the oxygen demand, hence a smaller respiratory minute volume might be sufficient for maintenance of metabolism. Thus, the behavioral actions can obscure the specific effect on respiration. Working in anesthetized animals the direct and indirect effects of a general anesthetic agent and its interaction with the opioids only further obscures the selective actions. Having mentioned these methodological considerations, the data available may be summarized as follows.

Administering β-EP to unanesthetized rodents distinct respiratory depression appears at relatively high dose levels,[59,60] i.e., in amounts beyond the analgesic ED_{50} values.

Tachypnoea and burst of panting were observed in conscious cats if giving β-EP directly into the third ventricle.[61,62] This "paradox" effect was obviously due to the general motor excitation and hyperthermic effect of the relatively small doses applied

Table 1
CARDIOVASCULAR EFFECTS OF OPIOID PEPTIDES AND OPIATE ANTAGONISTS IN VIVO

Treatment	Species	Anesthesia	Effect	Reversal by opiate antagonists	Ref.
Naloxone s.c. 0.1-3.0 mg/kg	Dog	No	Tachycardia	—	27
Naloxone i.c.v. 200 μg	Cat	Yes	No effect	—	4
Naloxone intracisternally 50-100 μg	Cat	Yes	No effect	—	4
Naltrexone s.c. 0.125-0.50 mg/kg	Rat	No	Diuresis	—	28
Met-E intracisternally 500 μg	Dog	Yes	No effect	—	1
β-EP intracisternally 50 μg	Dog	Yes	Hypotension and bradycardia	Partially	1
(D-Ala^2)-Met-E intracisternally 500 μg	Dog	Yes	Hypotension and bradycardia	Partially	1
Met-E, Leu-E i.c.v. 0.1-1.0 nmoles	Rat	Yes	Increase in urine outflow	Yes	12
(D-Ala^2, D-Leu^5)-E i.c.v. 20-200 μg	Cat	Yes	Tachycardia	Not mentioned	2
(D-Ala^2, D-Leu^5)-E i.c.v. 400 μg	Cat	Yes	Bradycardia and hypotension	Not mentioned	2
(D-Ala^2, D-Leu^5) intracisternally 10 μg	Cat	Yes	Bradycardia and hypotension	Not mentioned	2
(D-Ala^2, D-Leu^5)-E i.v. 0.03 mg/kg	Cat	Yes	Hypotension	Not mentioned	2
(D-Ala^2, D-Leu^5)-E s.c. 0.1-0.5 mg/kg	Cat	Yes	Hypotension, bradycardia	Not mentioned	2
Met-E i.c.v. 400 μg	Cat	Yes	No effect	—	2
β-EP i.c.v. 400-800 μg	Cat	Yes	Tachycardia	Yes	2
β-EP intracisternally 200-400 μg	Cat	Yes	Hypotension, bradycardia	Yes	2
Met-E intracisternally 18-180 nmoles	Rat	Yes	Hypertension then bradycardia	Yes	3

Table 1 (continued)
CARDIOVASCULAR EFFECTS OF OPIOID PEPTIDES AND OPIATE ANTAGONISTS IN VIVO

Treatment	Species	Anesthesia	Effect	Reversal by opiate antagonists	Ref.
Leu-E intracisternally 52-520 nmoles	Rat	Yes	Hypertension then bradycardia	Yes	3
α-endorphin intracisternally 5.7-57.0 nmoles	Rat	Yes	Hypertension then bradycardia	Yes	3
β-EP intracisternally 0.9-2.0 nmoles	Rat	Yes	Transient hypertension then hypotension and bradycardia	Yes	3
DALA intracisternally 17-33 nmoles	Rat	Yes	Hypotension and bradycardia	Yes	3
D-Met,Pro-EA intracisternally (dose not mentioned)	Rat	Yes	Hypotension and bradycardia	Not mentioned	3
β-EP i.v. 0.03-0.15 mg/kg	Untreated rat	Yes	Hypotension	Yes	5
β-EP i.v. 0.03-0.15 mg/kg	pCPA pretreated rat	Yes	No effect	—	5
DALA into n. ambiguus 0.01-1.0 μg	Dog	Yes	Bradycardia	Yes	33
Naloxone i.v. 10 mg/kg	Rat	No	Reversal of endotoxin caused hypotension and bradycardia	—	40
β-EP i.v. 0.085-0.85 mg/kg	Rat	Yes	Antidiuresis	Yes	7
β-EP i.c.v. 0.03-0.3 μmoles/kg	Rat	Yes	Antidiuresis	Yes	7
Leu-E i.v. 10-200 μg	Rat	Yes	Antidiuresis	Yes	8
β-EP i.v. 10-40 μg	Rat	Yes	Antidiuresis	Not mentioned	8
(D-Ala^2, D-Leu^5)-E i.v. 200 μg	Rat	Yes	Antidiuresis	Not mentioned	8

β-EP i.c.v. 0.1-3.0 μg	Rat	Yes	Antidiuresis, decrease in urine K^+ and Na^+ concentrations	Yes	9
DALA i.c.v. 0.3-3.0 μg	Rat	Yes	Antidiuresis, decrease in urine K^+ and Na^+ concentrations	Yes	9
Leu-E i.c.v. 50-200 μg	Rat	No	Inhibition of angiotensin induced vasopressin release and pressor response	Yes	48
Naloxone i.p. 1 mg/kg	Rat	No	No effect on water diuresis, increased Na^+ and Cl^- excretion	—	13
Naloxone i.v. 10-20 mg	Human	No	No effect on heart rate and blood pressure	—	49
Naloxone i.p. 0.2-2.0 mg/kg	Hypertensive rat	No	No effect alone, reversal of depressor effects of clonidine and α-methyl-DOPA	—	34 35
(−)-naloxone perfusion of ventricle IV	Dog	Yes	Reversal of halothane induced hypotension	—	47
(+)-naloxone perfusion of ventricle IV	Dog	Yes	No effect	—	47
Naloxone i.v. 1 mg/kg or continuously	Rat	No	Reversal of hypotonia in hypovolemic shock	—	42 50
Naloxone i.v. 0.2 mg/kg	Cat	Yes	Hypertension and tachycardia if sinus nerves are tied, vagi are cut, stellate ganglia and viscera are removed	—	30
Naloxone i.p. or i.v. 2-10 mg/kg	Rat	Yes	No effect on blood pressure in experimental hypertension	—	38
(−)-naloxone i.v. 10 mg/kg	Rat	Yes	Reversal of hypotension caused by spinal transection	—	51
(+)-naloxone i.v. 10 mg/kg	Rat	Yes	No effect	—	51
Naloxone ventriculocisternal perfusion	Dog	Yes	Reversal of endotoxin-induced shock	—	41

Table 1 (continued)
CARDIOVASCULAR EFFECTS OF OPIOID PEPTIDES AND OPIATE ANTAGONISTS IN VIVO

Treatment	Species	Anesthesia	Effect	Reversal by opiate antagonists	Ref.
Naloxone i.v. infusion	Normal and anephric cats	Yes	Reversal of hemorrhagic shock	—	43
Naloxone i.v. 0.2-0.6 mg	Young healthy males	No	Slight hypotension	—	29
β-EP i.v. ED_{50}: 0.07 μmoles/kg	Rat	Yes	Bradycardia	Not examined	6
Met-E i.v. ED_{50}: 1.3 μmoles/kg	Rat	Yes	Bradycardia	Not examined	6
Leu-E i.v. ED_{50}: 1.5 μmoles/kg	Rat	Yes	Bradycardia	Not examined	6
(D-Ala^2, D-Leu^5)-E i.v. ED_{50}: 0.45 μmoles/kg	Rat	Yes	Bradycardia	Not examined	6
FK 33-824 i.v. ED_{50}: 0.001 μmoles/kg	Rat	Yes	Bradycardia	Not examined	6
Naloxone i.p. 2.0 mg/kg	Normotensive rat	No	No effect on clonidine induced hypotension	—	35
Naloxone i.v. infusion	Cat	Yes	Reversal of clonidine induced hypotension	—	46
Naloxone i.v. infusion	Cat	Yes	Reversal of hemi-spinalization induced hypotension	—	46
Naltrexone i.v. 2.5-10.0 mg/kg	Dog	Yes	Reversal of hemorrhagic shock	—	44
Naloxone i.v. infusion	Dog	Yes	Reversal of hemorrhagic shock	—	45
Met-E i.c.v. 12-25 μg/kg	Goat	No	Antidiuresis	Yes	10
Leu-E i.c.v. 12-25 μg/kg	Goat	No	Antidiuresis	Yes	10
β-EP i.c.v. 2-200 ng/kg	Goat	No	Antidiuresis	Yes	10

in these experiments.[61,62] However, the same dose elicited a distinct decrease in tidal volume and respiratory frequency if it was administered to anesthetized cats.[63-65] In these experiments or in dogs[66] the depressant action was only further potentiated by the general anesthesia and the "specific" action was not concealed by the consequences of general motor excitation. Otherwise, in one study[63] the depressant action was preceded by short-lasting tachypnoea. Just as in the cases of vasodepressant effects, which also were introduced by a transient pressor effect (Chapter 4, Volume II). According to a recent study[65] injecting β-EP into the pontomedullary subarachnoid space no effect was seen, while Met-E and DALA elicited transient and lasting respiratory depression, respectively. Consequently, the distinct respiratory depressant effect of β-EP seems to be secondary to other actions on higher brain structures.

Central application of natural enkephalins has no durable effect on respiration and measuring the respiratory parameters only a very short-lasting depression was detected.[63,65] Interestingly just the latter group observed a rather strong and lasting decrease in respiratory frequency and tidal volume if Met-E was applied directly to the ventral surface of the brain stem.[64] Therefore, the inefficacy of enkephalins administered i.c.v. might be attributed to their incapability to reach the site of action (ostensibly due to rapid catabolism). The "stabilized" enkephalin analogues decrease the tidal volume and respiratory rate just as β-EP.[63,65,67] Thus, increased analgesic potency is accompanied by enhanced respiratory suppressant capacity. Florez et al.[63] found certain differences in the time-course of the effect of β-EP and those of enkephalins. β-EP and enkephalins might obviously have differential affinities to the different respiratory centers or simply due to the differences in their metabolic stabilities their relative concentrations in brain centers located close to or faroff the site of application may be different.

Applying the opioid peptides peripherally a basically different pattern of pharmacological activity may be observed.

Giving 30 mg β-EP to humans a slight decrease in respiratory frequency was recorded by Catlin et al.[68] However, no other group reported on any alteration of spontaneous breathing though it has been trialed in several laboratories (Chapter 3, Volume II). Similarly, the only enkephalin analogue FK 33-824 which has been repeatedly administered to humans, has not been found to alter the spontaneous respiration (Chapter 10, Volume III). This lack of any effect on breathing is definitely stated in the first clinical study with this analogue[69] and in the subsequent studies (Chapter 10, Volume III) no reference can be found on its contingent respiratory action. The absence of respiratory depression upon treatment with β-EP or the enkephalin analogue is probably not surprising considering that in humans these peptides have no analgesic effect either (if applying them peripherally) (Chapter 3, Volume II; Chapter 10, Volume III). Since several other mostly endocrine effects can be detected upon their peripheral administration (Chapter 5, Volume II), it is reasonable to suppose that in these cases they were not metabolized but did not reach (in sufficient amount) the brain areas mediating their respiratory depressant and analgesic actions.

These negative clinical findings are in good agreement with the corresponding animal studies. Catlin et al.[70] observed no apparent action in awake cats, if giving β-EP in the i.v. doses 0.01 to 1.00 mg/kg. This dose was not simply too low to elicit characteristic action since some very peculiar symptoms appeared (e.g., prolapse of nictitating membrane, stereotyped licking movements), but neither respiratory depression nor analgesia could be observed. In our experiments[71] another enkephalin analogue stronger than morphine upon i.v. application, D-Met, Pro-EA, also proved to be a rather weak respiratory depressant agent if giving it to conscious animals. We found the spontaneous respiration hardly influenced in cats and dogs in the doses of 1 to 30 mg/kg i.p. and 1 to 30 mg/kg i.v., respectively. In rats and in mice at toxic dose levels, i.e., above

100 mg/kg (s.c. or i.v.) the breathing was apparently inhibited.[71] Even in toxic doses the effects of morphine and D-Met, Pro-EA seemed to be different. The morphine treated animals showed irregular, gasping breathing, where the convulsive jerks and irregular respiratory movements were difficult to be distinguished and also cyanosis could be seen. However, upon D-Met, Pro-EA treatment the respiration became gradually superficial to such an extent that it was difficult to guess whether the animals did breathe at all. Of course, these experiments were made in unanesthetized animals.[71] Thus, opioid peptides do induce respiratory depression, but this effect appears in attenuated form if the substances are given to conscious animals and especially upon peripheral application.

The close correlation between analgesic and respiratory depressant actions is not a completely unexpected finding. Green[52] demonstrated that in the rat the ratio of analgesic to respiratory depressant potency was practically the same for morphine, methadone, codeine, and several other narcotics. Why are these actions of opiates correlated with each other? The opiate-induced respiratory depression is connected with elevation of the arterial carbon dioxide tension.[55] As arterial carbon dioxide is the main physiological stimulus of the respiratory centers, its elevated level in the blood points to a decreased sensitivity of neurones initiating the inspiratory movements. Accordingly, Moss and Friedman[66] observed that in dogs β-EP (given intracisternally) decreased the ventilatory response induced by carbon dioxide rebreathing. Conversely, the increase in respiratory minute volume due to carbogen inhalation was further augmented upon naloxone treatment in anesthetized rabbits.[72] Consequently, endogenous opioids might regulate the pain threshold and the sensitivity of respiratory neurones in similar ways. If opioids affect the respiration via suppression of sensory stimulation, the above-mentioned parallelism between analgesia and respiratory suppression is easy to understand.

Furthermore, the question should be answered just as in cases of other centrally mediated functions whether the endogenous opioids tonically inhibit the respiratoy functions or the observed depressant action is only an artefact due to the usage of unnaturally high doses. In the former case naloxone is expected to enhance the respiratory minute volume. In some studies naloxone had no effect when administered to waking healthy humans[73] or to rats.[67]

Of the numerous authors who examined the supposed therapeutic action of naloxone in the treatment of schizophrenia (Chapter 10, Volume III) nobody mentioned that it exerted any effect on the spontaneous respiration. However, it increased the activity of phrenic nerve in decerebrated cats[74] and facilitated the carbogen-induced ventilatory response in anesthetized rabbits.[72]

Recently Kumazawa et al.[75] reported that naloxone reversed the respiratory depression observed after cessation of thin-fiber muscular afferent stimulation in anesthetized dogs. In this relation probably the similar data on the cardiovascular effects of naloxone are worth recalling. As discussed in Chapter 4, Volume II naloxone had little if any effect on the cardiovascular parameters in waking healthy subjects, but the hypotension induced by anesthesia or different kinds of shock could be reversed by its i.v. application. As seen above, naloxone facilitated the respiration only in anesthetized animals,[72,74,75] which might indicate that endorphins are mobilized only in pathological or stressful situations.

Of course also in relation to the respiratory effects the matter arises whether the autonomic, i.e., respiratory depressant, and analgesic actions of opiates and opioid peptides are mediated by the same receptor population or not. For the time being only the first question can be answered. Using some classical morphine congeners as agonists and naloxone as antagonist, the apparent pA_2 values for the same agonist-antagonist combinations were significantly different in relation to analgesia and respiratory depression.[58] Upon repeated morphine treatment a smaller degree of tolerance developed to its respiratory depressant effect than to its analgesic action.[57] A single morphine treatment produced already a significant shift of naloxone's pA_2 value against morphine in relation to analgesia but not for respiratory depression.[57] Consequently, the receptors responsible for the antinociceptive action are surely different from those mediating this autonomic effect. Interestingly if comparing the prototypes of selective μ- and κ-receptor agonists (methadone and morphine vs. ketocyclazocine and ethylketocyclazocine, respectively), their respiratory depressant effects could equally be antagonized by naloxone.[76] Consequently, contrary to their analgesic actions, the respiratory depressant effects of μ- and κ-receptor agonists seem to be mediated by the same receptor population. However, according to other studies even the μ-receptor agonists may act by partially different mechanisms. Hassen et al.[53] reported that morphine selectively depressed the pontin apneustic respiratory neurones leaving the medullary units responsible for gasping respiration unaffected but meperidine depressed both types of neurones.

Considering the basically similar but in their time-course different patterns of respiratory depression elicited by β-EP, Met-E, and its analogues[63] and the above-mentioned[71] differences in breathing of animals treated with toxic doses of morphine and D-Met, Pro-EA, it is conceivable that these substances have selective actions on the pontin and medullary respiratory centers.

Summarizing the above-discussed data (see also Table 2) it may be stated that the endogenous opioid peptides have respiratory depressant action, which roughly correlates with their analgesic potency. The mechanism of their action is very likely a suppression of the central chemoreceptor functions. It is probable that the receptors mediating their respiratory depressant action are partially different from those responsible for the similar effect of opiates though in this aspect also the latter group seems to be heterogeneous. Their participation in the physiological tonic control of respiration in waking intact organism is questionable.

III. GASTROINTESTINAL EFFECTS

Surveying the literature on the in vivo gastrointestinal effects of opioid peptides, we are faced with several peculiar contradictions. First, our knowledge is rather scanty in this direction, though the most extensively used opiate sensitive in vitro preparation, the longitudinal muscle strip of guinea pig ileum, is an integral part of the gastrointestinal tract. Thus, we have rather poor information on the in vivo action of substances on an organ which is probably the most commonly used in their biological titration in vitro. Secondly, from a physiological point of view the response elicited by opiates in the isolated guinea pig ileum is anything but characteristic of the role of endogenous opioids in the digestive system. Opioids are generally known to increase the tone of the intestinal wall and not to relax it (see later). However, the endogenous opioids were detected via inhibition by them of the electrically induced contraction of this organ (Chapter 1, Volume I; Chapter 2, Volume III). Furthermore, application of the guinea pig ileum preparation in the arsenal of pharmacologists dealing with opiate-like substances substantially contributed to the discovery of enkephalins and endorphins in the brain. Nevertheless, several lines of evidence indicate that the physiological actions of the endogenous opioids are only partially mediated by the central opiate receptors, in relation to the gastrointestinal tract rather the peripheral opiate receptors and their locally released ligands count.

Otherwise, the constipant effect of opiates is an old clinical observation and this property of morphine congeners has been utilized in the symptomatic treatment of diarrhea for a long time. Therefore, let us stress here that contrary to the commonly

Table 2
RESPIRATORY (R) EFFECTS OF OPIOID PEPTIDES AND OPIATE ANTAGONISTS IN VIVO

Treatment	Species	Anesthesia	Effect	Reversal by opiate antagonists	Ref.
β-EP i.c.v. 3-15 nmoles	Rat	No	r depression	Yes	59
β-EP i.c.v. 10-20 μg	Cat	No	Tachypnoea, panting	Yes	61 62
D-Met, Pro-EA s.c. up to 30 mg/kg	Mouse, rat	No	No apparent effect	—	71
D-Met, Pro-EA i.v. up to 30 mg/kg	Dog	No	No apparent effect	—	71
D-Met, Pro-EA i.p. up to 30 mg/kg	Cat	No	No apparent effect	—	71
Met-E i.c.v. 900 μg	Cat	Yes	Transient drop in r frequency and tidal volume	Not examined	63 65
DALA i.c.v. 10 μg	Cat	Yes	Lasting decrease in r frequency and tidal volume	Not examined	63 65
β-EP i.c.v. 10 μg	Cat	Yes	After initial tachypnoea decrease in r frequency and tidal volume	Not examined	63 65
β-EP intracisternally 14.5 nmoles	Dog	Yes	Decrease in r frequency and tidal volume, decreased response to carbon dioxide rebreathing	Yes	66
β-EP i.v. 0.01-1.00 mg/kg	Cat	No	No visible effect	—	70
β-EP i.v. 30 mg	Human	No	Decrease in r frequency	Not examined	68
FK 33-824 i.m. 0.1-1.2 mg	Human	No	No effect	—	69
DALA i.c.v. 100 μg	Rat	No	Decrease in r frequency and tidal volume	Yes	67
β-EP i.c.v. 0.54-17.0 μg	Golden hamster	No	Decrease in r frequency	Yes	60
Naloxone i.v. 0.4 mg/kg	Decerebrated cat	—	Increase in the activity of the phrenic nerve	—	73

Naloxone i.v. 10-20 mg	Human	No	No effect	—	49
Naloxone i.v. 0.5-5.0 mg/kg	Rabbit	Yes	Enhanced r response to carbon dioxide	—	72
Met-E directly to the ventral surface of the brain stem	Cat	Yes	Decrease in r frequency and tidal volume	Yes	64
Met-E 900 μg into the ponto-medullary subarachnoid space	Cat	Yes	Immediate short-lasting r depression	Not mentioned	65
DALA 10 μg into the ponto-medullary subarachnoid space	Cat	Yes	Lasting r depression	Not mentioned	65
β-EP 10 μg into the ponto-medullary subarachnoid space	Cat	Yes	No significant effect	Not mentioned	65
Naloxone i.v. 0.2-2.0 mg/kg	Dog	Yes	Facilitation of activity of the phrenic nerve	—	75

used in vitro preparation of guinea pig ileum, opiates exert spasmogenic action in the different laboratory species, i.e., they enhance the intraluminar pressure, increase the tone of the intestinal wall, contract the biliary tract, the pylorus, and sphincter muscles, moreover, they decrease the gastric, pancreatic, and intestinal secretions.[77-79] The in vivo pharmacological model of this action is the inhibition of gastrointestinal propulsion. Green[80] demonstrated first the dose-dependent deceleration of transport of charcoal meal in the rat and also the close correlation between the analgesic, respiratory depressant, and constipating actions of different morphine congeners (administering them either peripherally or centrally). This correlation indicates that not only the antinociceptive and respiratory suppressant but also the antipropulsive effects of opiates might be centrally mediated; a rather surprising finding in view of the presence of enkephalins in the gut (Chapter 3, Volume III). Moreover, this correlation is not valid for the opioid peptides (see later) and for some newly developed morphine congeners either. Loperamide is a powerful inhibitor of intestinal propulsion;[81] it has rather strong affinity to the opiate receptors[82] but practically no central effect at all.[83] As loperamide is not absorbed from the bowels and does not penetrate into the brain, it may exert its opiate agonist action only locally. Consequently, any opiate specific in vivo effect on the gastrointestinal system may be mediated by at least two different receptor populations, one located in the gut the other in the brain. Therefore, the experimentally recorded activity is codetermined by the relative affinities to the peripheral and central receptor populations and by the distribution of the substance examined in the organism, i.e., by the relative concentrations in the brain and intestinal wall, respectively. Before reviewing the data some points deserve special consideration.

The gastrointestinal tract, contrary to many other peripheral organs, contains opiate receptors in substantial amount (Chapter 1, Volume III). Moreover, it is also rich in enkephalin releasing neurones[79,84,85] (and Chapter 3, Volume III). Thus, the gut is a target organ, which contrary to other peripheral tissues contain not only enkephalin releasing nerve terminals but cell bodies as well. Since the neuronal circuitry within the intestinal wall is partially independent (Chapter 3, Volume III), the existence of local autonomic reflex regulation may be supposed, where enkephalins are physiological transmitters.

Only certain gastrointestinal reflexes are centrally mediated, such as vomiting, fasting contractions of the stomach, defecation, etc., but the basic motor and secretory activities of the gut proceed more or less normally even after severing the central connections. Hence, it is easy to understand that the constipating action of peripherally administered morphine is maintained after decapitation, vagotomy, and/or administration of ganglionic blockers.[77,78]

Centrally acting drugs may influence the gastrointestinal activity also by alteration of the central sympathetic or parasympathetic tone, but these are only pharmacological phenomena hardly related to the physiological functions of endogenous opioids. This supposition seems to be valid also in the cases of centrally induced intestinal motility decreasing effects of opiates. Recently Schulz et al.[83] demonstrated that quaternary naloxone failed to antagonize the morphine-induced inhibition of intestinal propulsion if the alkaloid was given centrally or in relatively low i.v. doses. Increasing, however, the amount of i.v. injected morphine, the relative antagonist potency of quaternary naloxone was gradually increased.[83] On the other hand, tertiary naloxone erased the action of morphine administering it by any route.[83] Obviously, if opiates are given in relatively low peripheral dosage, they attain a brain level high enough to exert their parasympatholytic action as they readily cross the blood-brain barrier. Applying them in higher doses, their tissue levels in the intestinal wall will be high enough to exert also direct local effect on the gut. On the other hand, since opioid peptides are known to cross the blood-brain barrier only with great difficulty (Chapter 3, Volume III), the peripheral effects might prevail in relation to actions mediated both by centrally and peripherally located opiate receptors.

Accordingly, it is questionable whether the data on the gastrointestinal actions of centrally administered opioid peptides (Table 3) have any physiological significance. Cowan et al.[86] demonstrated first that the natural enkephalins similarly to morphine inhibited the gastrointestinal propulsion. In their work[86] Leu-E and Met-E were found to act in doses two to three orders of magnitude higher than morphine upon i.c.v. application. In this study[86] enkephalins (and morphine) elicited analgesia at about the dose level needed for the antidiarrheal action. Also Schulz et al.[83] stated that FK 33-824 produced antinociception and obstipation in approximately similar i.c.v. doses. These data[83,86] are in good agreement with the findings reported by Green,[80] who observed a close correlation of the relative analgesic, respiratory depressant, and antipropulsive potencies of different morphine derivatives upon intracisternal application.

Hence, the question emerges whether the central opiate receptors mediating the actions of opiates and opioid peptides moreover, their analgesic and gastrointestinal effects are the same or not. Probably the answer is no to both questions, but for the time being no definite conclusion is justified. Cowan et al.[86] found the antipropulsive action of centrally administered enkephalins more difficult to antagonize by naloxone than the similar action of morphine. Differences in their antagonizability point to differences in the receptorial mechanisms. Moreover, this observation correlates well with the in vitro data on the selective affinity of natural enkephalins to the δ- and that of the opiates to the μ-receptors (Chapter 1, Volume III).

As for the contingent identity of central opiate receptors responsible for constipation and analgesia, the data are conflicting. Parolaro et al.[77] found the pA_2 value of naloxone against morphine-induced antipropulsive action upon i.c.v. application 7.14. This value is very similar to those obtained in other experiments measuring the interaction in different analgesic assays (Table 1 in Chapter 2, Volume I). However, Takemori et al.[103] found the pA_2 value in relation to constipant action significantly lower than upon measuring it in an analgesic assay. The latter[103] data suggest that different receptor populations mediate these actions. In the latter experiment[103] morphine and naloxone were administered peripherally while in the former one[77] centrally. Consequently, the centrally located receptors mediating the analgesic and gastrointestinal actions of opiates are similar, in agreement with the suggestion of Green,[80] but the peripheral receptors coactivated upon peripheral drug application might be different.

Administering the opioid peptides i.v., an inhibition of the natural propulsion has been observed in the mouse,[83,95] just as upon i.c.v. administration. Recently intraarterially administered enkephalin was found to inhibit the transpyloric flow in anesthetized cats by Edin,[84] the action being reversible by naloxone. On the other hand, naloxone antagonized the pylorus contraction induced by vagal stimulation.[84] Chey et al.[97] recorded a distinct fall in pancreatic protein and bicarbonate secretions and in plasma secretin level upon i.v. infusion of Met-E. Consequently, again the much-discussed controversy reappears; how is it possible that natural enkephalins given i.v. induce recognizable autonomic actions without causing central effects (analgesia). In view of the existence of peripheral opiate receptors all these actions might be ascribed to the activation of opiate receptors located outside of the blood-brain barrier. (Also these observations on the effectivity of peripherally administered enkephalins prove that their weak biological activity cannot be attributed to the frequently claimed "metabolic instability," since they do induce different pharmacological actions but not centrally.)

Comparing the natural enkephalins to the synthetic ones, the antidiarrheal activity of the formers was found extremely weak, while β-EP and D-Met,Pro-EA have been found about three times weaker than morphine on molar basis.[95] The ED_{50} values of β-EP, D-Met,Pro-EA and morphine were found 1.42; 1.38, and 0.46 μmoles/kg, re-

Table 3
EFFECTS OF OPIOID PEPTIDES AND NARCOTIC ANTAGONISTS ON THE MOTOR AND SECRETORY ACTIVITIES OF THE GASTROINTESTINAL TRACT IN VIVO

Treatment	Species	Effect	Reversal by opiate antagonists	Ref.
Met-E i.c.v. 100-500 μg	Mouse	Inhibition of gastrointestinal propulsion	Yes	85
Leu-E i.c.v. 50-250 μg	Mouse	Inhibition of gastrointestinal propulsion	Yes	85
Met-E i.v. infusion	Fasted dog	Increase in myoelectrical (spike) activity of the small intestine	Yes	86
Met-E i.v. infusion	Fed dog	Decrease in myoelectrical (spike) activity of the small intestine	Yes	86
Met-E intraarterial infusion	Dog	Increase in histamine induced gastric acid and pepsin output but no effect on basal secretion, increase in gastric mucosal blood flow	Yes	87
Naloxone intraarterial infusion	Dog	No effect	—	87
Naloxone s.c. 0.1-2.0 mg/kg	Rat	No effect on stress-induced reduction in gastrointestinal activity	—	88
(D-Ala^2,Met^5)-E intraarterial infusion	Guinea pig	Inhibition of peristaltic reflex in isolated but vascularly intact ileum	Yes	89
FK 33-824 intraarterial infusion	Guinea pig	Inhibition of peristaltic reflex in isolated but vascularly intact ileum	Yes	89
Naloxone s.c. 100 mg/kg	Mouse	Inhibition of diarrhea induced by $PGF_{2\alpha}$, 5-HTP, metacholine	—	90
Naloxone s.c. 0.5 mg/kg	Rat	No effect on PGE_2 induced diarrhea	—	91
Naloxone s.c. 10 mg/kg	Rat	Potentiation of intestinal fluid secretion elicited by PGE_1 and vasoactive intestinal peptide	—	92
FK 33-824 s.c. 3.2-10.0 mg/kg	Rat	Inhibition of PGE_1 induced increase in intestinal fluid volume	Yes	93
FK 33-824 i.v. 1.0 mg/kg	Pithed rat	Inhibition of PGE_1 induced increase in intestinal fluid volume	Yes	93
Naloxone s.c. 10.0 mg/kg	Pithed rat	No effect on basal or PGE_1 stimulated intestinal fluid secretion	—	93
FK 33-824 i.c.v.	Mouse	Inhibition of gastrointestinal motility ED_{50} : 0.45 ng/kg	Not examined	82
FK 33-824 i.v.	Mouse	Inhibition of gastrointestinal motility ED_{50} : 8.0 / μg/kg	Not examined	82
Met-E i.v.	Mouse	Inhibition of 5-HTP induced diarrhea ED_{50} above 132 /μmoles/kg	Yes	94

Table 3 (continued)
EFFECTS OF OPIOID PEPTIDES AND NARCOTIC ANTAGONISTS ON THE MOTOR AND SECRETORY ACTIVITIES OF THE GASTROINTESTINAL TRACT IN VIVO

Treatment	Species	Effect	Reversal by opiate antagonists	Ref.
β-EP i.v.	Mouse	Inhibition of 5-HTP induced diarrhea ED_{50} about 1.42 μmoles/kg	Yes	94
D-Met, Pro-EA i.v.	Mouse	Inhibition of 5-HTP induced diarrhea ED_{50} about 1.38 /moles/kg	Yes	94
Met-E i.v. infusion	Guinea pig	Inhibition of cholecystokinin and secretin induced pancreatic flow and protein secretion	Yes	95
Met-E-amide intraarterially 0.2-1.0 mg/kg	Cat	Decrease in transpyloric flow	Yes	83
Met-E i.v. infusion	Dog	Decrease in pancreatic bicarbonate and protein secretion, fall in the plasma secretin level	Yes	96
Naloxone i.v. infusion	Dog	No effect on postprandial plasma secretin level	—	96
Naloxone i.v. infusion	Human	Decrease in basal and stimulation gastric acid secretion	—	97
FK 33-824 i.v. 0.005-0.1 μg/kg	Cat	Increase in the contractile activity of the colon	Not examined	98
FK 33-824 i.v. 0.001 μg/kg	Human	Increase in the phasic and tonic pressure activity in the sigmoid colon	Not examined	98
Met-E intraarterial infusion	Dog	Increase in intestinal blood flow, oxygen extraction and consumption, and motility	Yes	99
Met-E i.v. 1×10^{-5}-5.0 μg/kg	Cat	Contraction of the ileocecal sphincter	Yes	100
Leu-E i.v. 1×10^{-5}-1.0 μg/kg	Cat	Contraction of the ileocecal sphincter	Yes	100
Met-E intraarterially 0.125-4.0 μg/kg	Opossum	Contraction of the lower esophageal sphincter	Yes	101
α-endorphin intraarterially 0.10 ng/kg	Opossum	Relaxation of the lower esophageal sphincter	Yes	101
Naloxone intraarterially 4.0 mg/kg	Opossum	Relaxation of the lower esophageal sphincter	Yes	101

Note: 5-HTP = 5-hydroxytryptophan.

spectively.[95] In another study FK 33-824 elicited constipation in the i.v. dose of 8 μg/kg which roughly corresponds to a molar concentration of 0.02 μmoles.[83] In this study[83] the Sandoz analogue was found about 70 times more potent in inducing constipation than antinociception if administering it i.v., whereas upon i.c.v. application about the same amount was needed to elicit both effects. Calculating the corresponding ratio for D-Met,Pro-EA, the enkephalin analogue developed at the Institute for Drug Research (Hungary, Budapest), a very different relative potency ratio can be calculated. Also in mice its analgesic ED_{50} value upon i.v. application is about 1 mg/kg,[104] i.e. about 2 /μmoles/kg. This value is somewhat higher than its antidiarrheal ED_{50} measured in another laboratory.[95] As for β-EP, its analgesic ED_{50} was found about 0.3 /μmoles/kg[105] i.v. Thus, D-Met,Pro-EA and β-EP are roughly equiactive as analgesic and antidiarrheal agents, while the constipant action of FK 33-824 is far superior to its analgesic potency.

Of course rather erroneous conclusions may be drawn if data obtained in different laboratories and in different test systems are quantitatively compared. In view of the great differences measured it may probably be concluded that the relative analgesic and intestinal motility inhibiting potencies of opioid peptides do not correlate with each other. Consequently, the peripheral opiate receptors mediating the inhibition of gastrointestinal motility are surely different from those central ones, responsible for analgesia. Such a supposition is further confirmed by the differences in antagonizability as discussed above. Of course, the conclusion that the intestinal and central opiate receptors are different might appear trivial. Considering, however, that Met-E and Leu-E occur both in the brain and the bowel, and their biological activity is titrated mainly in vitro based on their activity at the intestinal receptors, the divergence of the central and peripheral receptors is probably not so self-evident.

How do the opioid peptides elicit this antipropulsive action? According to Edin[84] the morphine and enkephalin induced decrease of transpyloric flow was due to tonic contraction of the corresponding pyloric muscles and the tone of the whole gastric wall was enhanced, the action being naloxone sensitive. Opioid peptides contract not only the pyloric muscle but also the lower esophageal[102] and ileocecal[101] sphincters. As for the former, it was contracted only by the natural enkephalins, while α-endorphin, β-EP caused relaxation and morphine had inconsistent effect. Thus, Rattan and Goyal[102] supposed that enkephalins increased the tone by activation of the δ-receptors, while the others induced opposite action via μ-receptors.

Otherwise, Met-E, Leu-E, and FK 33-824 were found to increase the contractile activity of the colon in humans and in rats as well.[99,101] Increase was observed in the phasic and tonic activities alike, and also the myoelectrical spike frequency was enhanced.[99] The increased intestinal motility is accompanied by an increase in intestinal blood flow, oxygen extraction, and consumption, respectively.[100] These circulatory changes[100] are in all probability secondary consequences of the facilitated muscular activity. As shown in the Table 3, all these effects can be reversed by naloxone though in not the same degree. In one study[102] the effects of natural enkephalins could be antagonized only with relatively high doses of naloxone.

The antipropulsive activity of DALA (given i.c.v. or i.p.) could be prevented by subdiaphragmatic vagotomy,[106] but the contraction of ileocecal sphincter induced by Met-E and Leu-E was only partially blocked by atropine.[101] The contractile response of lower esophageal sphincter elicited by Met-E could be reversed by tetrodotoxin but the opposite effect of α-endorphin not.[102] Accordingly, the constipation may the consequence of a general increase in the gastrointestinal tone. Gillan and Pollock[107] showed that Met-E and Leu-E similarly to morphine inhibited the contraction of rat colon induced by electrical stimulation of the corresponding spinal nerves. Since this action was also observed in pithed animals,[107] the effect could only be peripherally mediated. However, as discussed above, morphine-like drugs are generally spasmogenic.[77,78]

Also in this study[107] enkephalins induced rhythmic contractions in the isolated rat colon preparation. Consequently, inhibition of the effect of electrical stimulation does not mean that opioids did not contract the intestinal preparation under study. In other words, opioids might elicit muscle relaxation via presynaptic inhibition and a direct spasmogenic action via postsynaptic mechanisms. This smooth muscle contracting effect might be connected with a direct action on the postsynaptic membrane as discussed in Chapter 2, Volume II, or it could be a consequence of serotonin or acetylcholine liberation both known to contract the intestinal wall. But neither atropine nor serotoninergic blocking drugs altered the spasmogenic action of opioids.[84,107]

Otherwise, the direct spasmogenic action of opioids is easier to demonstrate under in vitro conditions, which is only shortly discussed in this chapter destined for the in vivo data. β-EP and DALA caused dose-related phasic and tonic contractions in isolated and vascularly perfused segments of dog small intestine.[108] As both atropine and cinanserine inhibited these contractions, the effect might be correlated with cholinergic and serotoninergic mechanisms.[108]

Also in vitro Met-E potentiated the PGE_2 induced contractions of the rat colon, inhibited the similar response of chick rectum, but did not affect the rat colon.[109] Even more interesting is that the inhibitory action of β-EP was more sensitive to antagonism by naloxone than that of the enkephalin.[108] For this reason the opiate receptors mediating the direct intestinal effects of endogenous opioids should be heterogeneous. In a recent work Nijkamp and van Ree[110] demonstrated that DALA, Leu-E, α-endorphin, γ-endorphin, and β-LPH (80-91) contracted the isolated intestinal segments and colon of the rat but relaxed the gut of the guinea pig ileum. The effect was increased if examining the relatively more caudal segments of the bowels and decreased by applying relatively longer fragments of β-EP.[110] This contraction must be a direct myogenic action, as tetrodotoxin, a specific inhibitor of axonal conduction, potentiated the effect of Met-E.[110] As methysergide and cyproheptadine inhibited, while phentolamine potentiated the contraction, an interplay of noradrenergic and serotoninergic influences should determine the actual tone of smooth muscle.[110] Considering the much-discussed potentiation of serotoninergic functions by opioids, it is tempting to ascribe the myogenic action of opioids to the same interaction.

A further possibility is that the constipant action is independent of both the presynaptically mediated relaxation and postsynaptically induced myogenic actions, and it is brought about by suppressing specifically the peristaltic movements. Holzer and Lembeck[90] observed the inhibition of peristaltic reflex in the guinea pig ileum upon treatment with FK 33-824 and morphine. However, these observations[90] were made on guinea pig ileum, i.e., with a preparation which behaves ''irregularly'' as discussed above.

Actually, the corresponding in vitro observations were made by van Nueten et al.,[111] who found a dose-dependent inhibition of spontaneous peristaltic activity of the guinea pig ileum upon adding Met-E into the bathing solution. The same group[112] observed previously that the spontaneous peristalsis of guinea pig ileum distensed for prolonged period gradually decreased, which was regarded as a manifestation of exhaustion. However, adding naloxone to such an ''exhausted'' preparation, the peristaltic movements reappeared.[112] These studies[111,112] were confirmed by Kromer and Pretzlaff,[113] who also demonstrated the facilitation by naloxone of spontaneous peristalsis in guinea pig small intestine. However, this facilitatory action was independent of the previous prolonged work, i.e., ''exhaustion'', and it could be demonstrated only in the distal ileum but in the duodenum and jejunum not.[113] Enkephalin liberation might explain the gradual dissipation of peristalsis under certain physiological conditions.

In this aspect the observations of Thor et al.,[87] who examined the effect of enkephalins on the myoelectrical activity of the small intestines of conscious dogs are interesting. In this study[87] morphine, Met-E, and Leu-E given i.v. to fasted animals elicited an increase in spike activity and frequency of bursts of the interdigestive myoelectric complexes. Administering the same substances to previously fed dogs, a dose-dependent reduction in the postprandial spike activity was observed.[87] Thus, the electrographic data show that enkephalins may suppress the peristaltic activity but only postprandially.

Of course, the constipation inducing action might be a secondary phenomenon due to suppression of normal secretory activity in the gastrointestinal tract. Unfortunately the data are rather scanty in this relation too. In anesthetized dogs Met-E and morphine (given intraarterially) did not affect the basal secretion of gastric acid and pepsin but increased the gastric mucosal blood flow.[88] In the case of simultaneous histamine infusion the gastric acid and pepsin output was actually increased by opioids.[88] As this facilitation of gastric secretory activity was observed only in the presence of histamine and connected with an increase in the local blood flow,[88] it is reasonable to suppose that these effects were secondary phenomena due to vasodilatation and/or histamine release. On the other hand, naloxone reduced the basal gastric acid secretion and that induced by intragastric infusion of casein hydrolysate in healthy human subjects.[98] This experiment of Feldman and Walsh[98] indicates that endogenous opioids tonically facilitate the secretory activity of the stomach.

Otherwise, enkephalins have been found to inhibit the exocrine secretion of the pancreas, namely DALA and Met-E (given i.v.) drastically reduced the pancreatic flow, protein, and bicarbonate secretion and the plasma level of secretin.[96,97] In the same study[96] enkephalins did not inhibit the secretory activity of isolated pancreatic acini. Therefore, a nonendocrine explanation must be found. Probably the enkephalins contract the Oddi sphincter or other parts of the biliary tract as opiates do.[78] Therefore, it is interesting that the guinea pig ileum contractions elicited by cholecystokinin can be antagonized by morphine, β-EP and natural enkephalins.[114]

Considering the potent antidiarrheal activity, probably even more interesting are the observations of Beubler and Lembeck,[93,94] who found that morphine and FK 33-824 did not modify the basal intestinal fluid volume in intact unanesthetized fasted rats, but they dose-dependently inhibited the PGE_1 stimulated secretion of the intestinal fluid. The effect was reversed by naloxone, which alone had no effect but enhanced the stimulatory action of PGE_1.[93,94] Since the same inhibitory effect could be elicited in pithed rats,[94] and also loperamide[93] was active, i.e., an opiate which is not absorbed from the bowels,[81] obviously the local opiate receptors were involved, since the antidiarrheal action of loperamide also may be reversed by naloxone.[92]

The most attractive aspect of these experiments[93,94] is the correlation of antidiarrheal and antiprostaglandin activities. The prostaglandins are powerful stimulants of the intestinal secretion and they are suspected to be involved in the pathogenesis of diarrhea. As prostaglandins exert their effect via facilitation of adenylate cyclase and opioids do inhibit the activation by them of this enzyme (Chapter 2, Volume I and Chapter 1, Volume III), the antisecretory action of opioid peptides may be derived from one of their basic biochemical action.[115]

Needless to emphasize that the powerful antisecretory action may explain their antidiarrheal action too. Thus, their unclarified spasmogenic, muscle relaxant and antiperistaltic actions might be simply irrelevant as for their constipant effects. (For the time being it is difficult to interpret a report[91] according to the diarrhea induced by $PGF_{2\alpha}$, metacholine and 5-hydroxytryptophan can equally be reversed by naloxone.)

Some data indicate that endogenous opioids might mediate the inhibition of gastrointestinal activity in stressed animals. Bodycote and Chesher[116] observed that ilea, excised from guinea pigs previously stressed by forced swimming, reacted with contracture to naloxone and this contraction could be antagonized by atropine. Consequently, stress might inhibit the gastrointestinal activity via inhibition by endogenous opioids of acetylcholine release. However, in rats the ether stress-induced constipation was not reversed by naloxone.[89]

Based on the above-reviewed data, the role of the endogenous opioid peptides in regulation of gastrointestinal activity might be tentatively summarized as follows.

1. Contrary to their respiratory and cardiovascular actions the intestinal effects are primarily peripherally mediated.
2. Opioids inhibit the electrically induced contractions of different segments of gastrointestinl tract, this action is presumably due to a presynaptic inhibition of the transmitter release.
3. Opioids exert also a direct postsynaptically mediated spasmogenic effect especially on the caudalmost part of the intestinal tract. This action might be connected with their adrenolytic and serotonin receptor stimulant actions.
4. Endogenous opioids inhibit the spontaneous peristalsis in vivo mostly postprandially, in vitro upon exhausting the preparation.
5. Endogenous opioids decrease the pancreatic and intestinal fluid secretion. They do not modify the basal fluid secretion but inhibit that induced by prostaglandins.
6. Opioid peptides contract the pylorus, the esophageal and ileocecal sphincters as well.
7. The above-mentioned actions might explain the well known antipropulsive and anorexigenic effects of opioids as well. However, as for the latter action also centrally located opiate receptors must be counted with.
8. The intestinal opiate receptors responsible for these effects are at least partially different from those in the brain.

IV. MODULATION OF THE THERMOREGULATION

A. Effects of Opiates and Opioid Peptides on the Core Temperature: Structure-Activity and Dose-Response Relationships

Before reviewing the seemingly heterogeneous data concerning the effects of opioid peptides on thermoregulation it is worth mentioning that the relevant findings seem to be contradictory also in relation to morphine congeners. Upon systemic injection of morphine or its derivatives in several species (rabbit, dog, monkey, guinea pig) decrease in the body temperature was observed, while in other species rather hyperthermia was reported.[117-121] The more detailed studies revealed that the direction and magnitude of the effects observed depended on the doses administered and the route of application.[117-121] Moreover, certain components of opiate actions were found naloxone insensitive, generally the hyperthermic response.[120] Upon repeated treatment tolerance developed to the hypothermic action, while the hyperthermic response was not significantly modified.[117,120] This diversity of experimental data reminds one of those reported in relation to the locomotor activity, which also has been reported to be facilitated or suppressed depending on the dose, route of application, species examined, and on the eventual previous administration of the same drugs (Chapter 2, Volume I). Let us recall that a close examination of the data on spontaneous activity revealed a pattern, which seemed to be valid both for opiates (Chapter 2, Volume I) and opioid peptides (Chapter 3, Volume II) in almost all species examined. Applying opioids in subthreshold (subanalgesic) doses, a slight excitation could be observed. At moderate dose levels an initial depression was followed by hyperactivity observable also in the spontaneous electroencephalogram and further increase of the dose resulted in lengthening of the initial stupor and the subsequent hyperactivity as well. Upon repeated application tolerance developed primarily to the initial depressant effect (Chapter 2, Volume I and Chapters 2 and 3, Volume II).

If stating in advance that a similar biphasic pattern seems to emerge also in relation to thermoregulation, another point must be stressed beforehand, namely that only the pattern of dose-effect relationships might be similar, but there is no close correlation between the behavioral stupor and hypothermia or increased locomotion and hyperthermia, respectively (see below). Thus, the observed changes in the core temperature are not exclusively secondary epiphenomena of decreased or increased locomotion, since the receptors mediating the effects of opioids on thermoregulation are located in brain structures which have little to do with the analgesic action.

First Hermann[122] observed that in rats morphine increased the body temperature if giving it in doses below the 30 mg/kg dose level, while in higher dosage hypothermia could be observed. Gunne[117] demonstrated in a carefully designed study that increasing the dose of morphine, hypothermia replaced the hyperthermic response elicited by small amounts, which was later followed by the pyretic action. A further increment in the dose resulted in potentiation of the initial fall of body temperature and in postponement of subsequent phase of hyperthermia.[117] Upon repeated treatment tolerance developed primarily to the initial hypothermia, thus, the latency time of late hyperthermia gradually decreased.[117] This transformation of hypothermic reaction into a hyperthermic response upon repeated application could be seen also in other species.[118] The demonstration of an initial hypothermic reaction is missing only in cats.

Bearing in mind these findings, the data on the temperature effects of opioid peptides are already less inconsistent.

In rats β-EP, if giving it i.c.v. in doses smaller than 10 /μg, hyperthermia has been observed.[123] In higher doses the body temperature was rather reduced[123-126] and the hyperthermia ensued subsequently.[127] The hypothermic response could be reversed by naloxone,[124,125] while the data on the antagonism by naloxone of the hyperthermic reaction are less convincing. Upon repeated application of β-EP tolerance develops to its hypothermic action.[127] Unfortunately the tolerance liability of hyperthermic reaction, appearing after dissipation of the initial hypothermia or evolving alone if giving small dose, has not been studied yet.

In our own experiments[128] giving morphine to rats i.c.v. in doses of 10-30-100 nmole a monophasic hyperthermic reaction was recorded, the response being hardly dose-dependent (Figure 1). Upon repeated administration (on five consecutive days) this hyperthermic reaction seemed to be slightly attenuated, especially in animals treated with the smaller doses. Contrary to any expectation no initial hypothermia was seen in these experiments upon central administration of morphine. Giving β-EP in a similar way but in smaller doses (1 to 10 nmoles/animal), practically the same monophasic elevation of the core temperature was observed and practically no tolerance development upon the repeated treatments (Figure 2). Examining the effects of the potent enkephalin analogue, D-Met, Pro-EA, also a monophasic hyperthermic reaction was recorded, which gradually diminished during the five subsequent experimental days (Figure 3). Actually the most puzzling are the data obtained with β-EP. A close inspection of Figure 1 reveals that in animals repeatedly treated with 1 nmole simply no tolerance developed. However, those animals receiving 10 nmole displayed gradually increasing reaction, i.e., in these rats the hyperthermic reaction was rather increased upon the subsequent treatments. This peculiar phenomenon might be explained if supposing that a "latent" hypothermic action coinciding with the much stronger hyperthermic reaction results in a netto elevation of body temperature. Latter tolerance develops only (or primarily) to the hypothermic action resulting in enhancement of the netto hyperthermic reaction. However, no such pattern of changes could be observed in the cases of morphine or the enkephalin analogue.

These observations made in our laboratory[128] show that it is rather difficult to develop tolerance to the hyperthermic action of opioids. Of course five treatments, as applied in these experiments, might have been insufficient for tolerance development. Nevertheless, in another study of Miglécz et al.[129] a rapid attenuation of β-EP-induced analgesic effect was observed using practically the same treatment schedule. It is possible that continuing the daily administration of opioids tolerance could have been

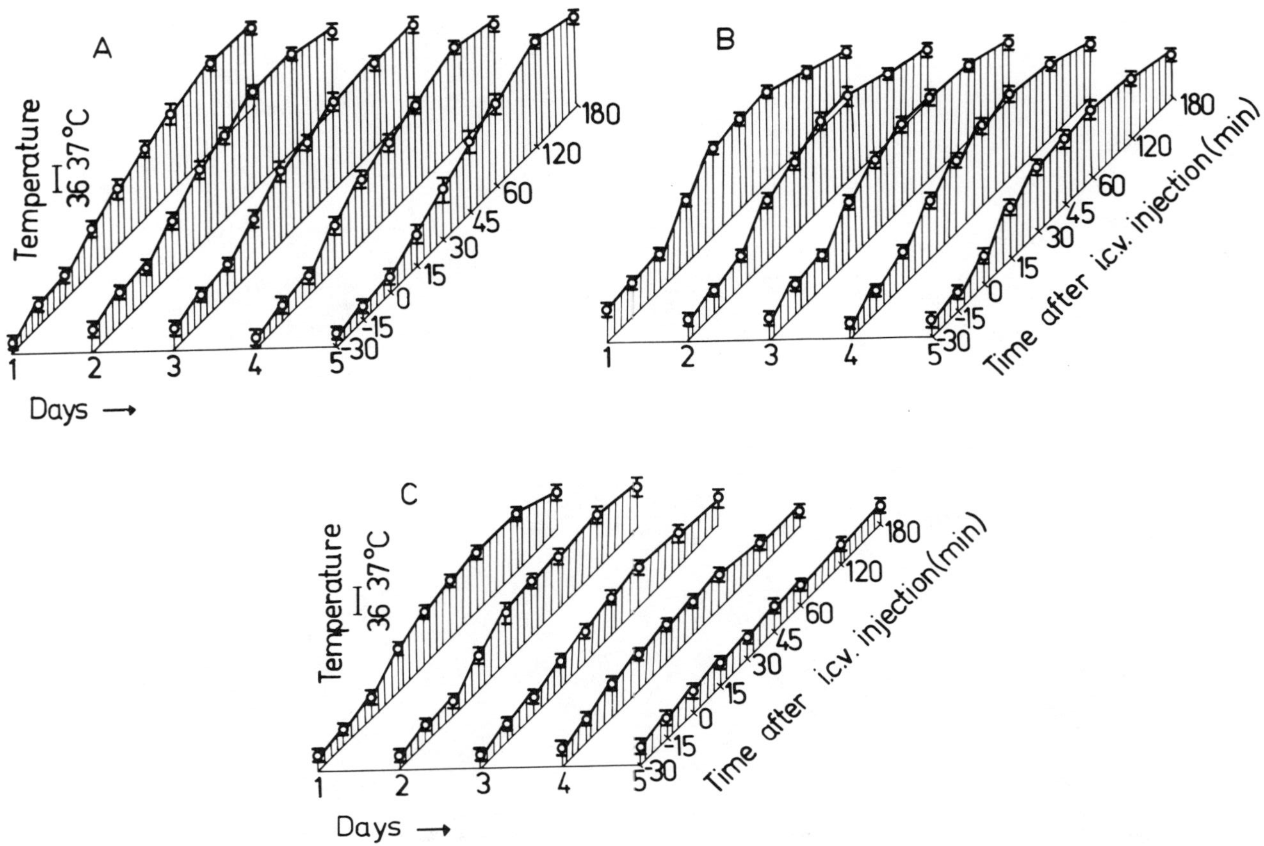

FIGURE 1. Effect of repeated i.c.v. morphine treatment on rectal temperature of rats. The same dose of morphine was administered on five consecutive days via permanently implanted cannula into the right lateral ventricle. Panel A: 10, panel B: 30, panel C: 100 nanomol 10 $\mu\ell$/animal, respectively. The rectal temperature was measured by Elab thermometer.

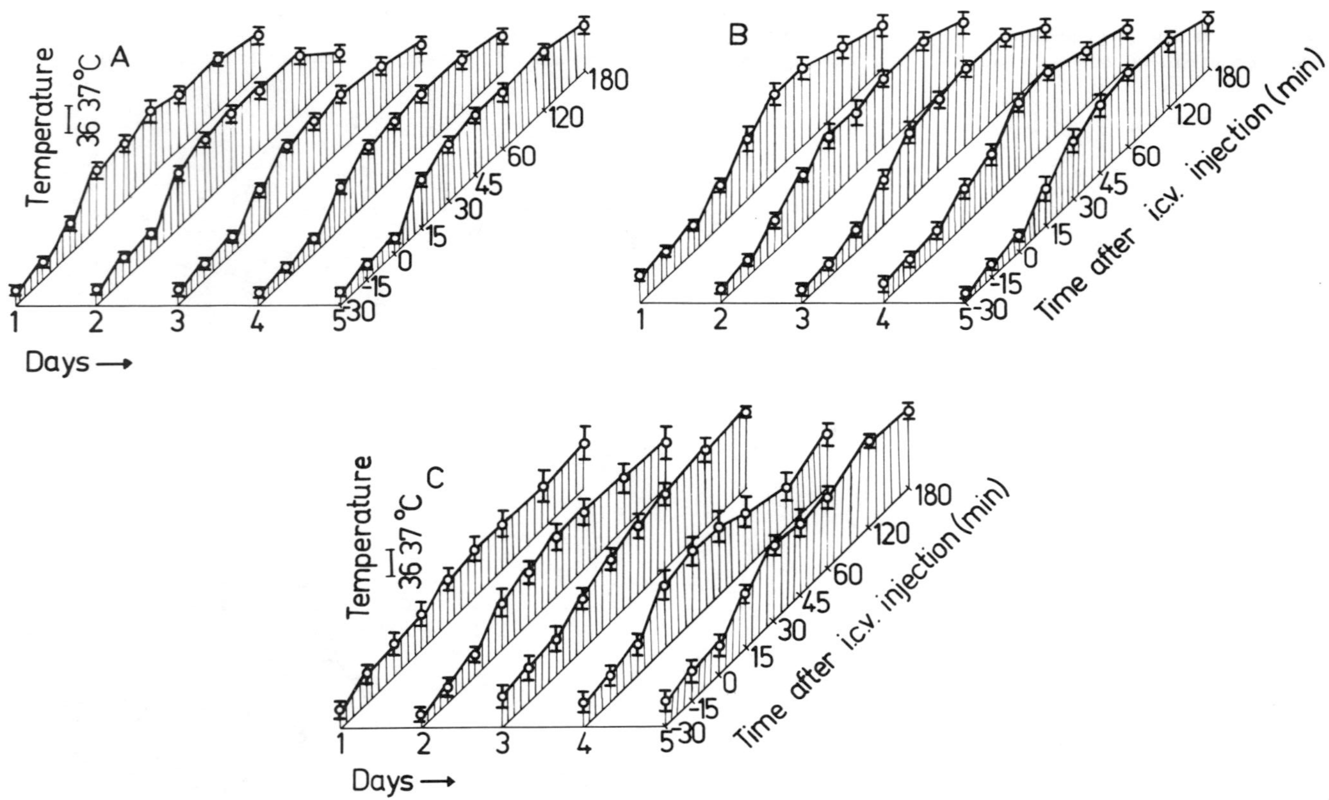

FIGURE 2. Effect of repeated i.c.v. β-EP treatment on rectal temperature of rats. Panel A: 1.0, panel B: 3.0, panel C: 1.0 nanomol/10 μℓ/animal, respectively. For further details see Figure 1.

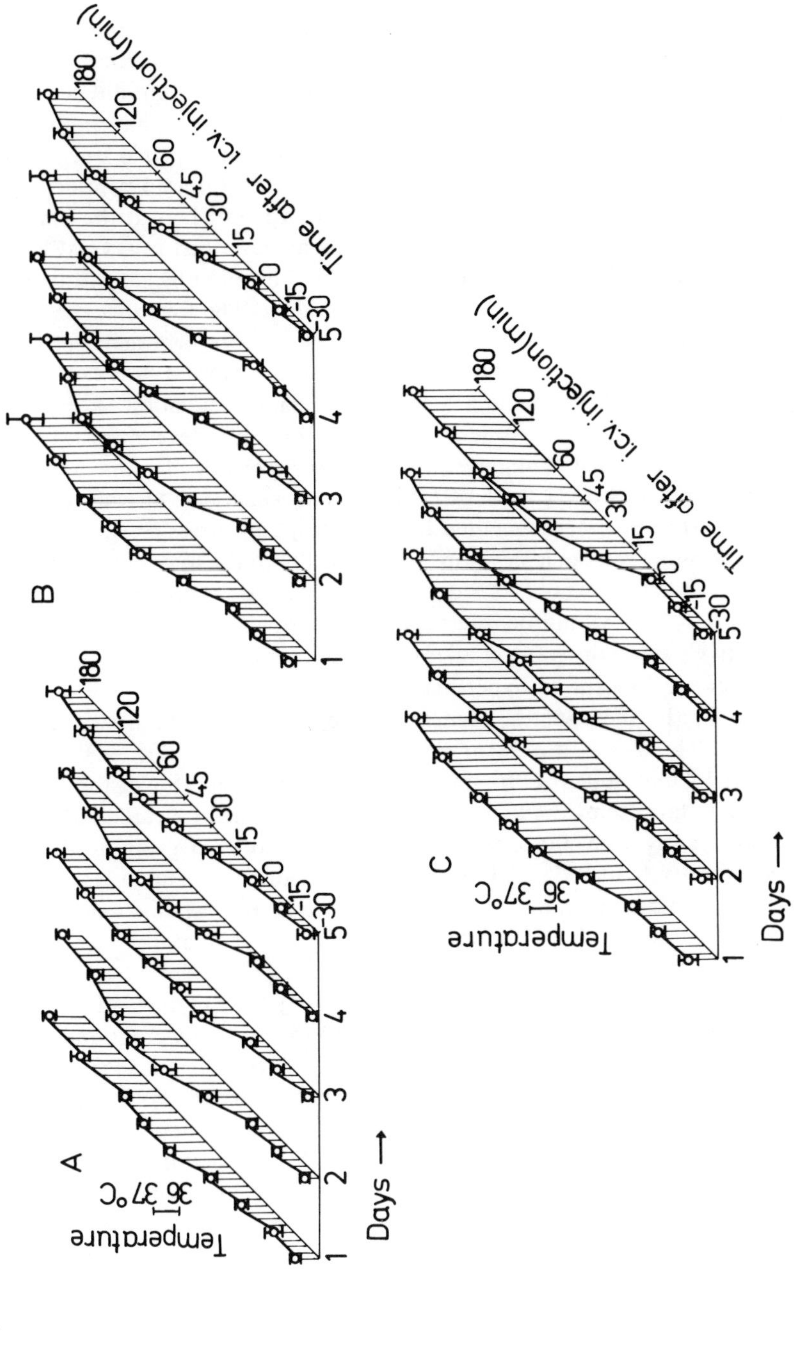

FIGURE 3. Effect of repeated i.c.v. D-Met, Pro-EA treatment on treatment on rectal temperature of rats. Panel A: 3.0, panel B: 1.0, panel C: 0.3 nanamol/10 $\mu\ell$/animal, respectively. For further details see Figure 1.

established in relation to the hyperthermic action, as in another laboratory applying higher morphine doses, i.e., implanting 6 pellets in 7 days, each containing 75 mg, tolerance developed to the hyperthermic action of the potent enkephalin analogue, FK 33-824.[130] Nevertheless, the great difference in tolerance development to the analgesic and hyperthermic effects of opioid peptides points to differences in the corresponding receptorial mechanisms, at least in rats.

In experiments of Holaday et al. 30 /μg β-EP elicited only a slight drop of temperature[123,126] or no effect at all[131] in sham operated control rats, while in hypophysectomized animals a marked hypothermic effect could be observed.[126,131] In these experiments[126] hypophysectomy potentiated the hypothermic action of β-EP but did not alter its analgesic and cataleptogenic effects. Hence the possibility emerges that the hyperthermic effect is mediated by the hypophysis (adenohypophysis) i.e., it is due to the stressor action of opioids, as the hyperthermic effect induced by opioid peptides and morphine reaches its peak generally 1 to 2 hr after the treatment, i.e., at the time of dissipation of their analgesic and cataleptogenic actions as described in several laboratories[125,126,132,133] and also shown in Figures 1 to 3. Thus, it might be easy to understand that the tolerance development to this action is so slow. The tolerance development to the ACTH mobilizing action of opioids must be a more protracted process than in relation to analgesia. In opiate addicts the adrenals are more or less exhausted (see the corresponding textbooks) proving a prolonged previous activation. According to Bläsig et al.[130] chronic morphine treatment resulted in tolerance development to the hyperthermic action of FK 33-824. The question arises whether this phenomenon was not due to the exhaustion of ACTH (and LPH) producing cells. The answer is no as in these animals a hyperthermic reaction could be elicited by cold stress.[130] Considering that in the case of morphine tolerance develops more rapidly to the hypothermic effect than to the hyperthermic action[117] and β-EP does not induce hyperthermia in hypophysectomized animals,[123,126,131] it would be interesting to know whether the β-EP-elicited hyperthermia is relatively more resistant to tolerance development.

Of course a further mechanism of hyperthermic effect independent of the specific influence exerted on the thermoregulatory centers may be an enhanced muscular activity, i.e., locomotion. Thus, it is not surprising that the hyperthermic reaction was found suppressed upon immobilizing the animals.[130] As discussed above, the hyperthermic action appears later than the analgesic response, it coincides rather with the subsequent behavioral activation. Since tolerance develops more sluggishly to the opiate-induced hypermotility than to its sedative effects (Chapter 2, Volume I), this action might even be potentiated as the cateleptic effects are already attenuated. Consequently, the relative resistance of hyperthermia might have an explanation independent of the adenohypophysis.

In cats only hyperthermic action was observed by Feldberg and Smyth[132] administering β-EP into the third ventricle. Moreover, no qualitative differences were found upon comparing β-EP with its shorter fragments or Met-E.[132-134] As in feline species locomotor excitation and increased activity of the sympathetic nervous system prevails upon opioid treatment, the consequences of hyperactivity of the pituitary-adrenal system might conceal the primary centrally mediated hypothermic action. However, in cats the β-EP-induced hyperthermia may easily be antagonized by naloxone.[132] Thus, the hyperthermic response might be centrally mediated in this species. Of course the question could be decided only by concomitant blockade of the hypophyseal and adrenal functions.

In mice again another pattern of effects appears. Huidobro-Toro and Way[133] recorded a hyperthermic response at the minute i.c.v. dose of 100 picogram/mouse, and increasing the dose of β-EP this elevation in body temperature was rather reduced, but no hypothermia has been observed. Others[136] found hypothermia in fairly high (i.c.v.)

doses. Bloom and Tseng[137] demonstrated that β-EP induced hyperthermia if given in small amounts, while hypothermia could be elicited by higher doses. The same authors[137] reported that no tolerance developed to the hyperthermic action and this response could not be antagonized by naloxone. Giving it i.v., no dose-related hyperthermia has been observed.[139] However, in another laboratory the hyperthermic action of β-EP was readily antagonized by naloxone even if the latter was given 2 hr after the peptide, i.e., at the time of full development of pyretic reaction.[133] Moreover, in morphine-tolerant animals the hyperthermic reaction was considerably attenuated.[133]

Interestingly in mice a single treatment with β-EP (giving at least 10 ng i.c.v.) resulted in tolerance development to its hyperthermic action.[138] A similarly rapid cross-tolerance development was demonstrated by Huidobro-Toro and Way[138] between β-EP and Leu-E, Met-E, or DALA, respectively.[138] It is worth recalling that Cox et al.[120] did not observe tolerance development in rats if morphine was applied in low doses eliciting only hyperthermia.

Thus, the tolerance development in mice to the hyperthermic effect of β-EP proceeds extremely rapidly,[138,139] while in rats the same process is difficult to detect or proceeds rather slowly whether β-EP,[128] morphine,[120] or the active enkephalin analogues[130] are examined.

The tolerance development, reversal by naloxone, and the observations that this rise in body temperature might be elicited also in subanalgesic dosage of β-EP or its fragments,[132,133] point to central (hypothalamic) mediation of hyperthermic reaction in cats and mice.

A close inspection of the data summarized in Table 4 reveals a similar pattern in relation to the natural enkephalins and their potent synthetic analogues. Met-E and Leu-E given centrally in small doses induce no measurable effect[136,145] or hyperthermia,[130,133,145] while in higher doses hypothermia or biphasic action can be detected in rats and in mice as well.

Nevertheless, of natural enkephalins much higher doses are needed to elicit any change in the core temperature than in the case of β-EP (Table 4). Reviewing the structure activity relationships two curious observations must be mentioned. Cowan et al.[135] found hypothermia upon i.c.v. application of Leu-E, while Met-E induced hyperthermia in cats. However, no other data indicate that the two natural enkephalins exert differential effects on thermoregulation. In an early study Bloom et al.[124] stated that β-EP induced hypothermia, while γ-endorphin induced a rise in the body temperature. This finding is in good agreement with the hypothesis on differential physiological functions of β-EP and γ-endorphin (Chapter 3, Volume II). However, considering the above-discussed data on the biphasic effect of opioids on body temperature in rodents, it cannot be excluded that the dose applied in this study was high enough to induce hypothermia in the case of β-EP, but the amount of γ-endorphin was relatively low for a similar action. As for the enkephalin analogues, DALA induced in chicks hypothermia just as β-EP did,[153] while in rodents their actions depend also on the dose applied[130,133] (Table 4).

B. Possible Mechanism of Alterations in Thermoregulation Induced by Opioids

As the above-outlined alterations in thermoregulation are not simply epiphenomena or motor manifestations of the much-discussed behavioral actions, the question arises that which brain regions and neurotransmitters mediate the hypo- and hyperthermic reactions.

As for the precise site(s) of action within the brain the study of Martin and Bacino[125] should be mentioned first. These authors[125] recorded hyperthermia if β-EP was injected directly into the preoptic area of hypothalamus or into the spinal subarachnoid space, while administering it into the lateral ventricles the body temperature decreased. A

Table 4
EFFECTS OF OPIOID PEPTIDES AND NARCOTIC ANTAGONISTS ON THERMOREGULATION

Treatment	Species	Effect	Reversal by opiate antagonists	Ref.
Naloxone s.c. 10 mg/kg	Normal or cold stressed rat	No significant effect	—	138
β-EP i.c.v. 7.4-14.9 × 10^{-9} moles	Rat	Hypothermia	Yes	122
γ-endorphin i.c.v. 2.8 × 10^{-7} moles	Rat	Hyperthermia	Not examined	122
Met-E i.c.v. 100-1000 μg	Cat	Hyperthermia	Yes	133
Leu-E i.c.v. 20-700 μg	Cat	Hypothermia	Yes	133
Naltrexone into anterior hypothalamus 10-50 μg	Morphine dependent rat	Hypothermia	—	139
Naloxone i.p. 10 mg/kg	Rat	Inhibition of conditioned hyperthermia	—	140
Naloxone s.c. 0.1-3.0 mg/kg	Dog	Slight hypothermia	—	140
Met-E i.c.v. 1-2 mg	Cat	Hyperthermia	Yes	132
β-EP i.c.v. 10 μg	Cat	Shivering, hyperthermia	Yes	130
LPH (61-78) i.c.v. 50-100 μg	Cat	Slight shivering	Yes	130
LPH (61-69) i.c.v. 100 μg	Cat	Slight shivering	Yes	130
Met-E i.c.v. 400 μg	Cat	Sometimes shivering	Yes	130
β-EP i.c.v. 94 μg	Rat	Hypothermia followed by hyperthermia	Yes	125
β-EP i.c.v. 2.5-20.0 μg	"Intact" rat	Slight hypothermia	Not examined	124
β-EP i.c.v. 2.5-20.0 μg	Hypophysectomized rat	Hypothermia	Not examined	124
Naloxone intrathecally 10 μg	Morphine dependent rat	Hypothermia	—	141
β-EP i.c.v. 7.5 μg	Rat	Hyperthermia	Not examined	121
β-EP i.c.v. 30.0 μg	Rat	Slight hypothermia	Not examined	121
β-EP i.c.v. 30.0 μg	"Intact" rat	No effect	—	129
β-EP i.c.v. 30.0 μg	Hypophysectomized rat	Hypothermia	Yes	129
Met-E i.p. 10-40 mg/kg	Rat	No effect	—	142
Met-E i.c.v. 100 μg	Rat	Hyperthermia	Yes	142
Met-E i.c.v. 400 μg	Rat	Hypothermia followed by hyperthermia	Yes	142
Met-E i.c.v. 100-400 μg	Morphine tolerant rat	No effect	—	142
β-EP intrathecally 1 μg	Rat	Hyperthermia	Not examined	143
(−)-naloxone i.c.v. 50 μg	Rat	Inhibition of emotional hyperthermia	—	144
(+)-naloxone i.c.v. 50 μg	Rat	No effect	—	144
Naloxone i.p. 10 mg/kg	"Intact" rat	Facilitation of escape from heated enclosure	—	145

Table 4 (continued)
EFFECTS OF OPIOID PEPTIDES AND NARCOTIC ANTAGONISTS ON THERMOREGULATION

Treatment	Species	Effect	Reversal by opiate antagonists	Ref.
Naloxone i.p. 10 mg/kg	Hypophysectomized rat	No effect on thermoregulatory behavior	—	145
β-EP intracisternally about 2-3 μg	Mouse	Hypothermia	Not examined	134
Met-E intracisternally about 0.3-0.4 μg	Mouse	No effect	—	134
Leu-E intracisternally about 0.3-0.4 μg	Mouse	No effect	—	134
FK 33-824 s.c. 0.33-5.0 mg/kg	Rat	Hyperthermia	Yes	128
FK 33-824 s.c. 20.0 mg/kg	Rat	Hypothermia	Yes	128
FK 33-824 i.c.v. 20-80 ng	Rat	Hyperthermia	Yes	128
FK 33-824 i.c.v. 320 ng	Rat	Hypothermia then hyperthermia	Yes	128
Met-E i.c.v. 100-500 μg	Rat	Hyperthermia	Not examined	128
β-EP into preoptic area of the hypothalamus 0.74-7.4 nmoles	Rat	Hyperthermia	Partially	123
β-EP i.c.v. 7.4 nmoles	Rat	Hypothermia	Yes	123
β-EP intrathecally 2.96 nmoles	Rat	Hyperthermia	Yes	123
Met-E i.c.v. 10 μg	Mouse	Hyperthermia	Not examined	131
Leu-E i.c.v. 10 μg	Mouse	Hyperthermia	Not examined	131
DALA i.c.v. 2-100 μg	Mouse	Hypothermia then hyperthermia	Not examined	131
β-EP i.c.v. 0, 1-200 ng	Mouse	Hyperthermia	Yes	131
β-EP i.c.v. 100-1000 ng	Morphine tolerant mouse	Hypothermia	Not examined	131
β-EP i.c.v. 50 μg	Rat	Hypothermia at cold or ambient temperature	Yes	146
Naloxone i.p. 2.5-10.0 mg/kg	Rat	Hypothermia; inhibition of noise stress induced hyperthermia	—	147
Naloxone i.v. 10-20 mg	Human	Hypothermia	—	148
β-EP s.c. 50 μg	2-7 day old rat pup	Increase in threshold of thermal stimuli	No	149
Naloxone s.c. 100 μg	2-7 day old rat pup	Increase in threshold of thermal stimuli	—	149
β-EP i.c.v. 0.03-1.0 μg	Mouse	Hyperthermia	No	135
β-EP i.c.v. 3.0-10.0 μg	Mouse	Hypothermia	Not mentioned	135
β-EP i.v. 2.6-25.5 mg/kg	Mouse	Not dose-related hyperthermia	Not examined	137
β-EP i.c.v. 1-12 μg	Chick	Hypothermia	Yes	150
DALA i.c.v. 2.5-20.0 μg	Chick	Hypothermia	Yes	150
Naltrexone i.p. repeatedly 25.0 mg/kg	Dormouse	Suppression of spontaneous hibernation	—	151
Naloxone i.p. 10 mg/kg	Rat	Reversal of inhibition of TSH release upon heat exposure	—	152

further interesting finding of this group[125] is that naloxone only partially attenuated the hyperthermic response induced by intrahypothalamic treatment, while the hypo- and hyperthermic actions of intraventricular and intrathecal β-EP injections were readily reversed by naloxone. These findings are consistent with the data obtained with morphine in similar experiments. Giving it into the preoptic area it also induced hyperthermia,[156] while naltrexone elicited a significant fall in body temperature of morphine-dependent animals via the same area.[141] Unfortunately the relevant data are not quite unambiguous. While Martin and Morrison[156] found the hyperthermic response elicited by intrahypothalamic morphine administration dose-dependent, according to Lotti et al.[118] morphine elicits via this region hyperthermia in low dose but hypothermia if given in higher doses or injecting it more caudally within the hypothalamus, i.e., into the area of the mamillary nuclei.

Considering the presence of β-EP in this area (Chapter 3, Volume III) endogenous opioids located in the anterior hypothalamus might be supposed to participate in the thermoregulation. Otherwise, Lotti et al.[118] and Cox et al.[120] postulated that opiates acted not directly on the heat gain or heat loss effector pathways, but via decreasing or increasing the hypothalamic thermoregulatory set point. Rats injected with a relatively small dose of morphine remained in the vicinity of radiant heat source significantly longer than the control ones.[118,120] As opiates suppress the nociceptive reactions in general, similar suppression of autonomic reactions induced by noxious heat stimuli would be a logical consequence. If the endogenous opioids acted similarly, suppression by them of thermoregulatory behavior were easy to interpret. Furthermore, it must be taken into consideration that hypothalamus contains not only β-EP but also enkephalins. The physiological role of the latter might be different from that of the β-EP. In this aspect it is worth mentioning that the time course of hyperthermia elicited by morphine was similar to that of the enkephalins but different from that shown by β-EP.[133] This region has, however, little to do with the analgesic actions (Chapter 2, Volume I) proving the dissociation of the two effects.

The simultaneous onset of hypothermia, catalepsy, and analgesia in rats and in mice if opioids are given into the ventricular system (Chapter 2, Volume I; Chapter 3, Volume II) indicates the existence of other thermoregulatory center(s) somewhere in the central gray or in a structure functionally closely related to it. Thus, in this brain area the analgesic, catalepsy inducing, and thermoregulatory effects of opioids might be mediated by the same receptor population.

A third site of action lies somewhere in the spinal cord. Stimulation of these opiate receptors by β-EP[125,146] or morphine[144] results in hyperthermia, which can easily be reversed by naloxone. Therefore, the spinal cord represents an area from where analgesia (Chapter 2, Volume I) and hyperthermia may be elicited but no behavioral action. The hyperthermic effect of intrathecally administered opioids seems to be difficult to interpret considering that the brain areas responsible for the physiological thermoregulation are located primarily in the hypothalamus and the spinal animals have no thermoregulatory capacity. Considering, however, that the thermal and nociceptive sensory information is mediated at least in part by the spino-thalamic pathways, the observed action might be attributed to the blockade of ascending pathways carrying information into the diencephalic thermoregulatory centers. Rudy and Yaksh[144] reported that the hyperthermic action of peripherally administered morphine could be only partially reversed by naloxone if giving it intrathecally. Consequently, the spinal cord cannot be the only site mediating the hyperthermic action of opioids.

Thus, considering the data on the changes in the core temperature, the antagonizability of these effects by naloxone, and the effectivity of local microinjections in certain cases, at least three sites of the brain must be supposed to mediate the effects of opioids on thermoregulation. Probably there are still further unexplored regions also participating in these actions. The multiplicity of brain areas mediating these effects may explain the diversity of experimental data (summarized in Table 4).

A further question is which neurotransmitters are involved in these effects of opioids. Of course, first of all serotonin has been suspected since potentiation of opiate actions by serotoninergic agonists is a well-known phenomenon in relation to analgesic and other behavioral actions (Chapter 2, Volume I). Already in 1972 Samanin et al.[121] observed an almost total abolition of both hypo- and hyperthermic effects of morphine in rats upon depletion of brain serotonin by lesioning of the raphe complex. In a recent study[149] the hypothermic effect of i.c.v. administered β-EP has been found attenuated in rats if the brain serotonin was depleted by *p*-chlorophenylalanine or 5-hydroxytryptamine. Not only its hypothermic action is mediated by serotonin. Methergoline inhibits the β-EP-induced hyperthermia if giving the former systemically and the latter into the preoptic area.[125] It is reasonable to suppose that other monoamines and acetylcholine are somehow involved in the modulation of thermoregulation by opiates or opioid peptides. As for the former of the many relevant publications, which actually lie outside of the scope of this book, let us mention only a few. French et al.[157] reported that the morphine-induced hyperthermia in cats was potentiated by the serotonin uptake blocker fluoxetine or inhibition of dopaminergic synapses by pimozide. Considering the antagonism between the serotonin and dopamine-mediated mechanisms, these findings[157] suggest that opiates act via upsetting their functional balance in favor of the former monoamine. As naloxone antagonizes the apomorphine-induced hypothermia and this antagonism may be abolished by methysergide,[158] endogenous opioids might act in a similar way. In rats the morphine-induced hypothermic action was found reversed by the α-adrenergic receptor blocker phentolamine,[159] which implicates the role of noradrenaline; a rather strange finding, since as discussed previously (Chapter 2, Volume I) adrenergic mechanisms do not seem to be involved in the analgesic action of opiates in this species.

As for acetylcholine, the data available are conflicting. In one study morphine-induced hyperthermia was found attenuated by physostigmine[160] while in another[161] by atropine. Regarding the acetylcholine release inhibitory action of opiates and opioid peptides (Chapter 2, Volume I; Chapter 1, Volume II), the latter finding is difficult to interpret.

Finally let us mention that recently prostaglandins have been implicated in the hyperthermic action of opiates. In rats the β-EP-elicited rise of temperature (upon its administration into the anterior hypothalamus) could only be partially antagonized by naloxone or naltrexone, but it was completely blocked by the prostaglandin synthesis inhibitor indomethacin.[125] In good agreement with this finding Scoto et al.[162] observed increased prostaglandin level in the brain of rats given a hyperthermic dose of Met-E. However, in mice the hyperthermic response of i.c.v. administered β-EP could not be antagonized by either indomethacin or acetylsalicylic acid.[133]

Finally, it should probably be recalled here that in two different laboratories[163,165] the pA_2 value of naloxone against morphine with regard to the hyperthermic effect of the latter was found smaller than in relation to analgesia. More naloxone is needed to antagonize the morphine-induced hyperthermia than its analgesic action. These observations prove that different opiate receptors are involved in the two actions. Unfortunately no such experiments have been made in relation to the hypothermic action.

C. Are the Endogenous Opioids Involved in the Physiological Thermoregulation?

If endogenous opioids are involved in regulation of normal body temperature, pure opiate antagonists alone should induce hypo- or hyperthermic reactions. In the only related human study Volavka et al.[151] reported recently that naloxone decreased the oral temperature in healthy young males. However, this effect was not dose-dependent and also the first animal studies gave conflicting results. Goldstein and Lowery[140] giving naloxone to "untreated" rats observed only a very slight decrease in the body temperature. (However, in dogs a slight but significant hypothermic effect has been observed upon naloxone treatment.[143]) Thus they[140] hesitantly concluded that endorphins did not play any significant role in thermoregulation at least in intact animals. Later Stewart and Eikleboom[150] repeated this experiment taking special care to habituate the animals to the potentially stressful experimental conditions (handling, insertion of the rectal probe, etc.). In these animals naloxone elicited a significant and dose-dependent reduction in temperature and effectively prevented the development of hyperthermia to noise stress.[150] It is reasonable to suppose that in the above-mentioned experiment[140] and also in many other in vivo studies reviewed in this book the failure of naloxone to modify the normal functioning of the organism was due to the stressful experimental conditions, which resulted in activation of the endogenous opioid mechanisms and thus obscured the specific actions of exogenous opiate antagonists. In a related study Bläsig et al.[147] found that the first encounter with a new environment and handling of the animals resulted in hyperthermia and a dramatic elevation of plasma β-EP level. Pretreatment with naloxone, however, was able to counteract to a certain extent the stress-induced hyperthermia.[147] The receptors mediating this action ought to have been opiate receptors, as only (−)-naloxone was effective its strereoisomer (+)-naloxone not.[147]

Nevertheless, the finding that previous familiarization to the experimental environment "disclosed" the specific action of opiate antagonists[150] might indicate tolerance development to the mobilization of endogenous opioids. Since noise elicited the stress response even after having habituated the animals to the new surroundings,[150] it is reasonable to suppose that actually the stimulus lost its significance, but the organism did retain its ability to mobilize endorphins under stressful conditions.

It is a further question whether the stress-induced rise in temperature is due to the peripheral effect of β-EP or the concomitantly liberated ACTH or the former acts centrally by recirculating into the brain. Considering the relative inefficacy of much higher amounts of β-EP if injected i.v. (Chapter 3, Volume II), this latter possibility should be rejected. Moreover, the stress-induced increase in plasma β-EP concentration is not accompanied by a similar increase in its brain level.[165] The role of the concomitantly released ACTH is shown by the experiments of Holaday et al.,[126] who found that a certain dose of s.c. injected morphine elicited a slight hyperthermia in intact mice but a dose-dependent drop of temperature in adrenalectomized animals.

Otherwise, the morphine-induced hyperthermia can be conditioned by pairing its injection with olfactory stimulation.[142] In this experiment naloxone antagonized the hyperthermic effect of conditional stimulus.[142] Thus, the conditioned response was already elicited by liberation of the endogenous opioids. In other words, the liberation of endogenous opioids may be conditioned. Obviously, such a process might take place under natural conditions.

The physiological significance of endogenous opioids is demonstrated also by the experiments of Sandman et al.,[152] who found that neonatal administration of β-EP or naloxone resulted in chronic insensitivity to thermal stimuli. Unfortunately in this very interesting work[152] no information is given on the spontaneous temperature of desensitized animals. The importance of endorphins in the physiological regulation of body temperature is also shown by the observation[154] according to naloxone that inhibits the spontaneous hibernation in the garden dormouse.

It must be emphasized that these effects might be at least in part secondary consequences of altered TRH secretion as discussed in Chapter 5, Volume II. Sharp et al.[155] demonstrated that naloxone abolished the customary decrease in serum TSH concentration observable upon exposure to hot (37°C) environment.

D. Summary

Based on the above-reviewed findings some tentative conclusions might be put forward.

1. β-EP and possibly other endogenous opioids are involved in normal physiological thermoregulation.
2. Due to the complexity of their effects and to the multiplicity of the brain sites mediating their actions the data on the influence of opioids on thermoregulation are seemingly conflicting, especially if the experimental conditions are not carefully planned and analyzed (Table 4).
3. Opioids induce hyperthermia by elevating the thermoregulatory set point of the organism's thermostat in the anterior hypothalamus resulting in inhibition of the heat escaping behavior.
4. The opiate receptors mediating the hyperthermic action cannot be identical with those responsible for analgesia. They are located in brain areas hardly participating in the antinociceptive action, the hyperthermic effect in rats is less prone to tolerance development, and the pyretic reaction is more difficult to be antagonized by naloxone than the opposite effect (Table 4).
5. Opioids also induce a hypothermic action via brain areas close to the caudal ventricular system. This hypothermic effect is accompanied by catalepsy and analgesia. Consequently the hypothermia might be an unspecific phenomenon due to the cessation of locomotor activity. The hypothermic action disappears with the abolishment of analgesia upon naloxone treatment and tolerance development.
6. Stressful events cause hyperthermia via mobilization of the adenohypophyseal β-EP and ACTH. This action is also mediated by opiate receptors. The decreased ability of repeated stress to mobilize β-EP from the pituitary reflects rather habituation than true tolerance development at the level of opiate receptors.
7. Opioids are capable to elicit hyperthermia also by receptors located in the spinal cord. This hyperthermic action might be due to the blockade of the sensory input into the higher brain centers, but the experimental control of this supposition is still missing.
8. For the time being no data are available on the specific roles of different subtypes of opiate receptors in these phenomena.

V. EFFECTS ON THE EYE

Some actions of opiates and opioid peptides are so characteristic that they deserve special attention, as the pharmacological analysis of these events may contribute to the understanding of endogenous opioid mechanisms in general.

A. Effect on Pupillary Size

Opiates have long been known to induce miosis in humans, dogs, and rabbits but mydriasis in rats, mice, monkeys, and cats.[166-169] As both the miotic and mydriatic actions can be reversed by naloxone or other narcotic antagonists,[166,168,170] both effects should be mediated by opiate receptors. Interestingly the miotic effect of opiates is especially strong in humans,[170] which means that in experimental animals doses generally higher than the analgesic ED_{50} are needed to elicit measurable effect on pupillary size, while in humans morphine congeners induce miosis already in therapeutic doses. The specificity of this action is further substantiated by tolerance development to the miotic effect of opiates in humans[171] and the strong mydriatic action of conjuctivally instilled naloxone in opiate addicts proving the development of receptor supersensitivity. According to the data available opioid peptides exert similar action.

In cats mydriasis accompanies the other characteristic behavioral and autonomic symptoms induced by i.c.v. injected β-EP.[172-174] The pupillary dilatation could be reversed by naloxone, similarly to the other vegetative effects.[172-174] Interestingly Catlin et al.[175] found no apparent change in the pupillary diameter (and respiratory rate) upon i.v. injection of β-EP in freely moving cats. Let us mention here that the behavioral and autonomic effects of β-EP are generally different upon central and peripheral application in this species (Chapter 4, Volume II). For the time being no explanation can be given for these differences. Feldberg and Smyth[172,173] found the shorter fragments of β-EP inactive, while Clark observed mydriasis if Met-E was applied in the heroic i.c.v. dose of 0.5 to 2.0 mg.[176] In our experiments[177] D-Met, Pro-EA, applied i.p. in doses sufficient to induce analgesia and deep sedation, did not significantly modify the pupillary diameter. However, it must be added, that in these experiments[177] the nictitating membrane was prolapsed to such a degree, that the pupils were hardly visible rendering the observation unreliable. In cats the opioid peptides seem to mimic the mydriatic action of opiates only upon i.c.v. application but not when given peripherally.

Opioid peptides similarly to morphine induce mydriasis in rodents. We found mydriatic effect upon s.c. application of the potent enkephalin analogue D-Met, Pro-EA in mice. This effect was dose-related (Table 5) and it could be reversed by naloxone.[179] A similar mydriatic effect was observed by Korczyn et al.[180] in mice upon i.v. treatment with Met-E, Leu-E, and (D-Ala^2, D-Leu^5)-E, the latter being somewhat stronger than the natural enkephalins.

We have found distinct mydriatic action also in rats applying morphine, β-EP, or D-Met, Pro-EA whether centrally (i.c.v.) or peripherally.[177] On the other hand, Tortella et al.[181] observed a significant increase in pupillary diameter upon i.p. injection of Met-E and Leu-E in doses of 10 to 20 mg/kg. Leu-E even potentiated the similar action of morphine.[181]

Thus, we are faced with a perplexing contradiction. β-EP in cats induces mydriasis upon central administration,[172-174] but upon peripheral administration not.[175,176] The potent enkephalin analogue D-Met, Pro-EA also fails to induce mydriasis upon i.p. injection, though it does elicit analgesia.[176] The incapability of β-EP to induce mydriasis might be connected with its ineffectiveness as analgesic also upon peripheral application (Chapter 3, Volume II). Nevertheless, it is difficult to explain that the natural enkephalins have mydriatic action if injected i.p.,[181] though they are very weak analgesics even upon intracerebral application. This controversy is a bit similar to that observed in relation of the cardiovascular actions. As discussed in Chapter 4, Volume II, enkephalins elicit only a short-lasting pressor effect if applying them centrally. This effect though naloxone sensitive may be regarded unspecific due to its short duration, rapid onset, and its similarity to the initial vasocontrictor action of potent enkephalin analogues which also elicit a subsequent phase of long-lasting dose-dependent hypotension. On the other hand, giving the natural enkephalins peripherally, they were capable to induce bradycardia and antidiuresis, the latter in very low doses.

Recently Wallenstein[182] demonstrated that morphine did not induce mydriasis in adrenalectomized cats. On the contrary, in these animals a dose-related pupillary constriction was observed.[182] Otherwise, in adrenalectomized cats its action on blood pressure became monotonous, i.e., in place of the customary biphasic action a dose-related decrease in heart rate and blood pressure was recorded.[182] Considering the related data of the same laboratory[183] that the morphine-induced mydriasis could be antagonized by phenoxybenzamine whether giving it systemically or locally into the eye, the conclusion is simple. The mydriatic action of opioids is due to their catecholamine mobilizing

Table 5
MYDRIATIC EFFECT OF MORPHINE AND D-MET, PRO-EA IN MICE PUPILLARY DIAMETER (mm)

	Morphine s.c.			D-Met, Pro-EA		
Dose mg/kg	Control	Test	P	Control	Test	P
1.25	0.33 ± 0.013	0.65 ± 0.090	<0.01	0.33 ± 0.017	0.52 ± 0.026	<0.001
2.50	0.34 ± 0.014	0.74 ± 0.056	<0.001	0.33 ± 0.013	0.76 ± 0.061	<0.001
5.00	0.35 ± 0.020	1.22 ± 0.089	<0.001	0.32 ± 0.010	1.08 ± 0.078	<0.001
10.00	0.33 ± 0.018	1.74 ± 0.058	<0.001	0.30 ± 0.017	1.20 ± 0.092	<0.001

Note: Pupillary diameter examined according to Pulewka.[176]

effect just as the pressor response. The mydriatic effect might be exerted even directly on the muscle fibers with noradrenergic innervation, since the action could be reversed by topical application of the α-adrenergic blocking agent phenoxybenzamine.[183] (Nevertheless, the effect of phenoxybenzamine may also be attributed to its affinity to the opiate receptors as discussed in Chapter 2, Volume I.)

Consequently, this pupillary dilatation is only a secondary phenomenon due to adrenergic mobilization, which obscures the primary centrally mediated miosis-induction visible in several species and also in humans. Supposing that the catecholamines are primarily mobilized from the adrenal medulla it is obvious why found Korczyn et al.[167] the morphine-induced mydriasis in mice essentially unaltered upon pretreatment with propranolol, phentolamine, haloperidol, *p*-chlorophenylalanine, or in animals with cervical sympathetic denervation.

As for the mechanism of miotic action in dog, Lee and Wang[170] showed that morphine had no effect if applying it directly into the anterior chamber of the eye. Giving it, however, i.v. the pupillo-constrictor neurones in the visceral part of the oculomotor nuclear complex responded with increased frequency of discharges.[170] Consequently, an increased activity of parasympathetic neurones may explain the miotic response.

In our laboratory D-Met, Pro-EA induced mydriasis in dogs upon i.v. application.[177] In a preliminary experiment, this enkephalin analogue induced mydriasis also upon intracisternal application (unpublished).

Finally it must be emphasized that contrary to the classical opiates, opioids more precisely β-EP and FK 33-824 have not been reported to induce miosis in humans. More exactly one group[184] explicitly stated that the above-mentioned potent enkephalin analogue did not alter the pupillary diameter in healthy humans if giving it in a relatively low dose, while in another study[185] naloxone also failed to alter the pupillary size. Otherwise, though many reports deal with the effects of β-EP this enkephalin analogue and naloxone in humans, nowhere is there any allusion to modification of pupillary diameter in either of those cases where β-EP was applied intracerebrally (Chapter 10, Volume III). Of course these studies were not devoted primarily to the examination of eventual effects on the pupillary size. The lack of any explicit reference shows that no striking action has been observed. Moreover, in animal experiments Korczyn et al.[180] failed to detect any change in the pupillary diameter if giving naloxone alone to mice.[180]

The data available are difficult to interpret. Opiates induce miosis or mydriasis depending on the species used but independently of the route of application in rodents. The specific effect might be the miosis, which is frequently obscured by catecholamine mobilization presumably from the adrenals. Centrally administered β-EP and potent enkephalin analogues induce mydriasis in rodents and in cats, i.e., in species responding in similar way to morphine. However, upon peripheral application they fail to elicit the expected mydriasis in cats[175,177] and miosis in humans.[184] Moreover, they induce mydriasis also in dogs,[177] i.e., in a species known to react with miosis to opiates. Thus far nobody observed miosis upon treatment with opioid peptides, though this is the typical, i.e., centrally mediated specific effect of opiates.

As opioid peptides cross the blood-brain barrier only with great difficulty (Chapter 3, Volume III), their mydriatic action might be attributed to the prevalence of peripheral actions. However, nobody reported on miotic effect either in those human studies where β-EP administered intracerebrally (Chapter 10, Volume III) or upon i.c.v. administration in animal experiments. Therefore, the inability of opioid peptides to induce miosis represents one of the in vivo demonstration of opiate receptor multiplicity.

B. Relaxation of Nictitating Membrane

A further interesting action of opioid peptides is the relaxation of nictitating membrane in cats. Catlin et al.[175] and ourselves[177] observed a distinct and dose-dependent relaxation of nictitating membrane in unanesthetized cats applying β-EP or D-Met, Pro-EA i.v. and i.p., respectively. The effect is naloxone-reversible.[175,177] No such relaxation was observed upon central application of opioid peptides to cats.[172,173] Neither morphine congeners have been found to elicit such relaxation only etorphine.[186]

Yen et al.[187] observed the inhibition of electrically induced contractions of the nictitating membrane in cats anesthetized with pentobarbitone upon i.v. application of (D-Ala^2, Met^5)-E. In the same experiment this enkephalin analogue failed to antagonize the noradrenaline-induced contraction.[187] For this reason the authors concluded that opioid peptides inhibited the contractions via presynaptic inhibition of noradrenaline release.[187]

The in vitro inhibition by opiates of electrically induced contractions of isolated nictitating membrane is a well-known phenomenon (Chapter 1, Volume I). Henderson et al.[188] demonstrated that this effect was due to the presynaptic inhibition of noradrenaline release from sympathetic nerve terminals innervating the smooth muscle of nictitating membrane.

The above-discussed observations indicate that peripherally located opiate receptors mediate the effect. The high potency of β-EP[175,177] (contrary to morphine) in this test suggests that the inhibition of contraction might be mediated by receptors selectively sensitive to β-EP. It must, however, be added that in vitro morphine and β-EP are nearly equiactive in inhibition of the contraction of nictitating membrane, while D-Met, Pro-EA is stronger than any of the two (Chapter 1, Volume I). Thus, either separate subgroup of opiate receptors is located in this organ which is specifically sensitive to this enkephalin analogue or D-Met, Pro-EA has a certain adrenolytic action unrelated to its opioid agonist potency. Further experiments are needed to clarify this question.

C. Eyelid Retraction and Induction of Lenticular Opacity

Exophthalmos is a well-known effect of opiates in rodents just as the Straub phenomenon. Exophthalmos, i.e., retraction of the eyelids has been observed also upon i.c.v. application of β-EP in rats. This effect was specifically stated by Bloom et al.,[189] not mentioned but visible on the figures published by Jacquet and Marks,[190] and it has obviously been observed in other laboratories that the analgesic action of β-EP was accompanied by the typical autonomic effects characteristic of opiates.

We have also observed exophthalmos in rats and in mice upon i.c.v. application of morphine, D-Met, Pro-EA and β-EP or on peripheral administration of the formers.[177] A similar widening of palpebral fissure can be observed in cats upon i.c.v. injection of β-EP and morphine[172,173] but upon peripheral (i.v., i.p.) application no such effect has been detected.[175,177]

To understand the mechanism of this interesting phenomenon again the competent studies on the similar action of morphine may be helpful. First Weinstock et al.[191] detected that opiates elicited lenticular opacity in mice in doses somewhat higher than the corresponding analgesic ED_{50} values. This effect was mediated by opiate receptors as psychoactive drugs of other types did not elicit this phenomenon.[191] It could be reversed by opiate antagonists[191,192] and upon repeated application tolerance developed to this action.[193] In a subsequent study Weinstock and Marshall[194] demonstrated that the action was due to the retraction of the eyelids since numerous drugs and experimental procedures inducing ptosis prevented this effect of opiates. A prolonged opening of the eye might allow the evaporation of a sufficient amount of water through the cornea rendering the aqueous humor hypertonic, which is reflected by clouding of the eye.[194] Consequently, exophthalmos in rodents, widening of the palpebral fissure in cats, and lenticular opacity in mice reflect the same process, the opiate receptor mediated retraction of the eyelids.

This lenticular opacity inducing action of opioid peptides has only been studied in our laboratory.[179] As shown in Table 6, D-Met, Pro-EA is capable of eliciting this phenomenon just as is morphine. These data also show that the action of the enkephalin analogue is much stronger than that of morphine, i.e., in the case of the peptide the ED_{50} of lenticular opacity inducing action is lower than that of morphine especially if comparing the ED_{50} values to the corresponding LD_{50} value. In other words, morphine induced lenticular opacity only in nearly lethal amounts, while the enkephalin analogue in amounts well below the toxic level (Table 6).[179]

Thus, the questions that should be answered are where are the opiate receptors responsible for lenticular opacity, are they similar to those mediating the analgesic action, and the most interesting one, are the receptors mediating this action of morphine identical with those responsible for the effect of D-Met, Pro-EA or not.

As for the first question, levorphanol given intracerebrally induced lenticular opacity in such small doses which were insufficient to elicit the phenomenon upon i.p. application.[195] Consequently, the receptors mediating the action should be centrally located. However, topical application of noradrenaline facilitated the lenticular opacity inducing action of i.p. administered levorphanol.[193] Moreover, lenticular opacity did not develop in mice in which the cervical sympathetic nerves were severed.[194] Therefore, also peripheral adrenergic mechanisms are involved. Moreover, catecholamines are capable to elicit this phenomenon in themselves the potency order being adrenaline > noradrenaline > isopropylnoradrenaline.[193] Accordingly, the centrally administered opiates might act by peripheral adrenergic mechanisms. As the cataractogenic action of levorphanol was enhanced by catecholamines injected i.c.v.,[196] both central and peripheral adrenergic mechanisms should take part in the action.

The opiate receptors responsible for opacity might be similar to those mediating the analgesic effects. Determining the pA_2 values of naloxone and nalorphine against morphine in hot plate test and in relation of lenticular opacity, very similar values have been measured.[192,197]

Due to the almost total lack of competent data it is difficult to answer whether the same receptors are involved in induction of lenticular opacity upon treatment with opiates and opioid peptides. The data summarized in Table 6 suggest that at least partially different receptors mediate the action. In mouse hot plate and writhing tests the s.c. ED_{50} values of morphine and D-Met, Pro-EA were 3.10; 2.10, and 7.20; 2.45 mg/kg, respectively, in a previous experiment.[177] In this study[179] (Table 6) morphine induced lenticular opacity in mice in a dose of 140 mg/kg (ED_{50}), which is about 60 times higher than the analgesic ED_{50} and 2.5 times smaller than the LD_{50}. In the case

Table 6
LENTICULAR OPACITY PRODUCING ACTIVITY OF MORPHINE AND D-MET, PRO-EA AND THEIR ACUTE TOXICITY IN MICE

Treatment	ED_{50}(s.c.)	LD_{50}(s.c.)
Morphine	140 (90-217) mg/kg	340 (281-411) mg/kg
D-Met, Pro-EA	36 (26-50) mg/kg	1710 (1326-2206) mg/kg

Note: Lenticular opacity examined according to Cox and Weinstock.[190]

of D-Met, Pro-EA the ED_{50} for lenticular opacity was 34 mg/kg and its LD_{50} 1710 mg/kg. Consequently, there is hardly any correlation between the analgesic and cataractogenic potencies if comparing the two substances, which points to differences in the receptorial mechanisms mediating these actions. Let us recall here that certain qualitative differences were found also in relation to their effects on the pupillary diameter and on the tone of nictitating membrane as well. Of course, if the opioid peptides are more potent liberators of adrenal catecholamines, these differences were easy to interpret. However, so far no data are available on the effect of adrenalectomy on development of lenticular opacity elicited by opiates or opioid peptides.

Summarizing the above-reviewed data it might be stated that opioid peptides induce their characteristic effects on the eye (myosis or mydriasis, retraction of the eyelids, and relaxation of the nictitating membrane) in at least four or five different ways.

Stimulation of the central parasympathetic tone may explain the miotic action in several species, which is one of the most sensitive indications of opiate receptor stimulation in humans.

A peripherally mediated sympatholytic effect might count for the relaxation of the nictitating membrane.

Exophthalmos, i.e., retraction of the eyelids is a symptom of adrenergic receptor stimulation, which is at least in part centrally mediated. This action is very difficult to interpret in view of the much discussed parasympathomimetic influence of opioids.

Finally, mobilization of the adrenal catecholamines surely contributes to the mydriatic action observed in several species. However, it is questionable whether opioids also facilitate certain peripheral adrenergic pathways contributing to the development of lenticular opacity and mydriasis. As local adrenaline instillation potentiates the levorphanol induced opacity, it cannot be excluded that certain adrenergic pathways are facilitated. This would mean that in cats upon central administration of opiates or opioid peptides the sympathetic stimulatory action and catecholamine mobilization prevails resulting in exophthalmos and mydriasis. Upon peripheral application opiates would act in a similar way but opioid peptides such as β-EP and D-Met, Pro-EA might elicit a preferential inhibition of peripheral adrenergic activity resulting in relaxation of nictitating membrane.

REFERENCES

1. **Laubie, M., Schmitt, H., Vincent, M., and Remond, G.,** Central cardiovascular effects of morphinomimetic peptides in dogs, *Eur. J. Pharmacol.,* 46, 67, 1977.
2. **Feldberg, W. and Wei, E.,** Central cardiovascular effects of enkephalins and C-fragment of lipotropin, *J. Physiol. (London),* 280, 18P, 1978.
3. **Bolme, P., Fuxe, K., Agnati, L. F., Bradley, R., and Symthies, J.,** Cardiovascular effects of morphine and opioid peptides following intracisternal administration in chloralose-anesthetized rats, *Eur. J. Pharmacol.,* 48, 319, 1978.
4. **Feldberg, W. and Wei, E.,** The central origin and mechanism of cardiovascular effects of morphine as revealed by naloxone in cats, *J. Physiol. (London),* 272, 99P, 1977.
5. **Lemaire, I., Tseng, R., and Lemaire, S.,** Systemic administration of β-endorphin: potent hypotensive effect involving a serotonergic pathway, *Proc. Natl. Acad. Sci. USA,* 75, 6240, 1978.
6. **Wei, E. T., Lee, A., and Chang, J. K.,** Cardiovascular effects of peptides related to the enkephalins and β-casomorphin, *Life Sci.,* 26, 1517, 1980.
7. **Tseng, L.-F., Loh, H. H., and Li, C. H.,** β_h-endorphin: antidiuretic effects in rats, *Int. J. Peptide Protein Res.,* 12, 173, 1978.
8. **Bisset, G. W., Chowdrey, H. S., and Feldberg, W.,** Release of vasopressin by enkephalin, *Br. J. Pharmacol.,* 62, 370, 1978.
9. **Huidobro-Toro, J. P., Huidobro, F., and Croxatto, R.,** Effects of β-endorphin and D-alanine²-enkephalinamide on urine production and urinary electrolytes in the rat, *Life Sci.,* 24, 297, 1979.
10. **Vandeputte van Messom, G. and Peeters, G.,** Effects of intracerebroventricular administration of opioid peptides on urinary function in the conscious goat, *Arch. Int. Pharmacodyn.,* 243, 304, 1980.
11. **Weitzman, R. E., Fisher, D. A., Minick, S., Ling, N., and Guillemin, R.,** β-endorphin stimulates secretion of arginin vasopressin *in vivo, Endocrinology,* 101, 1643, 1977.
12. **Fujimoto, S., Kato, T., and Hisada, S.,** Changes in urine outflow with intracerebroventricular administration of morphine and its related compounds in rats, *Jpn. J. Pharmacol.,* 28, 170P, 1978.
13. **Huidobro, F. and Huidobro-Toro, J. P.,** Antidiuretic effect of morphine and vasopressin in morphine tolerant and non-tolerant rats, differential effects on urine composition, *Eur. J. Pharmacol.,* 59, 55, 1979.
14. **Priano, L. L. and Vatner, S. F.,** Effects of morphine on cardiac output and regional blood flow distribution in conscious dogs, *Anesthesiology,* 51, 140S, 1979.
15. **Simantov, R., Childers, S. R., and Snyder, S. H.,** (^{3}H)-opiate binding: anomalous properties in kidney and liver membranes, *Mol. Pharmacol.,* 14, 69, 1978.
16. **Schurig, J. E., Cavanagh, R. L., and Buynski, J. P.,** Effect of butorphanol and morphine on pulmonary mechanics, arterial blood pressure and venous plasma histamine in the anesthetized dog, *Arch. Int. Pharmacodyn.,* 233, 296, 1978.
17. **Evans, A. G. J., Nasmyth, P. A., and Stewart, H. C.,** The fall of blood pressure caused by intravenous morphine in the rat and the cat, *Br. J. Pharmacol.,* 7, 542, 1952.
18. **Knoll, J.,** Neuronal peptide (enkephalin) receptors in the ear artery of the rabbit, *Eur. J. Pharmacol.,* 39, 403, 1976.
19. **Rónai, A. Z. and Berzétei, I.,** Similarities and differences of opioid receptors in different isolated organs, in *Endorphins'78,* Gráf, L., Palkovits, M., and Rónai, A. Z., Eds., Publishing House of the Hungarian Academy of Science, Budapest, Hungary, 1978, 237.
20. **Dressler, W. E., D'Alonzo, G., Rossi, V. G., and Orzechowski, R. F.,** Modification of certain vascular responses to dopamine by morphine, *Eur. J. Pharmacol.,* 28, 108, 1974.
21. **Hanko, J. H. and Hardebo, J. E.,** Enkephalin-induced dilatation of pial arteries *in vitro* probably mediated by opiate receptors, *Eur. J. Pharmacol.,* 51, 295, 1978.
22. **Moore, A. F.,** The interaction of enkephalins with substance P on vascular smooth muscle, *Res. Commun. Chem. Pathol. Pharm.,* 23, 233, 1979.
23. **Konishi, S., Tsunoo, A., and Otsuka, M.,** Enkephalins presynaptically inhibit cholinergic transmission in sympathetic ganglia, *Nature (London),* 282, 515, 1979.
24. **Di Giulio, A. M., Yang, H.-Y. T., Lutold, B., Fratta, W., Hong, J., and Costa, E.,** Characterization of enkephalin-like material extracted from sympathetic ganglia, *Neuropharmacology,* 17, 989, 1978.
25. **Kumakura, K., Karoum, F., Guidotti, A., and Costa, E.,** Modulation of nicotinic receptors by opiate receptor agonists in cultured adrenal chromaffin cells, *Nature (London),* 283, 489, 1980.
26. **Grundy, H. F.,** Cardiovascular effects of morphine, pethidine, diamorphine and nalorphine on the cat and rabbit, *Br. J. Pharmacol.,* 42, 159, 1971.
27. **Jacob, J. J. and Michaud, G. M.,** Production par la naloxone d'effets inverses de ceux de la morphine chez le chien éveillé, *Arch. Int. Pharmacodyn.,* 222, 332, 1976.
28. **Khazan, U. and Moreton, J. E.,** Naltrexone blockade of morphine-induced-antidiuresis in the rat, *Pharmacologist,* 19, 156, 1977.

29. **Zilm, D. H.**, Naloxone response in non-dependent man: effect on six physiological variables, *Neuropharmacology,* 19, 591, 1980.
30. **Dashwood, M. R. and Feldberg, W.**, Release of opioid peptides in anesthetized cats? *Br. J. Pharmacol.,* 68, 697, 1980.
31. **Laubie, M., Schmitt, H., and Drouillat, M.**, Central sites and mechanisms of the hypotensive and bradycardic effects of the narcotic analgesic agent fentanyl, *Naunyn-Schmiedeb. Arch. Pharmacol.,* 296, 255, 1977.
32. **Rockhold, R. W. and Caldwell, R. W.**, Effects of lesions of the nucleus tractus solitarii on the cardiovascular actions of clonidine in conscious rats, *Neuropharmacology,* 18, 347, 1979.
33. **Laubie, M., Schmitt, H., and Vincent, M.**, Vagal bradycardia produced by microinjections of morphine-like drugs into the nucleus ambiguus in anesthetized dogs, *Eur. J. Pharmacol.,* 59, 287, 1979.
34. **Farsang, C. and Kunos, G.**, Naloxone reverses the antihypertensive effect of clonidine, *Br. J. Pharmacol.,* 67, 161, 1979.
35. **Farsang, C., Ramirez-Gonzalez, M. D., Mucci, L., and Kunos, G.**, Possible role of an endogenous opiate in the cardiovascular effects of central *alpha* adrenoceptor stimulation in spontaneously hypertensive rats, *J. Pharmacol. Exp. Ther.,* 214, 203, 1980.
36. **Laubie, M., Schmitt, H., Canellas, J., Roquebert, J., and Demichel, P.**, Centrally mediated bradycardia and hypotension induced by narcotic analgesics: dextromoramide and fentanyl, *Eur. J. Pharmacol.,* 28, 66, 1974.
37. **Di Giulio, A. M., Yang, H.-Y. T., Fratta, W., and Costa, E.**, Decreased content of immunoreactive enkephaline-like peptide in peripheral tissues of spontaneously hypertensive rats, *Nature (London),* 278, 646, 1979.
38. **Zamir, N., Simantov, R., and Segal, M.**, Pain sensitivity and opioid activity in genetically and experimentally hypertensive rats, *Brain Res.,* 184, 299, 1980.
39. **Zamir, N. and Segal, M.**, Hypertension-induced analgesia: changes in pain sensitivity in experimental hypertensive rats, *Brain Res.,* 160, 170, 1979.
40. **Holaday, J. W. and Faden, A. I.**, Naloxone reversal of endotoxin hypotension suggests role of endorphins in shock, *Nature (London),* 275, 450, 1978.
41. **Janssen, H. F. and Lutherer, L. O.**, Ventriculocisternal administration of naloxone protects against severe hypotension during endotoxin shock, *Brain Res.,* 194, 608, 1980.
42. **Faden, A. I. and Holaday, J. W.**, Opiate antagonists: a role in the treatment of hypovolemic shock, *Science,* 205, 317, 1979.
43. **Feuerstein, G., Ailam, R., and Bergmann, F.**, Reversal by naloxone of hemorrhagic shock in anephric cats, *Eur. J. Pharmacol.,* 65, 93, 1980.
44. **Gurll, N. J., Lechner, R., Vargish, T., and Reynolds, D. G.**, Dose-dependent naltrexone improvement of cardiovascular function and survival suggests opiate receptors involved in pathophysiology of canine hemorrhagic shock, *Clin. Res.,* 28, 17A, 1980.
45. **Gurll, N. J., Lechner, R., Vargish, T., and Reynolds, D. G.**, Endorphins and opiate receptors are involved in myocardial depression of canine hemorrhagic shock, *Clin. Res.,* 28, 8A, 1980.
46. **Ailam, R., Bergmann, F., and Feurstein, G.**, Naloxone antagonizes the hypotensive effect of clonidine. Naloxone prevents the fall of blood pressure in certain cases of spinal shock, *Isr. J. Med. Sci.,* 16, 220, 1980.
47. **Arndt, J. O. and Freye, E.**, Opiate antagonist reverses the cardiovascular effects of inhalation anesthesia, *Nature (London),* 277, 399, 1979.
48. **Summy Long, J. Y., Keil, L. C., Deen, K. C., and Severs, W. B.**, Leu-enkephalin inhibition of central angiotensin. II. Effects, *Pharmacologist,* 21, 198, 1979.
49. **Volavka, J., James, B., Reker, D., Pollock, V., and Cho, D.**, Electroencephalographic and other effects of naloxone in normal men, *Life Sci.,* 25, 1267, 1979.
50. **Faden, A. I. and Holaday, J. W.**, Naloxone treatment of endotoxin shock: stereospecificity of physiologic and pharmacologic effects in the rat, *J. Pharmacol. Exp. Ther.,* 212, 441, 1980.
51. **Holaday, J. W. and Faden, A. I.**, Naloxone acts at central opiate receptors to reverse hypotension, hypothermia and hypoventilation in spinal shock, *Brain Res.,* 189, 295, 1980.
52. **Green, A. F.**, Comparative effects of analgesics on pain threshold, respiratory frequency and gastrointestinal propulsion, *Br. J. Pharmacol.,* 14, 26, 1959.
53. **Hassen, A. H., St. John, W. M., and Wang, S. C.**, Selective respiratory depressant action of morphine compared to meperidine in the cat, *Eur. J. Pharmacol.,* 39, 61, 1976.
54. **Kallos, T. and Caruso, F. S.**, Respiratory effects of butorphanol and pethidine, *Anaesthesia,* 34, 633, 1979.
55. **White, W. J. and Zagon, I. S.**, Acute and chronic methadone exposure in adult rats: studies on arterial blood gas concentrations and pH, *J. Pharmacol. Exp. Ther.,* 209, 451, 1979.
56. **Tress, K. H. and El Sobky, A. A.**, Responses to intravenous heroin in dependent and non-dependent volunteers, *Neuropharmacology,* 16, 530, 1977.

57. **McGilliard, K. L. and Takemori, A. E.**, Effect of acute and chronic morphine pretreatment on morphine-induced respiratory depression and analgesia, *Pharmacologist*, 18, 174, 1976.
58. **McGilliard, K. L. and Takemori, A. E.**, Antagonism by naloxone of narcotic-induced respiratory depression and analgesia, *J. Pharmacol. Exp. Ther.*, 207, 494, 1978.
59. **Bloom, F., Segal, D., Ling, N., and Guillemin, R.**, Endorphins: profound behavioral effects in rats suggest new etiological factors in mental illness, *Science*, 194, 630, 1976.
60. **Tseng, L.-F., Ostwald, T. J., Loh, H. H., and Li, C. H.**, Behavioral activities of opioid peptides and morphine sulfate in golden hamsters and rats, *Psychopharmacology*, 64, 215, 1979.
61. **Feldberg, W. and Smyth, D. G.**, The C-fragment of lipotropin-a potent analgesic, *J. Physiol. (London)*, 260, 30P, 1976.
62. **Feldberg, W. and Smyth, D. G.**, C-fragment of lipotropin-an endogenous potent analgesic peptide, *Br. J. Pharmacol.*, 60, 445, 1977.
63. **Flórez, J., Mediavilla, A., and Pazos, A.**, Differential effects on respiration of β-endorphin, D-Ala²-Met-enkephalinamide and Met-enkephalin, in *Proc. Int. Narcotic Res. Conf.*, van Ree, J. M. and Terenius, L., Eds., Elsevier/North-Holland, Amsterdam, 1978, 167.
64. **Flórez, J. and Mediavilla, A.**, Respiratory and cardiovascular effects of Met-enkephalin applied to the ventral surface of the brain stem, *Brain Res.*, 138, 585, 1977.
65. **Flórez, J., Mediavilla, A., and Pazos, A.**, Respiratory effects of β-endorphin, D-Ala²-Met-enkephalinamide and Met-enkephalin injected into the lateral ventricle and the pontomedullary subarachnoid space, *Brain Res.*, 199, 197, 1980.
66. **Moss, I. R. and Friedman, E.**, β-endorphin: effects on respiratory regulation, *Life Sci.*, 23, 1271, 1978.
67. **Isom, G. E., Allerton, H. W., and Zobrist, R. H.**, Central respiratory effects of D-Ala²methionine-enkephalinamide in rats, *Pharmacologist*, 21, 198, 1979.
68. **Catlin, D. H., Hui, K. K., Loh, H. H., and Li, C. H.**, β-endorphin: subjective and objective effects during acute narcotic abstinence in man, in *Advances in Biochemical Psychopharmacology*, Vol. 18, Costa, E. and Trabucchi, M., Eds., Raven Press, New York, 1978, 341.
69. **Graffenfried, von B., del Pozo, E., Roubicek, J., Krebs, E., Pöldinger, W., Burmeister, P., and Kerp, L.**, Effects of the synthetic enkephalin analogue FK 33-824 in man, *Nature (London)*, 272, 729, 1978.
70. **Catlin, D. H., George, R., and Li, C. H.**, β-endorphin: pharmacological and behavioral activity in cats after low intravenous doses, *Life Sci.*, 23, 2147, 1978.
71. **Székely, J. I., Dunai-Kovács, Z., Miglécz, E., and Tarnawa, I.**, The *in vivo* pharmacology of (D-MET², PRO⁵)-enkephalinamide, in *Endorphins'78*, Gráf, L., Palkovits, M., and Rónai, A. Z., Eds., Publishing House of the Hungarian Academy of Science, Budapest, Hungary, 1978, 319.
72. **Beubler, E.**, Naloxone increases carbon dioxide stimulated respiration in the rabbit, *Naunyn-Schmiedeb. Arch. Pharmacol.*, 311, 199, 1980.
73. **Volavka, J., James, B., Reker, D., Pollock, V., and Cho, D.**, Electroencephalographic and other effects of naloxone in normal men, *Life Sci.*, 25, 1267, 1979.
74. **Lawson, E. E., Waldrop, T. G., and Eldridge, F. L.**, Naloxone enhances respiratory output in cats, *J. Appl. Physiol.*, 47, 1105, 1979.
75. **Kumazawa, T., Tadaki, E., and Kim, K.**, A possible participation of endogenous opiates in respiratory reflexes induced by thin-fiber muscular afferents, *Brain Res.*, 199, 244, 1980.
76. **Metcalf, G., Rees, J. M. H., and Ward, S. J.**, *In vivo* antagonism of analgesia and respiratory depression induced by proposed μ- and κ-opiate agonists, *Br. J. Pharmacol.*, 66, 473P, 1979.
77. **Parolaro, D., Sala, M., and Gori, E.**, Effect of intracerebroventricular administration of morphine upon intestinal motility in rat and its antagonism with naloxone, *Eur. J. Pharmacol.*, 46, 329, 1977.
78. **Ambinder, R. F. and Schuster, M. M.**, Endorphins: new gut peptides with a familiar face, *Gastroenterology*, 77, 1132, 1979.
79. **Polak, J. M., Sullivan, S. N., Bloom, S. R., Facer, P., and Pearse, A. G. E.**, Enkephalin-like immunoreactivity in the human gastrointestinal tract, *Lancet*, 1, 972, 1977.
80. **Green, A. F.**, Comparative effect of analgesics on pain threshold, respiratory frequency and gastrointestinal propulsion, *Br. J. Pharmacol.*, 14, 26, 1959.
81. **Van Nueten, J. M., Janssen, P. A., and Fontaine, J.**, Loperamide a novel type of antidiarrheal agent, *Arzneim.-Forsch. (Drug Res.)*, 24, 1641, 1977.
82. **Stahl, K. D., van Bever, W., Janssen, P., and Simon, E. J.**, Receptor affinity and pharmacological potency of a series of narcotic analgesic, anti-diarrheal and neuroleptic drugs, *Eur. J. Pharmacol.*, 46, 199, 1977.
83. **Schulz, R., Wüster, M., and Herz, A.**, Centrally and peripherally mediated inhibition of intestinal motility by opioids, *Naunyn-Schmiedeb. Arch. Pharmacol.*, 308, 255, 1979.
84. **Edin, R.**, Evidence for vagal enkephalinergic neural control of the feline pylorus and stomach, *Gastroenterology*, 78, 492, 1980.

85. **Uddman, R., Alumets, J., Hakanson, R., Sundler, F., and Walles, B.,** Peptidergic (enkephalin) innervation of the mammalian esophagus, *Gastroenterology,* 78, 732, 1980.
86. **Cowan, A., Doxey, J. C., and Metcalf, G.,** A comparison of pharmacological effects produced by leucine-enkephalin, methionine-enkephalin, morphine and ketocyclazocine, in *Proc. Int. Narcotic Res. Club Meeting,* Kosterlitz, H. W., Ed., North-Holland, Amsterdam, 1976, 95.
87. **Thor, P. K., Krol, R., Siebers, R., Konturek, S. J., Coy, D. H., and Schally, A. V.,** Influence of methionine-enkephalin and morphine on the myoelectrical activity of the small bowel, *Gastroenterology,* 74, 1148, 1978.
88. **Konturek, S. J., Pawlik, W., Walus, K. M., Coy, D. H., and Schally, A. V.,** Methionine-enkephalin stimulates gastric secretion and gastric mucosal blood flow, *Proc. Soc. Exp. Biol. Med.,* 158, 156, 1978.
89. **Howd, R. A., Adamovics, A., and Palekar, A.,** Naloxone and intestinal motility, *Experientia,* 34, 1310, 1978.
90. **Holzer, P. and Lembeck, F.,** Effect of neuropeptides on the efficiency of the peristaltic reflex, *Naunyn-Schmiedeb. Arch. Pharmacol.,* 307, 257, 1979.
91. **Dajani, E. Z., Woods, E. M., Roge, E. A. W., Bertermann, R. E., and Schweingruber, F. L.,** Naloxone reversal of drug-induced diarrhea in mice, *Arch. Int. Pharmacodyn,* 240, 340, 1979.
92. **Piercey, M. F. and Ruwart, M. J.,** Naloxone inhibits the anti-diarrhoeal activity of loperamide, *Br. J. Pharmacol.,* 66, 373, 1979.
93. **Beubler, E. and Lembeck, F.,** Inhibition of stimulated fluid secretion in the rat small and large intestine by opiate agonists, *Naunyn-Schmiedeb. Arch. Pharmacol.,* 306, 113, 1979.
94. **Lembeck, F. and Beubler, E.,** Inhibition of PGE_1 induced intestinal secretion by the synthetic enkephalin analogue FK 33-824, *Naunyn-Schmiedeb. Arch. Pharmacol.,* 308, 261, 1979.
95. **Dairman, W., Kappell, B., and Smolen, M.,** Anti-diarrheal activity of endorphins in the mouse, *Pharmacologist,* 21, 198, 1979.
96. **Gorelick, F., Jamieson, J. D., Kinder, B., Binder, H. J., and Dobbins, J. W.,** Enkephalins inhibit pancreatic exocrine secretion, *Clin. Res.,* 27, 266A, 1979.
97. **Chey, W. Y., Coy, D. H., Konturek, S. J., Schally, A. V., and Tasler, J.,** Enkephalin inhibits the release and action of secretin on pancreatic secretion in the dog, *J. Physiol. (London),* 298, 429, 1980.
98. **Feldman, M. and Walsh, Jr. H.,** Endogenous opiates facilitate gastric acid secretion in man, *Gastroenterology,* 78, 1164, 1980.
99. **Wienbeck, M., Dünzen, R., Körner, M., and Gerges, W.,** Enkephalins affect colonic motility, *Gastroenterology,* 78, 1290, 1980.
100. **Walus, K. M., Pawlik, W., Fondacaro, J. D., and Jacobson, E. D.,** Effects of Met-enkephalin on intestinal circulation, oxygen consumption and motility, *Gastroenterology,* 78, 1287, 1980.
101. **Ouyang, A., Snape, W. J., Jr., and Cohen, S.,** Opiate receptor properties of the feline ileocecal sphincter, *Gastroenterology,* 78, 1231, 1980.
102. **Rattan, S. and Goyal, R. K.,** Effect of morphine and endogenous opiates on the opossum lower esophageal sphincter, *Gastroenterology,* 78, 1241, 1980.
103. **Takemori, A. E., Kupferberg, H. J., and Miller, J. W.,** Quantitative studies of the antagonism of morphine by nalorphine and naloxone, *J. Pharmacol. Exp. Ther.,* 169, 39, 1969.
104. **Székely, J. I., Dunai-Kovács, Z., Miglécz, E., and Tarnawa, I.,** The *in vivo* pharmacology of (D-Met^2,Pro^5)-enkephalinamide, in *Endorphins'78,* Gráf, L., Palkovits, M., and Rónai, A. Z., Eds., Publishing House of the Hungarian Academy of Science, Budapest, Hungary, 1978, 319.
105. **Tseng, L.-F., Loh, H. H., and Li, C. H.,** β-endorphin as a potent analgesic by intravenous injection, *Nature (London),* 263, 239, 1976.
106. **Burks, T. F. and Galligan, J. J.,** Centrally mediated inhibition of intestinal transit: role of endogenous opioid peptides, *Gastroenterology,* 78, 1146, 1980.
107. **Gillan, M. G. C. and Pollock, D.,** Acute effects of morphine and opioid peptides on the motility and responses of rat colon to electrical stimulation, *Br. J. Pharmacol.,* 68, 381, 1980.
108. **Burks, T. E.,** Endorphin actions in dog intestine, *Pharmacologist,* 21, 175, 1979.
109. **Abraham, G. J. S., Dhume, V. G., Diniz, R. S., and Pereira, J. S.,** Effect of Met-enkephalin on the responses of rat fundus and colon and chick rectum to PGE_2 and its release from rat aortic strip, *Pharm. Res. Commun.,* 12, 131, 1980.
110. **Nijkamp, F. P. and van Ree, J. M.,** Effects of endorphins on different parts of the gastrointestinal tract of rat and guinea-pig in vitro, *Br. J. Pharmacol.,* 68, 599, 1980.
111. **Van Nueten, J. M., van Ree, J. M., and van Houtte, P. M.,** Inhibition by Met-enkephalin of peristaltic activity in the guinea pig ileum, and its reversal by naloxone, *Eur. J. Pharmacol.,* 41, 341, 1977.
112. **Van Nueten, J. M., Janssen, P. A. J., and Fontaine, J.,** Unexpected reversal effects of naloxone on the guinea pig ileum, *Life Sci.,* 18, 803, 1976.

113. **Kromer, W. and Pretzlaff, W.**, In vitro evidence for the participation of intestinal opioids in the control of peristalsis in the guinea pig small intestine, *Naunyn-Schmiedeb. Arch. Pharmacol.*, 309, 153, 1979.
114. **Zetler, G.**, Antagonism of cholecystokinin-like peptides by opioid peptides, morphine or tetrodotoxin, *Eur. J. Pharmacol.*, 60, 67, 1979.
115. **Kimberg, D. V., Field, M., Johnson, J., Henderson, A., and Gershon, E.**, Stimulation of intestinal mucosal adenyl cyclase by cholera enterotoxin and prostaglandins, *J. Clin. Invest.*, 50, 1218, 1971.
116. **Bodycote, I. J. and Chesher, G. B.**, Naloxone-induced contracture of ileum from stressed guinea pigs, *Eur. J. Pharmacol.*, 57, 259, 1979.
117. **Gunne, L.-M.**, The temperature response in rats during acute and chronic morphine administration; a study of morphine tolerance, *Arch. Int. Pharmacodyn*, 129, 416, 1960.
118. **Lotti, V. J., Lomax, P., and George, R.**, Temperature responses in the rat following intracerebral microinjection of morphine, *J. Pharmacol. Exp. Ther.*, 150, 135, 1965.
119. **Banerjee, U., Feldberg, W., and Lotti, V. J.**, Effect on body temperature of morphine and ergotamine injected into the cerebral ventricles of cats, *Br. J. Pharmacol.*, 32, 523, 1968.
120. **Cox, B., Ary, M., Chesarek, W., and Lomax, P.**, Morphine hyperthermia in the rat: an action on the central thermostats, *Eur. J. Pharmacol.*, 36, 33, 1976.
121. **Samanin, R., Kon, S., and Garattini, S.**, Abolition of the morphine effect on body temperature in midbrain raphe lesioned rats, *J. Pharm. Pharmacol.*, 24, 374, 1972.
122. **Hermann, J. B.**, The pyretic action on rats of small doses of morphine, *J. Pharmacol.*, 76, 309, 1942.
123. **Holaday, J. W., Loh, H. H., and Li, C. H.**, Unique behavioral effects of β-endorphin and their relationship to thermoregulation and hypothalamic function, *Life Sci.*, 22, 1525, 1978.
124. **Bloom, F., Segal, D., Ling, N., and Guillemin, R.**, Endorphins: profound behavioral effects in rats suggest new etiological factors in mental illness, *Science*, 194, 630, 1976.
125. **Martin, G. E. and Bacino, C. B.**, Action of intracerebrally injected β-endorphin on the rat's core temperature, *Eur. J. Pharmacol.*, 59, 227, 1979.
126. **Holaday, J. W., Law, P.-Y., Tseng, L.-F., Loh, H. H., and Li, C. H.**, β endorphin: pituitary and adrenal glands modulate its action, *Proc. Natl. Acad. Sci. USA*, 74, 4628, 1977.
127. **Tseng, L.-F., Loh, H. H., and Li, C. H.**, Human β-endorphin: development of tolerance and behavioral activity in rats, *Biochem. Biophys. Res. Commun.*, 74, 390, 1977.
128. **Miglécz, E. and Székely, J. I.**, unpublished data, 1978.
129. **Miglécz, E., Székely, J. I., and Dunai-Kovács, Z.**, Comparison of tolerance development and dependence capacities of morphine, β-endorphin, and (D-Met[2],Pro[5])-enkephalinamide, *Psychopharmacology*, 62, 29, 1979.
130. **Bläsig, J., Bäuerle, U., and Herz, A.**, Endorphin-induced hyperthermia: characterization of the exogenously and endogenously induced effects, *Naunyn-Schmiedeb. Arch. Pharmacol.*, 309, 137, 1979.
131. **Holaday, J. W., Tseng, L. F., Loh, H. H., and Li, C. H.**, Thyrotropin releasing hormone antagonizes β-endorphin hypothermia and catalepsy, *Life Sci.*, 22, 1537, 1978.
132. **Feldberg, W. and Smyth, D. G.**, C-fragment of lipotropin — an endogenous potent analgesic peptide, *Br. J. Pharmacol.*, 60, 445, 1977.
133. **Huidobro-Toro, J. P. and Way, E. L.**, Studies on the hyperthermic response of β-endorphin in mice, *J. Pharmacol. Exp. Ther.*, 211, 50, 1979.
134. **Clark, W. G.**, Emetic and hyperthermic effects of centrally injected methionine-enkephalin in cats, *Proc. Soc. Exp. Biol. Med.*, 154, 540, 1977.
135. **Cowan, A., Doxey, J. C., and Metcalf, G.**, A comparison of pharmacological effects induced by leucine-enkephalin, methionine-enkephalin, morphine and ketocyclazocine, in *Proc. Int. Narcotic Res. Club Meeting*, Kosterlitz, H. W., Ed., North-Holland, Amsterdam, 1976, 95.
136. **Nemeroff, C. B., Osbahr, A. J., III, Manberg, P. J., Ervin, G. N., and Prange, A. J., Jr.**, Alterations in nociception and body temperature after intracisternal administration of neurotensin, β-endorphin, other endogenous peptides, and morphine, *Proc. Natl. Acad. Sci. USA*, 76, 5368, 1979.
137. **Bloom, A. and Tseng, L.-F.**, 1979 cit post 23.
138. **Huidobro-Toro, J. P. and Way, E. L.**, Rapid development of tolerance to the hyperthermic effect of β-endorphin, and cross-tolerance between the enkephalins and β-endorphin, *Eur. J. Pharmacol.*, 65, 221, 1980.
139. **Tseng, L.-F. and Li, C. H.**, β-endorphin: hyperthermia in mice by intravenous injection, *Int. J. Peptide Protein Res.*, 15, 471, 1980.
140. **Goldstein, A. and Lowery, P. J.**, Effect of the opiate antagonist naloxone on body temperature in rats, *Life Sci.*, 17, 927, 1975.
141. **Baizman, E. R.**, An investigation of the central mechanisms of morphine- and narcotic antagonist-induced hypothermia in the naive and dependent rat, *Dissertation Abstr. Intern.*, B 37, 718, 1976.

142. **Lal, H., Miksic, S., and Smith, N.**, Naloxone antagonism of conditioned hyperthermia: an evidence for release of endogenous opioid, *Life Sci.*, 18, 971, 1976.
143. **Jacob, J. J. and Michaud, G. M.**, Production par la naloxone d'effets inverses de ceux de la morphine chez le chien éveillé, *Arch. Int. Pharmacodyn.*, 222, 332, 1976.
144. **Rudy, T. A. and Yaksh, T. L.**, Hyperthermic effects of morphine: set point manipulation by a direct spinal action, *Br. J. Pharmacol.*, 61, 91, 1977.
145. **Ferri, S., Reina, R. A., Santagostino, A., Scoto, G. M., and Spadaro, C.**, Effects of Met-enkephalin on body temperature of normal and morphine-tolerant rats, *Psychopharmacology*, 58, 277, 1978.
146. **Yaksh, T. L. and Henry, J. L.**, Antinociceptive effects of intrathecally administered human β-endorphin in the rat and cat, *Can. J. Physiol. Pharmacol.*, 56, 754, 1978.
147. **Bläsig, J., Höllt, V., Bäuerle, U., and Herz, A.**, Involvement of endorphins in emotional hyperthermia of rats, *Life Sci.*, 23, 2525, 1978.
148. **Holaday, J. W., Wei, E., Loh, H. H., and Li, C. H.**, Endorphins may function in heat adaptation, *Proc. Natl. Acad. Sci. USA*, 75, 2923, 1978.
149. **Lin, M. T., Chern, Y. F., Chen, F. F., and Su, C. Y.**, Serotonergic mechanisms of beta-endorphin-induced hypothermia in rats, *Eur. J. Physiol.*, 382, 87, 1979.
150. **Stewart, J. and Eikelboom, R.**, Stress masks the hypothermic effect of naloxone in rats, *Life Sci.*, 25, 1165, 1979.
151. **Volavka, J., James, B., Reker, D., Pollock, V., and Cho, D.**, Electroencephalographic and other effects of naloxone in normal men, *Life Sci.*, 25, 1267, 1979.
152. **Sandman, C. A., McGivern, R. F., Berka, C., Walker, J. M., Coy, D. H., and Kastin, A. J.**, Neonatal administration of β-endorphin produces "chronic" insensitivity to thermal stimuli, *Life Sci.*, 25, 1755, 1979.
153. **Nistico, G., Rotiroti, D., Naccari, F., de Sarro, G. B., and Marmo, E.**, Effects of intraventricular β-endorphin and D-Ala^2-methionine-enkephalinamide on behavior, spectrum power of electrocortical activity and body temperature in chicks, *Res. Comm. Chem. Pathol. Pharm.*, 28, 295, 1980.
154. **Kromer, W.**, Naltrexone influence on hibernation, *Experientia*, 36, 581, 1980.
155. **Sharp, B., Gordon, J., Morley, J., Carlson, H., and Hershman, J. M.**, Endogenous opioid peptides modulate the decrease of rat serum thyrotropin related to heat exposure, *Clin. Res.*, 28, 26A, 1980.
156. **Martin, G. E. and Morrison, J. E.**, Hyperthermia evoked by the intracerebral injection of morphine sulphate in the rat: the effect of restraint, *Brain Res.*, 145, 127, 1978.
157. **French, E. D., Vasquez, S. A., and George, R.**, Potentiation of morphine hyperthermia in cats by pimozide and fluoxetine hydrochloride, *Eur. J. Pharmacol.*, 48, 351, 1978.
158. **Moon, B. H., Feigenbaum, J. J. I., Weiner, W., Klawans, H., Carson, P. E., and Davis, J.**, Naloxone altered hypothermic responses to apomorphine: role of serotonergic mechanism(s), *Clin. Res.*, 26, 292A, 1978.
159. **Burks, T. F. and Rosenfeld, G. C.**, Neurotransmitter mediation of morphine hypothermia in rats, *Life Sci.*, 24, 1067, 1979.
160. **Sharkawi, M.**, Morphine hyperthermia in the rat: its attenuation by physostigmine, *Br. J. Pharmacol.*, 44, 544, 1972.
161. **Oka, T. and Negishi, K.**, Effect of neurohumoral modulators on the morphine-induced hyperthermia in non-tolerant rats, *Eur. J. Pharmacol.*, 42, 225, 1977.
162. **Scoto, G. M., Spadaro, C., Spampinato, S., Arrigo-Reina, R., and Ferri, S.**, Prostaglandins in the brain of rats given, acutely and chronically a hyperthermic dose of Met-enkephalin, *Psychopharmacology*, 60, 217, 1979.
163. **McGilliard, K. L., Tulunay, F. C., and Takemori, A. E.**, Antagonism by naloxone of morphine- and pentazocine-induced respiratory depression and analgesia and of morphine-induced hypothermia, in *Proc. Int. Narcotic Res. Club Meeting*, Kosterlitz, H. W., Ed., North-Holland, Amsterdam, 1976, 281.
164. **Tulunay, F. C.**, Pharmacological characterization of narcotic receptors for hyperthermia in rats, in *Proc. Int. Narcotic Res. Conf.*, van Ree, J. M. and Terenius, L., Eds., Elsevier/North-Holland, Amsterdam, 1978, 495.
165. **Rossier, J., French, E. D., Rivier, C., Ling, N., Guillemin, R., and Bloom, F. E.**, Foot-shock induced stress increases β-endorphin levels in blood but not brain, *Nature (London)*, 270, 618, 1977.
166. **Tallarida, R. J., Kramer, M. S., Roy, J. W., Kester, R. A., Murray, R. B., and Adler, M. W.**, Miosis and fluctuation in the rabbit pupil: effects of morphine and naloxone, *J. Pharmacol. Exp. Ther.*, 201, 587, 1977.
167. **Korczyn, A. D., Boyman, R., and Shifter, L.**, Morphine mydriasis in mice, *Life Sci.*, 24, 1667, 1979.
168. **Klemfuss, H., Tallarida, R. J., Adler, C. H., and Adler, M. W.**, Morphine-induced mydriasis and fluctuation in the rat: time and dose relationships, *J. Pharmacol. Exp. Ther.*, 208, 91, 1979.
169. **Fanciullacci, M., Boccuni, M., Pietrini, U., and Sicuteri, F.**, The naloxone conjunctival test in morphine addiction, *Eur. J. Pharmacol.*, 61, 319, 1980.

170. **Lee, H. K. and Wang, S. C.**, Mechanism of morphine-induced miosis in the dog, *J. Pharmacol. Exp. Ther.*, 192, 415, 1975.
171. **Tress, K. H. and El Sobky, A. A.**, Pupil responses to intravenous heroin (diamorphine) in dependent and nondependent humans, *Br. J. Clin. Pharmacol.*, 7, 213, 1979.
172. **Feldberg, W. and Smyth, D. G.**, C-fragment of lipotropin—an endogenous potent analgesic peptide, *Br. J. Pharmacol.*, 60, 445, 1977.
173. **Feldberg, W. and Smyth, D. G.**, The C-fragment of lipotropin — a potent analgesic, *J. Physiol. (London)*, 260, 31P, 1976.
174. **Meglio, M., Hosobuchi, Y., Loh, H. H., Adams, J. E., and Li, C. H.**, β-endorphin: behavioral and analgesic activity in cats, *Proc. Natl. Acad. Sci. USA*, 74, 774, 1977.
175. **Catlin, D. H., George, R., and Li, C. H.**, β-endorphin: pharmacological and behavioral activity in cats after low intravenous doses, *Life Sci.*, 23, 2147, 1978.
176. **Clark, W. G.**, Emetic and hyperthermic effects of centrally injected methionine-enkephalin in cats, *Proc. Soc. Exp. Biol. Med.*, 154, 540, 1977.
177. **Székely, J. I., Dunai-Kovács, Z., Miglécz, E., and Tarnawa, I.**, The *in vivo* pharmacology of (D-Met^2, Pro^5)-enkephalinamide, in *Endorphins'78*, Gráf, L., Palkovits, M., and Rónai, A. Z., Eds., Publishing House of the Hungarian Academy of Science, Budapest, Hungary, 1978, 319.
178. **Pulewka, P.**, Das Auge der weissen Maus als pharmakologisches Testobjekt, *Naunyn-Schmiedeb. Arch. Exp. Pathol. Pharmacol.*, 168, 307, 1932.
179. **Forgács, O. and Székely, J. I.**, unpublished data, 1979.
180. **Korczyn, A. D., Eshel, Y., and Keren, O.**, Enkephalin mydriasis in mice, *Eur. J. Pharmacol.*, 65, 285, 1980.
181. **Tortella, F. C., Cowan, A., Koenig, D., and Adler, M. W.**, Pupillary effects of i.p. administered methionine enkephalin and leucine enkephalin in the rat, *Pharmacologist*, 21, 199, 1979.
182. **Wallenstein, M. C.**, Biphasic effects of morphine on cardiovascular system of the cat, *Eur. J. Pharmacol.*, 59, 253, 1979.
183. **Wallenstein, M. C. and Wang, S. C.**, Mechanism of morphine-induced mydriasis in the cat, *Am. J. Physiol.*, 236, 292, 1979.
184. **Stubbs, W. A., Delitala, G., Jones, A., Jeffcoate, W. J., Edwards, C. R. W., Ratter, S. J., Besser, G. M., Bloom, S. R., and Alberti, K. G. M. M.**, Hormonal and metabolic responses to an enkephalin analogue in normal man, *Lancet*, 2, 1225, 1978.
185. **Volavka, J., James, B., Reker, D., Pollock, V., and Cho, D.**, Electroencephalographic and other effects of naloxone in normal men, *Life Sci.*, 25, 1267, 1979.
186. **Herz, A.**, personal communication, 1978.
187. **Yen, M.-H., Ku, P.-Y., and Lee, H.-K.**, The effect of (D-Ala^2)-Met-enkephalin on the contraction of nictitating membrane in cats, *Eur. J. Pharmacol.*, 63, 213, 1980.
188. **Henderson, G., Hughes, J., and Kosterlitz, H. W.**, The effects of morphine on the release of noradrenaline from the cat isolated nictitating membrane and the guinea-pig ileum myenteric plexus-longitudinal muscle preparation, *Br. J. Pharmacol.*, 53, 505, 1975.
189. **Bloom, F., Segal, D., Ling, N., and Guillemin, R.**, Endorphins: profound behavioral effects in rats suggest new etiological factors in mental illness, *Science*, 194, 630, 1976.
190. **Jacquet, Y. F. and Marks, N.**, The C-fragment of β-lipotropin: an endogenous neuroleptic or antipsychotogen? *Science*, 194, 632, 1976.
191. **Weinstock, M., Stewart, H. C., and Butterworth, K. R.**, Lenticular effect in mice of some morphine-like drugs, *Nature (London)*, 182, 1519, 1958.
192. **Cox, B. M. and Weinstock, M.**, Quantitative studies of the antagonism by nalorphine of some of the actions of morphine-like analgesic drugs, *Br. J. Pharmacol.*, 22, 289, 1964.
193. **Smith, A. A., Karmin, M., and Gavitt, J. A.**, Interaction of catecholamines with levorphanol and morphine in the mouse eye, *J. Pharmacol. Exp. Ther.*, 151, 103, 1966.
194. **Weinstock, M. and Marshall, A. S.**, The influence of the sympathetic nervous system on the action of drugs on the lens, *J. Pharmacol. Exp. Ther.*, 166, 8, 1969.
195. **Smith, A. A., Karmin, M., and Gavitt, J. A.**, Central origin of the lenticular opacities induced in mice by opiates, *J. Pharm. Pharmacol.*, 18, 545, 1966.
196. **Smith, A. A.**, Potentiation of opioid-induced cataracts by catecholamines injected into the mouse brain, *Psychopharmacologia (Berl.)*, 16, 313, 1970.
197. **Smith, A. A., Albin, R., and Crofford, M.**, Heterogeneity of receptors for analgesia respiration and lenticular effect, in *Proc. Int. Narcotic Res. Club Meeting*, Kosterlitz, H. W., Ed., North-Holland, Amsterdam, 1976, 289.

Chapter 5

OPIOID PEPTIDES AND THE NEUROENDOCRINE FUNCTIONS

J. I. Székely

TABLE OF CONTENTS

I. Introduction 208

II. Effects on Prolactin and Growth Hormone (GH) Secretion 208

III. Effects on LHRH, LH, and FSH Secretion 217

IV. Effects on ACTH and TSH Secretion 220

V. Effects on Vasopressin and Oxytocin Secretion 222

VI. Other Hormonal Effects 223

References 224

I. INTRODUCTION

Upon discovering the opioid peptides, their potential role in regulation of the pain threshold and certain CNS functions were studied. However, more and more data indicate that these substances concomitantly take part in physiological modulation of neuroendocrine processes. Of course, there is hardly any complex central function of the organism which would not be modified by opioids. Substances which alter the normal brain activity and also have distinct circulatory, respiratory, and other autonomic actions obviously modify many other functions of the organism. However, as it will be discussed below, certain neuroendocrine effects of opioids seem to be primarily related to certain yet unclarified physiological functions of endogenous opioids and they cannot be qualified as secondary epiphenomena of other processes. The long-known neuroendocrine effects of exogenous opiates are mediated mainly by the hypothalamus, i.e., by a structure rich in opiate receptors (Chapter 1, Volume III) and endogenous opioids (Chapter 3, Volume III). Some characteristic neuroendocrine actions of opioids appear at very low dose levels, especially in vivo; in several cases their endocrine effects may be detected in smaller doses than necessary for analgesia. Thus, sometimes the question arises whether the psychotropic or endocrine actions are the more specific ones.

II. EFFECTS ON PROLACTIN AND GROWTH HORMONE (GH) SECRETION

Increase in serum levels of prolactin and GH are well-known effects of the opiates and also the reversal of this action by narcotic antagonists.[1,2] The specificity of these actions is further confirmed by a recent study[3] in which partial agonists were found to inhibit the morphine-induced GH release, while giving them alone, the serum level of GH was elevated. As for the prolactin, its level is decreased by the pure opiate antagonists (Table 1). The serum prolactin lowering or elevating activities of morphine derivatives correlate with their opiate agonist or antagonist character as determined in receptor binding assay.[3] Thus, the naloxone reversible elevation of serum prolactin level is a reliable indicator of the opiate agonist potency.

Of the opioid peptides, first the natural enkephalins were shown to elevate the serum prolactin level by Lien et al.[4] This finding was quickly extended to endorphins[5,6,9,17,18,28,35,49] and to the synthetic enkephalin analogues.[8,11,14-16,19,20-23,26,27,29,37,41,42,46] As shown in Table 1, in many studies the GH level was found elevated. Nevertheless, the data on the elevation of GH level upon treatment with opioids are less convincing. In all studies summarized in Table 1 the increase in serum GH concentration was of smaller magnitude and/or was elicited only by higher doses of opioids than the similar effect on serum prolactin. In one study[39] just decrement was observed. A decrease in the serum GH level was also found in acromegalic humans upon i.v. treatment with DALA.[50] In healthy men only the most active enkephalin analogue, FK 33-824, increased the serum GH level,[21,22] but β-EP (given i.v. or intracerebrally) or (D-Ala^2,D-Nle^5)-E-amide decreased it or their effect was variable both in humans and in monkeys.[37]

This increase by opioid peptides in serum prolactin or GH levels could be antagonized by naloxone (Table 1). Only Cocchi et al.[7] reported that in newborn rats the Leu-E-induced elevation of GH could be antagonized by naloxone but the similar increase in serum prolactin could not. Another interesting exception has been reported by Tache et al.,[36] who observed that the increase in serum prolactin elicted by stress could not be modified either by naloxone or β-EP. In other studies[10,12] naloxone inhibited the stress-induced prolactin release, but in a recent study Rossier et al.[43] reported

Table 1
EFFECTS OF OPIOID PEPTIDES AND NALOXONE ON SERUM PROLACTIN AND GROWTH HORMONE (GH) LEVELS

Treatment	Species	Prolactin	GH	Reversal by opiate antagonists	Ref.
Met-E i.v., s.c.	Rat	Increase	Not examined	Not examined	4
Leu-E i.v., s.c.	Rat	Increase	Not examined	Not examined	4
β-EP i.v., i.c.v.	Rat	Increase	Increase	Yes	5
α-endorphin i.v.	Rat	No effect	Not examined	—	5
Met-E i.v.	Rat	No effect	Not examined	—	5
β-EP i.c.v.	Rat	Increase	Increase	Not examined	6
Met-E i.c.v.	Rat	Increase	Increase	Not examined	6
Leu-E i.c.v.	Rat	Increase	Increase	Inhibition of GH release but no effect on prolactin liberation	7
DALA s.c.	Rat	Increase	Increase	Yes	8
(D-Ala²)-Leu-E s.c.	Rat	Increase	Increase	Yes	8
Met-E-amide s.c.	Rat	No effect	No effect	—	8
Naloxone s.c.	Rat	Decrease	Decrease	—	8
β-EP i.c.v.	Rat	Not examined	Increase	Not examined	9
Met-E i.c.v.	Rat	Not examined	Increase	Not examined	9
Naloxone i.p.	Intact rat	Decrease	Not examined	—	10
Naloxone i.p.	Rat with hypothalamic deafferentation	Decrease	Not examined	—	10
DALA i.c.v.	Rat	Increase	Increase	Not examined	11
Met-E-ethylamide i.c.v.	Rat	Increase	Increase	Not examined	11
N-acetyl-DALA i.c.v.	Rat	No effect	No effect	—	11
(D-Tyr¹)-Met-E i.c.v.	Rat	No effect	No effect	—	11
Naloxone i.p.	Intact or stressed rat	Decrease	Not examined	—	12
Met-E i.p.	Rat	Increase	Increase	Yes	13
Naloxone i.p.	Rat	Decrease	Decrease	—	13
Tyr-D-Ala-Gly-Phe-Met-$NH-C_3H_7$ i.v.	Rat	Increase	Not examined	Yes	14
Tyr-D-Ala-Gly-NH-C_2H_4-C_6H_5 i.v.	Rat	Increase	Not examined	Yes	14
Met-E i.v.	Rat	No effect	Not examined	—	14
(D-Ala²,D-Leu⁵)-E-amide s.c.	Rat	Increase	Increase	Yes	15
Enkephalin analogues with substitutions: N-Me Tyr¹, D-Ala², D-Trp⁵, D-Leu⁵ i.c.v.	Rat	Increase	Not examined	Yes	16
Enkephalin analogues with substitutions: D-Trp¹, D-Phe⁴, D-Met⁵, Phe¹, D-Ala¹ i.c.v.	Rat	No effect	Not examined	—	16
Met-E i.c.v.	Rat	Increase	Increase	Yes	17
α-endorphin i.c.v.	Rat	Increase	Increase	Yes	17
γ-endorphin i.c.v.	Rat	Increase	Increase	Yes	17
β-EP i.c.v.	Rat	Increase	Increase	Yes	17
β-EP i.c.v.	Rat	Increase	Increase	Not examined	18
Tyr-D-Ser-Gly-Phe-Met-*O*-methyl i.v.	Goat	Increase	Increase	Yes	19
Tyr-Gly-Gly-Phe-Met-*O*-methyl i.v.	Goat	Increase	Increase	Yes	19

Table 1 (continued)
EFFECTS OF OPIOID PEPTIDES AND NALOXONE ON SERUM PROLACTIN AND GROWTH HORMONE (GH) LEVELS

Treatment	Species	Prolactin	GH	Reversal by opiate antagonists	Ref.
Naloxone i.v., i.p.	Goat	Decrease	No effect	—	19
Met-E i.c.v.	Rat	Increase	No effect	Not examined	20
α-endorphin i.c.v.	Rat	Increase	No effect	Not examined	20
γ-endorphin i.c.v.	Rat	Increase	No effect	Not examined	20
DALA i.c.v.	Rat	Increase	Increase	Not examined	20
β-EP i.c.v.	Rat	Increase	Increase	Yes	20
FK 33-824 i.m.	Human	Increase	Increase	Not examined	21
FK 33-824 i.v.	Human	Increase	Increase	Yes	22
FK 33-824 i.v.	Monkey	Increase	Not examined	Yes	23
Naloxone s.c.	Morphine tolerant rat	No effect	Not examined	—	24
Naloxone i.v.	Human	Not examined	Increase	—	25
D-Met,Pro-EA i.v.	Rat	Increase	Increase	Yes	26
D-Met,Pro-EA i.v.	Rat	Increase	Not examined	Yes	27
β-EP i.c.v.	Rat	Increase	Not examined	Not examined	28
D-Met,Pro-EA i.v.	Rat	Increase	Increase	Yes	29
Naloxone i.v.	Human	Decrease	Not examined	—	30
Naloxone i.v.	Monkey	Decrease	Not examined	—	31
					32
Naloxone i.p.	Rat	Decrease	Not examined	—	33
Naloxone i.v.	Rat	No effect	No effect	—	34
Naloxone i.v.	Human	No effect	No effect	—	34
β-EP i.c.v.	Rat	Increase	Not examined	Yes	35
β-EP i.c.v.	Stressed rat	No effect	Not examined	—	36
Naloxone i.p.	Stressed rat	No effect	Not examined	—	36
Naloxone s.c.	Rat	Decrease	No effect	—	3
β-EP i.c.v.	Human	Increase	Decrease	Not examined	37
β-EP i.v.	Human	Increase	No effect	Not examined	37
(D-Ala^2,D-Nle^5)-E-amide i.v.	Baboon	Increase	Variable	Not examined	37
Naloxone i.v.	Monkey	Decrease	Not examined	—	38
β-EP intracerebrally	Rat	Not examined	Decrease	Not examined	39
Naloxone i.v.	Rat	Not examined	No effect	—	40
Naltrexone i.v.	Rat	Not examined	No effect	—	40
FK 33-824 i.v.	Dog	Not examined	Increase	Yes	41
					42
Naloxone s.c.	Rat	Decrease	Not examined	—	43
Naloxone i.v.	Human	Facilitation of TRH induced elevation	No effect	—	44
Naloxone i.p.	Rat	Decrease	Not examined	—	45
FK 33-824 i.v.	Human	Increase	Not examined	Yes	46
Naloxone i.v.	Human	No effect	Slight increase	—	47
Naloxone s.c.	Female rats	Decrease but only during estrus	Not examined	—	48

that naloxone only partially antagonized the stress-induced elevation of serum prolactin without affecting the concomitant increase in serum β-EP level. Therefore, they[43] suggested that the stress-induced elevation of plasma prolactin was at least in part due to the mobilization of endogenous β-EP, which was not antagonized by naloxone. It might be supposed that the receptors mediating the β-EP-induced prolactin liberation are relatively insensitive to naloxone.

The serum prolactin and GH releasing potencies of opioids correlate well with their other opiate-like activities. Thus, Met-E was found more potent than Leu-E.[4] Comparing the potencies of different opioid peptides, β-EP was repeatedly found to be more active than the natural enkephalins or α- and γ-endorphins.[5,6,9,17]

As for the synthetic enkephalin analogues, the increased activities of "enzymatically stabilized" derivatives is also well known (Table 1). Conversely, some analogues which are inactive in receptor binding assay or vas deferens preparation proved to be ineffective in relation to prolactin releasing activity (*N*-acetyl-Met-E; (D-Tyr^1)-Met-E).[17] The most detailed study on the structure activity relationship of enkephalin analogues in relation to the prolactin release was carried out by Meltzer et al.[16] Comparing their potencies in receptor binding assay and in prolactin release (upon i.c.v. application), generally good correlation has been found.[16] Of course, in such experiments the eventual deviations may be more interesting. (D-Ala^2,D-Leu^5-NH-ethyl)-E is an effective prolactin releaser but rather weak in receptor binding assay, while (D-Ala^2,D-Met^5)-E is rather weak as prolactin releaser,[16] though the analgesic activity of the latter is well known (Chapters 5 and 6, Volume III).

These differences in the analgesic and prolactin releasing activities are easy to explain if different receptors mediate the two biological actions. As the multiplicity of opiate receptors has been solidly established in the in vitro experiments (Chapter 1, Volume III), such an explanation of the divergence of in vivo potencies might be supposed. Moreover, considerable differences may be detected if the relative potencies of opioid peptides determined in the in vitro receptor binding assay are compared with the activities measured in any in vivo opiate specific test. The in vivo activity depends not only on the receptor agonist potency but also on the metabolic stability and relative distribution in the living organism. The extremely high in vivo efficacy of etorphine as compared it to morphine is only partially explained by its enhanced affinity to the opiate receptors. Increased lipid solubility is a further factor explaining its exceptional potency in the living organism.[51] As the lipid solubilities of different opioid peptides are presumably different, while this factor hardly plays any role in their receptor binding capacities, these deviations are easy to explain. In the cases of the opioid peptides, however, further factors might distort the correlation between the in vitro and in vivo potencies, such as their capacity to pass the blood-brain barrier and their in vivo metabolic stability. These problems are discussed in Chapter 5, Volume III, but let us recall here that the lack of analgesic activity in the case of natural enkephalins is generally attributed to their susceptibility to enzymatic degradation, especially if they are administered systemically, i.e., outside of the blood-brain barrier. Just from this point of view it is rather surprising that the natural enkephalins have been shown to elevate the serum prolactin and GH levels upon i.v. even s.c. and i.p. administration.[1,4,5,13] On the other hand, it has been repeatedly demonstrated that upon i.c.v. application β-EP and some synthetic enkephalin analogues are several orders of magnitude stronger prolactin and GH releasers than the natural enkephalins.[5,6,9,17,20] The enhanced activity of longer fragments of LPH (β-EP) and synthetic enkephalin derivatives is generally thought to be the consequence of their enhanced metabolic stability (Chapter IV/2). However, if this latter hypothesis were true, we would be faced with a rather strange situation, in which metabolic stabilization of opioid peptides is more important upon central administration than in the case of peripheral application.

In view of this apparent contradiction the data of Locatelli et al.[26] deserve special attention. These authors found the potent enkephalin analogue, D-Met,Pro-EA to elicit a sixfold elevation of serum prolactin level in rats in the i.v. dose of 0.2 mg/kg. Obviously, the minimally efficient dose (e.g., the dose only doubling the baseline concentration) is lower. Hermann et al.[27] found a nearly tenfold elevation of the prolactin level in the dose of 0.3 mg/kg i.v. In our experiments the analgesic ED_{50} of this substance, also in rats upon i.v. application, is somewhat higher than 0.2 mg/kg (Chapter 3, Volume II). Consequently, the prolactin releasing activity of this enkephalin analogue appears at lower dose level than its analgesic effect.

Similar contradictions arise upon comparing the data on the analgesic and endocrine actions of β-EP. Rivier et al.[5] reported that it elicited a significant elevation of serum prolactin level in the i.v. doses 0.03 to 0.3 mg/kg. However, the Guillemin group[52] reported in the same year, that β-EP in rats, contrary to observations of others[53] made in mice, did not elicit analgesia if applying it i.v. up to the dose of 8.0 mg/kg i.v.

Upon i.c.v. administration Kato et al.[17] observed prolactin liberation already in the dose of 10 ng, which is definitely less than needed for analgesia (Chapter 3, Volume II), though generally at least 0.5 to 2.0 /μg was given in the studies on its prolactin releasing activity.[5,6,9,28,35] The above-discussed statements are valid also for the opiates. In one of the above-mentioned works[26] morphine released prolactin already in the i.v. dose of 0.1 mg/kg but for analgesia in our laboratory at least 0.2 to 0.4 mg/kg was needed (Chapter 3, Volume II).

Summarizing these seemingly heterogeneous data probably the following conclusions may be drawn.

1. The in vitro potencies of opioid peptides (measured in receptor binding assay or in the isolated organ preparations) correlate more or less with their prolactin releasing capacities in vivo, upon i.c.v. application. This correlation is rather poor, but probably not worse than in relation of the analgesic effect, also upon i.c.v. administration.
2. Using the same route of administration, somewhat lower doses are needed to elevate the prolactin concentration than to elicit analgesia.
3. If applying them peripherally (i.v., s.c., i.p.), their prolactin releasing capacity uncomparably surpasses their analgesic effect, the latter being virtually nonexistent.

These findings strongly suggest that the opiate receptors mediating the prolactin (and probably the GH) releasing action(s) of opioid peptides and those responsible for their antinociceptive effect are different. Moreover, different receptors should mediate their hormonal actions upon central and peripheral application, respectively. Since opioids, especially natural enkephalins, are relatively more active upon peripheral than central application, the receptors mediating the effects upon systemic application cannot be situated beyond the blood-brain barrier. Moreover, as naloxone was found to decrease the serum prolactin level in rats with hypothalamic deafferentation,[10] the site of action of opioids cannot be located more proximally than the hypothalamus.

The rather low density of opiate receptors in the anterior hypophysis (Chapter 1, Volume III) indicates that it cannot be the site of action, since adenohypophysis contains only negligible amount of enkephalins (Chapter 3, Volume III). Its β-EP content is fairly high, but apart from the recent study of Rossier et al.,[43] there is no experimental proof that it would take part in the local regulation of release of other hormones from this lobe. The anterior hypophyseal β-EP is released together with ACTH to peripheral hormonal and metabolic stimuli and in stress situations (Chapters 3 and 4, Volume III).

As for the exact site of action, the in vitro experiments might be more informative (Table 2). Unfortunately rather inconsistent data have been obtained upon working with isolated organ cultures. First Lien et al.[4] observed stimulation by natural enkephalins of spontaneous prolactin release in pituitary cell culture. The effect was naloxone sensitive.[4] This finding has not been confirmed in the subsequent studies, i.e., neither β-EP nor natural and synthetic enkephalins stimulated the spontaneous release of prolactin[5,8,45,60] and GH[5,8] if incubating the adenohypophysis in vitro. Only an in vitro maintained pituitary tumor tissue (GH 3) showed this effect.[61]

Moreover, neither naloxone[4,45] nor naltrexone[10] modified the spontaneous prolactin release in vitro. These observations speak against a hypophyseal site of action of opioid peptides. Thus, the authors[5,8,10,45,60] concluded on a hypothalamic point of attack. This supposition is only further substantiated by the high density of opiate receptors and endogenous opioids in the mediobasal hypothalamus (Chapters 1 and 3, Volume III). Consequently, opioids might exert their action via the hypothalamic releasing factors which regulate the liberation of hypophyseal hormones. Since the activity of the adenohypophysis is regulated by hypothalamic releasing factors, the above-mentioned data may be simply irrelevant. In the living organism the hypophyseal prolactin secretion is inhibited by dopamine.[69] Thus, the question emerges whether opioids modify the dopamine-mediated inhibition of prolactin secretion in vitro. According to Login and McLeod[60] Met-E and (D-Ala^2,Met^5)-E did not modify the suppressant action of dopamine. However, in the experiments of Enjalbert et al.[55,59] Met-E and β-EP and morphine reversed the dopamine-mediated suppression of prolactin release and this effect could be abolished by naloxone. As both groups[55,59,60] worked with in vitro incubated hypophyseal tissue, the data are squarely contradictory. Consequently, for the time being it is difficult to find out whether the antidopamine action of opioid peptides is realized exclusively at hypothalamic level or also through the opiate receptors of the hypophysis.

Unfortunately, the contingent effect of the same interaction between dopamine and opioid peptides has not been examined in the same studies in relation to GH. Nevertheless, the GH release is primarily regulated by somatostatin. Consequently, the data on the effect of opioids on somatostatin release or on its action in the adenohypophysis might be more relevant. Really, somatostatin was reported to antagonize the morhine-induced increase of GH level.[1] As for the opioid peptides, Drouva et al.[65] observed a naloxone-sensitive inhibition of K^+-induced somatostatin release from the mediobasal hypothalamus without alteration of the spontaneous release. Hence, stimulation of GH release by opioids could be attributed to a decrease in somatostatin liberation from the hypothalamus. In a similar experiment Hollander et al.[56] observed facilitation of somatostatin release. In this study[56] morphine was inactive and the antagonizability by naloxone of the effect of β-EP was not examined. Furthermore, the inhibition by morphine and β-EP of somatostatin liberation has been observed also in isolated perfused pancreas.[70]

It is reasonable to suppose that at least in part the increase in plasma concentration of GH is the consequence of attenuated somatostatin liberation from the hypothalamus. Even this cautious conclusion can be challenged. In experiments of Chihara et al.[20] the GH releasing response to endorphins, natural, and synthetic enkephalins was not suppressed by treating the animals with antiserum to somatostatin, though in these animals the basal level of GH was elevated. Accordingly, opioids are capable to increase the serum GH level also in the absence of endogenous somatostatin.

Actually, also the antidopamine actions of opioids seem to be primarily mediated at hypothalamic level. As dopamine released from the hypothalamus tonically inhibits the prolactin secretion,[1] opioid peptides are expected to inhibit its liberation. In fact, a decrease in dopamine turnover has been observed upon treatment with morphine, Met-E or β-EP.[28,71]

Table 2
EFFECTS OF OPIOID PEPTIDES AND OPIATE ANTAGONISTS ON HORMONE RELEASE IN VITRO

Treatment	Assay	Effect	Reversal by opiate antagonists	Ref.
Met-E	Pituitary cell culture	Increase in spontaneous P and GH release	Yes	4
Leu-E	Pituitary cell culture	Increase in spontaneous P and GH release	Yes	4
Naloxone	Pituitary cell culture	No effect	—	4
Met-E	Pituitary cell culture	No effect on spontaneous P and GH release	—	5
α-endorphin	Pituitary cell culture	No effect on spontaneous P and GH release	—	5
β-EP	Pituitary cell culture	No effect on spontaneous P and GH release	—	5
DALA	Cultured anterior pituitary tissue	No effect on spontaneous P and GH release	—	8
(D-Ala^2,D-Leu^5)-E-amide	Cultured anterior pituitary tissue	No effect on spontaneous P and GH release	—	8
Naltrexone	Cultured pituitary tissue	No effect on spontaneous P release	—	10
β-EP	Isolated neural lobe	No effect on vasopressin release	—	54
Met-E	Cultured pituitary tissue	No effect alone but reversal of dopamine induced suppression of P release	Yes	55
β-EP	Cultured pituitary tissue	No effect alone but reversal of dopamine induced suppression of P release	Yes	55
β-EP	Rat hypothalamus organ culture	Facilitation of SRIF and TRH release, no effect on GHRH liberation	Not examined	56
Met-E	Incubated mediobasal hypothalamic slice	No effect on spontaneous LHRH release but suppression that induced by dopamine	Yes	57
Naloxone	Incubated mediobasal hypothalamic slice	No effect on spontaneous and dopamine induced LHRH release	—	57
β-EP	Isolated neural lobe	Facilitation of oxytocin release	Yes	58
Naloxone	Isolated neural lobe	Inhibition of oxytocin release	—	58
Naloxone	Cultured pituitary tissue	No effect on basal P release, inhibition of reversal by morphine of dopamine induced suppression of prolactine release	—	59
Met-E	Cultured anterior pituitary tissue	No effect on spontaneous P release and its suppression by dopamine	—	60
DALA	Cultured anterior pituitary tissue	No effect on spontaneous P release and its suppression by dopamine	—	60
Met-E	GH3 pituitary tumor cells	Facilitation of spontaneous P release	Yes	61
Met-E,Leu-E,(D-	Rat pancreatic mono-	Inhibition of insulin and	Yes	62

Table 2 (continued)
EFFECTS OF OPIOID PEPTIDES AND OPIATE ANTAGONISTS ON HORMONE RELEASE IN VITRO

Treatment	Assay	Effect	Reversal by opiate antagonists	Ref.
Ala²)-Met-E	layer culture	glucagon release		
Naloxone	Rat pancreatic monolayer culture	No effect	—	62
β-EP	Adrenal cell culture	Facilitation of steroid synthesis	Not examined	63
Met-E	Adrenal cell culture	No effect on spontaneous and $ACTH_{1-24}$ induced steroidogenesis	—	64
Naloxone	Adrenal cell culture	Inhibition of $ACTH_{1-24}$ induced steroidogenesis	—	64
Naltrexone	Adrenal cell culture	Inhibition of $ACTH_{1-24}$ induced steroidogenesis	—	64
β-EP	Cultured anterior pituitary	No effect on spontaneous P and LH release	—	45
Met-E	Cultured anterior pituitary	No effect on spontaneous P and LH release	—	45
Leu-E	Cultured anterior pituitary	No effect on spontaneous P and LH release	—	45
DALA	Cultured anterior pituitary	No effect on spontaneous P and LH release	—	45
Naloxone	Cultured anterior pituitary	No effect on spontaneous P and LH release	—	45
Met-E	Incubated mediobasal hypothalamic slice	No effect on spontaneous LHRH release but suppression of that induced by K^+	Yes	65
Met-E	Incubated mediobasal hypothalamic slice	No effect on spontaneous SRIF release but suppression of that induced by K^+	Yes	65
Naloxone	Incubated mediobasal hypothalamic slice	No effect on spontaneous or K^+ induced LHRH and SRIF release, resp.	—	65
β-EP	Isolated neural lobe and pituitary stalk	Inhibition of vasopressin release induced by electrical stimulation	Yes	66
(D-Ala²,D-Leu⁵)-E	Isolated neural lobe and pituitary stalk	Inhibition of vasopressin release induced by electrical stimulation	Yes	66
Leu-E	Rat islet cell culture	Inhibition of insulin and glucagon release	Yes	67
Met-E	Rat islet cell culture	Inhibition of insulin and glucagon release	Yes	67
(D-Ala²)-Leu-E	Rat islet cell culture	Inhibiion of insulin and glucagon release	Yes	67
(D-Phe⁴)-Leu-E	Rat islet cell culture	Inhibition of insulin and glucagon release	Yes	67
Naloxone	Rat islet cell culture	No effect	—	67
LPH	Rat adrenal capsular cell culture	Increase in aldosterone production	Not examined	68
LPH	Rat adrenal decapsular cell culture	No effect on corticosterone production	—	68

Note: P = prolactin; GH = growth hormone.

Actually, Ferland et al.[71] demonstrated that Met-E injection into the lateral ventricles of the rat decreased the dopamine turnover in the median eminence and at the same time elevated the serum prolactin level. As opiates and opioid peptides increase the dopamine turnover in other brain areas (Chapter 2, Volume I and Chapter 1, Volume II) these observations were probably not easily acknowledged in the literature. Nevertheless, the decreased rate of dopamine metabolism is only an indirect indicator of its reduced liberation. Therefore, the work of Gudelsky and Porter[69] is of special interest in which the dopamine concentration in pituitary stalk plasma of rats was examined. In this study[69] morphine, β-EP, and DALA reduced the dopamine concentration and this effect could be reversed by naloxone.

These data are, however, not surprising if considering that the serum prolactin level is decreased and increased by dopamine agonists and antagonists, respectively. Moreover, that the prolactin level decreases upon treatment with drugs inhibiting the dopamine reuptake and also this action can be antagonized by morphine.[28]

Dopamine itself is considered to enhance the GH liberation.[1,2] However, according to Takahara et al.[18] the β-EP-induced release of prolactin, GH and LH could equally be reversed by 2-Br-α-ergocryptine, a dopamine agonist. Probably the contradictions might be explained by supposing that opioid peptides increase the plasma GH level by inhibiting the somatostatin liberation from the hypothalamus as suggested by Drouva et al.[65] and exert an opposite action via decreasing the dopamine liberation.

A further question to be answered is whether the opioids act directly on the hypothalamic dopamine receptors. The answer is definitely no. The dopamine elicited decrease in serum prolactin level could be reversed by haloperidol, but not by morphine[24] and conversely, the haloperidol-induced increase in prolactin level could be abolished by apomorphine but not by naloxone.[24] The same authors reported that the morphine-induced elevation of serum prolactin level was antagonized by both naloxone and apomorphine.[72] This finding indicates that opiates and opioid peptides act behind the postsynaptic dopamine receptors. Accordingly, opioids would act presynaptically, i.e., they should exert their action via presynaptic inhibition of dopamine release. Okajima et al.[73] demonstrated that the β-EP-induced prolactin release was diminished in rats pretreated with 6-hydroxydopamine. Since this agent poisoning selectively catecholamine producing cells acts on the presynaptic side, also this observation[73] shows that opioids do not exert their antidopamine action by a direct effect on the postsynaptic dopamine receptors. Lal et al.[74] observed enhanced inhibition of prolactin release upon apomorphine injection in animals previously chronically treated with morphine. This finding indicates a hyperreactivity of tubero-infundibular dopaminergic system in the opiate-treated animals. The receptor supersensitivity might also be a presynaptically mediated phenomenon. Accordingly opioids might act in both the nigro-striatal and the hypothalamic tubero-infundibular systems via presynaptic inhibition of dopamine release. In the nigro-striatal system this presynaptic action is accompanied by a considerable increase in the dopamine turnover (Chapter 2, Volume I and Chapter 1, Volume II). However, in the hypothalamus no such seemingly secondary (compensatory) increase of dopamine turnover develops. Probably the differences in the neuronal circuitries, i.e., the existence of a striato-nigral feedback in the extrapyramidal system and the lack of a corresponding feedback pathway in the hypothalamic dopaminergic area explains this difference in their function.

Otherwise, the brain dopaminergic system is modulated by serotoninergic and cholinergic elements. Considering the well-known mutual antagonism between the serotonin and dopamine receptor stimulants, the data of Spampinato et al.[29] are easy to understand. In their experiments the prolactin release induced by D-Met,Pro-EA was antagonized by pharmacological manipulations which blocked the serotoninergic system (treatment with methysergide, metergoline, or 5,6-dihydroxytryptamine), while facilitation was observed upon blocking the dopaminergic transmission (by domperidone) or by rendering the serotonin receptors supersensitive (by depleting brain serotonin with *p*-chlorophenylalanine).[29] More or less similar conclusions may be drawn from the experiments on pharmacological modulation of morphine-induced prolactin release.[33] As for the GH releasing action, it was stimulated by *p*-chlorophenylalanine pretreatment, i.e., supersensitivity of the serotonin receptors resulted in facilitation of GH surge induced by β-EP, Met-E.[75] In another laboratory[41,42] the FK 33-824 induced GH mobilization could be reversed also by atropine, proving that not only the dopaminergic and serotoninergic systems mediate this action of opioid peptides. The β-EP and morphine-induced prolactin liberation could be reversed also by pilocarpine, while the latter did not modify the similar effect of haloperidol.[35] This effect of pilocarpine could be abolished by atropine but not by mecamylamine.[35] Thus, muscarinergic and not nicotinergic influences may modify the effects of opioids on prolactin secretion. This cholinergic action cannot be exerted directly on the postsynaptic dopaminergic membrane as pilocarpine did not modify the effect of haloperidol.[35] Consequently, either cholinergic nerve terminals impinge upon cells secreting opioids or both are connected to the input of dopamine producing neurones.

Having reviewed the data on the effects of opioids on serum prolactin and GH levels, again the much-discussed question emerges: do they act tonically or phasically? Obviously this problem may be clarified by administration of opiate antagonists to otherwise intact animals. As shown in Table 1, most authors recorded a decrease in the serum prolactin level upon naloxone treatment. Only Martin et al.[34] reported that naloxone failed to modify the nocturnal GH and prolactin surges in healthy women and the similar periodic release in rats. Similarly Volavka et al.[47] failed to detect any effect in normal male volunteers. These authors concluded that the endogenous opioid did not exert continuous tonic inhibitory action only mediate the stress-induced liberation of prolactin.[34] As for the GH level upon naloxone treatment, as shown in Table 1, the data are totally confusing, increase, decrease, or no effect have been recorded.

Therefore, though the elevation of serum prolactin level is probably the most sensitive in vivo indicator of opioid mechanisms, further experiments are needed to clarify their site of action, whether exclusively hypothalamic or even the hypophyseal action must be counted with. The participation other brain transmitters in their actions and their role in the physiological regulation also require additional experiments.

III. EFFECTS ON LHRH, LH, AND FSH SECRETION

The inhibition by chronic opiate treatment of sexual and reproductive functions in humans is an old clinical experience. Male opiate addicts have depressed serum testosterone level, and female addicts were repeatedly reported to have amenorrhea, infertility, and spontaneous abortion.[1,76] There is no indication that opiates would directly interact with the peripheral sexual organs but their serum LH decreasing action is well known[76,77] and also the castration induced rise in serum LH was antagonized by morphine.[1] A 3-day exposure of male rats to morphine resulted in significant atrophy of the prostate and the seminal vesicles, obviously due to a dramatic reduction in the serum testosterone level.[78] In females a single morphine injection given at the critical period resulted in inhibition of ovulation, an effect which could be reversed by naloxone.[79,80] Recently Pang et al.[81] showed that administering morphine during proestrus, the preovulatory surge of plasma LH and FSH did not appear, which explains the inhibition of normal ovulation. Moreover, the morphine-induced suppression of ovulation could be surmounted by LHRH.[81] Therefore, the attention has been focused on the effects of opioids on LHRH secretion from the hypothalamus.

Incubating the LHRH containing parts of the hypothalamus (the pallisadic zone of median eminence and the region of the infundibular sulcus), the dopamine-induced release was inhibited by morphine[82] and Met-E[57] as well. This effect of opioids was reversible by naloxone.[57,82] Opioids inhibited only the dopamine-induced release of LHRH, leaving the basal rate of secretion unaffected.[57,82] The dopamine-induced liberation could be inhibited also by pimozide and haloperidol, but not by naloxone.[57] Thus, the similar effects of dopamine antagonists and opioids are mediated by different receptors.

Accordingly, the question emerges whether the opiate receptors are located on the LHRH secreting cells themselves, which would mean that these neurones are directly inhibited by opioids, or the endogenous opioid peptides act via presynaptic inhibition of dopamine release. For the time being, in view of the scarcity of competent experimental data, no definite answer can be given to this question, unless supposing that the basic mechanism of action exerted by them is the same everywhere. If it is really true, then just as in the case of prolactin, an indirect effect might be proposed, i.e., a presynaptic inhibition of dopamine release would be responsible for the decreased LHRH secretion in the hypothalamus as suggested by the in vitro experiments.[57,82]

These in vitro experiments[57,82] can explain the repeatedly observed in vivo decrease in plasma LH level upon treatment with Met-E[1,13] or its potent analogues.[22] Interestingly, Takahara et al.[18] observed a significant increase in serum LH level upon β-EP administration, a finding, which is difficult to interpret just as the lack of any significant action on serum FSH level in the above-mentioned animal studies.[1,13] However, in a recent study of Kinoshita et al.[83] the i.c.v. administered β-EP was found to suppress the pulsatile discharge of LH into the plasma in castrated and intact male rats. The inconsistency of these data[18,83] concerning the effect of β-EP on serum LH level was attributed to the urethane anesthesia,[83] which is known to exert a profound effect on the hypothalamo—pituitary axis. If accepting the latter data[83] as relevant information on the physiological action of endogenous opioids, the suppression by them of LHRH release may probably be considered experimentally proven.

On the other hand, it is rather difficult to explain that in the above-mentioned studies[1,13] only the serum LH level was reduced, while the FSH concentration did not change. If the opioids exert their action exclusively via alteration of the LHRH secretion, the serum LH and FSH levels should change simultaneously. In humans both the FSH and LH levels were decreased upon treatment with FK 33-824.[22]

Finally the basic issue, is the LHRH production tonically inhibited by endogenous opioid peptides? Considering that naloxone treatment was found to increase the plasma LH[1,2,13,44,45,47,48,76,84-87] and FSH[1,2,44] levels, the answer might be yes. This LH releasing action of naloxone could be counteracted by 5-hydroxytryptophan, i.e., this effect was brought about by reducing the central serotoninergic tone just as in the case of prolactin.[85] The role of endogenous opioids in physiological regulation is further substantiated by the recent works of Blank et al.[48,88] In their experiments naloxone significantly increased the serum LH concentration in prepubertal female rats but did not affect it in adult females only during stress, while in males naloxone increased the hormone level after having reached the age of puberty.[88] Moreover, the naloxone-induced LH surge could be antagonized by prior injection of estradiol, but the similar effect on the prolactin level could not.[48,88] These data have been confirmed also by others.[85] Consequently, opioids might modulate the LH and prolactin secretion by partially different mechanisms. Moreover, due to the ability of estradiol to modify the action of naloxone,[88] the effect of the endogenous opioids might be dependent on the serum levels of gonadal steroids. However, it must be asked where are the receptors sensitive both to the opioids and peripheral gonadal hormones? Recently Cicero et al.[87] demonstrated that the depressive effect of testosterone on serum LH concentration could be competitively antagonized by naloxone. Hence the theoretical possibility arises according to the LHRH secreting neurones should have two inhibitory inputs, i.e., two different types of receptors, one of them activated by opioids while the other by peripheral gonadal hormones. Another possibility was that certain receptors are sensitive to both types of ligands. LaBella et al.[89] working with rat whole brain homogenate demonstrated that certain estrone derivatives were active in displacing ^{3}H-naloxone in opiate receptor binding assay. The displacement curves for active steroids were parallel to those of the opiate drugs,[89] which indicates true competitive antagonism and an affinity to the opiate receptors. Otherwise, allosteric inhibition also could explain this effect. On the other hand, in some recent works[84,90] neither naloxone nor morphine affected the dihydrotestosterone binding in hypothalamus and limbic brain homogenate. Thus, only a subfraction of opiate (or steroid) receptors might be sensitive both to opioids and gonadal hormones. If certain gonadal steroids bind to brain opiate receptors, it is easy to understand that why Hahn and Fishman[91] have found the saturable binding of ^{3}H-naltrexone significantly increased in the brain of castrated rats (comparing them to intact animals). Since the number of binding sites, as determined by Scatchard analysis, was found increased,[91] the lacking androgens might have had an agonist potency. The extrahypothalamic location of these receptors have been directly demonstrated by Maughan et al.,[92] who could block the ovulation in rats by microinjecting morphine into the dorsal and median raphe nuclei. Recently Parvizi and Ellendorf[93] reported that microinjecting β-EP into the amygdala resulted in a decrease in plasma LH level, while administering it into the mediobasal hypothalamus no such effect was seen. Thus, these receptor might be located among others also in the amygdala. In the experiments of Cicero et al.[87] naloxone did not compete with testosterone for the latter's cytosol binding proteins in the hypothalamus. Accordingly, they supposed that opioid-containing neurones mediated the inhibitory feedback effect of peripheral gonadal hormones on the LHRH secreting hypothalamic cells. Let us mention here that in aging male rats naloxone failed to facilitate the LH release.[86] Since in these animals the Met-E content of the hypothalamus was found concomitantly elevated, these authors[86] supposed that the age-related decrease in gonadal activity was due to enhanced inhibition of LHRH release by the endogenous opioid peptides.

It is probably reasonable to review the data concerning the role of endogenous opioid peptides in *sexual behavior* here. The inhibition by morphine of normal male sexual behavior has been unambiguously confirmed by several groups.[94] The same effect has been observed by Meyerson and Terenius[95] in rats upon giving 1 μg β-EP i.c.v. (This dose, corresponding about to 0.3 nanomoles, is smaller than the minimal analgesic ED_{50}, as seen in Chapter 3, Volume II.) As observed by the authors,[95] these animals displayed normal precopulatory behavior, i.e., they actively pursued the hormone treated receptive females and made contact with them, but the subsequent phases of the intercourse were more or less omitted, just as observed in morphine-treated males.[94] The authors themselves stressed that this action of β-EP appeared in such a low dose which was not sufficient to elicit catalepsy, rigidity, or other alterations in the normal behavior.[95] For this reason the inhibition of copulation cannot be attributed to unspecific general motor depressant effect. The specific inhibition of male sexual behavior was observed also upon i.c.v. application of DALA.[96]

Opioids inhibit the sexual behavior in females too. As morphine suppresses the lordosis induced by estrogen-progesterone treatment, and this inhibition can be reversed by naltrexone, also the potent enkephalin analogue FK 33-824 dose-dependently inhibited the LHRH-induced lordosis in estrogen + progesterone primed females.[97]

The sexual behavior due to its complex nature is obviously organized by to date unspecified higher brain structures. As even the most cranial areas of the brain contain steroid and opiate receptors as well, the presumable suppressant effect of endogenous opioids might be mediated either directly by opiate receptors or indirectly by lowering the plasma level of peripheral gonadal hormones. Considering, however, the prompt inhibition of sexual behavior upon treatment with opioids, the first mechanism seems to be more probable, at least in acute experiments.

Thus, the data available strongly suggest that endogenous opioid mechanisms inhibit the normal sexual behavior. Nevertheless, if they play such an important role in the sexual functions, opiate antagonists should facilitate them. Actually naloxone has been reported to increase the frequency of mounting, intromission, and ejaculation in sexually inactive rats[96] and to shorten the ejaculation latency in adult experienced rats.[98,99] The latter finding indicates that endorphins are primarily involved in regulation (physiological inhibition) of ejaculation. Unfortunately all these experiments were carried out in rats and no animal data are available concerning other species. Nevertheless, if these conclusions are valid for humans, opiate antagonists might be therapeutically useful drugs for treatment of certain kinds of impotence.

IV. EFFECTS ON ACTH AND TSH SECRETION

Release of ACTH from the pituitary upon acute opiate treatment is a well-known phenomenon and the consecutive elevation of serum corticoid level as well. Upon chronic treatment rather a lowering of plasma ACTH and corticosterone concentration has been found.[47,100,101] In humans also ACTH and adrenal cortical hormone release can be seen upon acute opiate administration,[2] but in opiate addicts the similar action of methadone was found attenuated.[102]

The consequences of both acute and chronic opiate treatments are generally attributed to their stressor effects. Eisenberg and Sparber[101] have shown that naloxone reversed the corticosterone releasing action of levorphanol if giving them together, while it elicited the same response if giving it alone after dissipation of the levorphanol-induced action. However, administering naloxone to drug naive animals, no increment was seen.[101] Thus, it seems that the naloxone-elicited increase in plasma corticosteroid level might be rather due to the reversal of opiate-induced actions than to direct effects elicited by the narcotic antagonist. Obviously, also the spontaneous dissipation of opiate-induced euphoria might be stressful and the repeated opiate administration might result in exhaustion of the pituitary-adrenal system. However, in the subsequent work of the same author[103] naloxone was found to induce a dose dependent increase in the plasma corticosterone level both in drug naive animals, and also in those repeatedly pretreated with this pure opiate antagonist. Moreover, naloxone increases the serum cortisol concentrations in normal healthy humans.[44] Probably the displacement of endogenous opioids is stressful for the organism, but there are some alternative possibilities too.

Gibson et al.[104] reported that the serum corticosterone level was elevated upon treatment with naloxone, morphine, or stressing the animals with ether.[104] The ether stress-induced elevation of corticosterone level was potentiated by normorphine but inhibited by naloxone and other narcotic antagonists.[104] They supposed the existence of an ACTH release inhibitory mechanism which was tonically suppressed by the endogenous opioids. Consequently, opiate antagonists should suppress the ether stress-induced ACTH mobilization via liberating the endogenous ACTH inhibitory system from the control of endogenous opioids.

To date the opioid peptides themselves have hardly been examined. An increase in plasma corticosterone level was observed upon injecting β-EP into different areas of the limbic system and hypothalamus in rats.[39] Administering the potent enkephalin analogue D-Met,Pro-EA to rats i.v. in a moderate dose (0.4 mg/kg), the serum ACTH level increased within 3 to 8 min and this elevation could be antagonized by simultaneously given naloxone.[105] Unfortunately Reina et al.[105] did not examine the action of naloxone in itself, but the rapid onset of ACTH mobilization excludes the possibility that it would have been due simply to the dissipation of the euphoric effect. In humans the other potent enkephalin analogue FK 33-824 was reported to reduce the serum levels of ACTH and cortisol,[22] while naloxone increased their levels in another study[47] also in healthy male volunteers[47] Leu-E decreased the cortisol level in humans.[106] In these studies, both Leu-E[106] and its analogue[22] were given in rather low i.v. doses (0.25 mg) which surely did not elicit significant central psychic effects but elevated the blood concentration of prolactin and GH.[22] Interestingly, Marton et al.[107] found the ACTH level increased or decreased upon i.c.v. administration of D-Met,Pro-EA in awake and anesthetized rats, respectively. Thus, the action of opioids also depends on the activity level.

The above-mentioned human[22,106] studies indicate that peripherally given opioid peptides, which hardly reach the receptors behind the blood-brain barrier, suppress the pituitary-adrenal system just as suggested by Gibson et al.[104] Thus, the opposite effect of peripherally given naloxone[44,47,103] is easy to interpret. Probably this physiological inhibitory effect is obscured if opioids are given in much higher doses sufficient to elicit euphoria and its dissipation exerts a stressor action.

A further mechanism of action might be elicited via alteration of vasopressin release, which also is a physiological regulator of ACTH mobilization (see later). A fourth possibility is a direct action of opioid peptides on the steroid synthesis in the adrenals. Of course, in this relation the contingent effect of β-EP (LPH) might be of interest, since they are present in the plasma (Chapter 3, Volume III). β-EP increases the corticosterone synthesis in isolated cells of zona fasciculata, but this action appears in a concentration three orders of magnitude higher than the similar effect of ACTH and actions of the two are not additive.[63] Accordingly, it is questionable if this weak action of β-EP has any physiological significance. If it does, the role of β-EP might be rather an allosteric modulation of ACTH-induced steroidogenesis. All the more as the $ACTH_{1-24}$-induced steroid production was uninfluenced by Met-E but somewhat inhibited by naloxone and naltrexone.[64]

Recently Matsuoka et al.[68] observed a dose-dependent facilitation of aldosterone synthesis in rat adrenal capsular cell culture, i.e., in glomerulosa cells upon LPH treatment. The same treatment did not modify the corticosterone production in fasciculata, reticularis, and medullary cells.[68] Thus, the possibility arises that LPH were the natural and selective aldosterone-stimulating factor of the pituitary. As the intact LPH molecule does not bind to the opiate receptors (Chapter 5, Volume III), this hormonal effect should be a nonopiate-like action.

Morphine decreases the serum TSH level and this effect can be reversed by naloxone.[1,2] The similar effect of peripherally given Met-E also has been demonstrated.[1,13] In a further study[39] intracerebrally given β-EP had no effect on the serum TSH level and Met-E-methyl esther did not modify the serum thyroxin concentration.[19] In humans FK 33-824 increased the TSH level.[22] Considering that dopamine inhibits, while adrenalin facilitates the TSH liberation,[2] the action of endogenous opioids on the thyroid function might be rather complex.

V. EFFECTS ON VASOPRESSIN AND OXYTOCIN SECRETION

The antidiuretic effect of opiates has been repeatedly demonstrated.[54,108-110] This action was generally related to an increase in vasopressin release. Such a hypothesis was only further substantiated by the recent findings on the high enkephalin and opiate receptor contents of the neurohypophysis (Chapters 1 and 3, Volume III). A more thorough survey of data, however, reveals a different pattern of relationships.

Opiates, given whether peripherally or centrally, induced dose-dependent oliguria with concomitant reduction in urinary Na^+, K^+, and Cl^- excretion in the experiments of Huidobro and Huidobro-Toro.[111] The arginine vasopressin-induced oliguria was, however, accompanied by increased secretion of monovalent ions.[111] Consequently, if the opiates acted via vasopressin liberation, they ought to have induced saline diuresis, similar to that observed upon direct application of this neurohypophyseal hormone. Since in this study[111] the endogenous creatinine clearance was decreased upon morphine treatment, primarily the glomerular filtration might have been affected. This action could, however, be the consequence of reduced renal blood flow and decreased blood pressure, i.e., circulatory changes characteristic of opioids (Chapter 4, Volume II). Consequently, the antidiuretic effect does not necessarily mean an enhancement of the vasopressin liberation.

In vivo Weitzman and his co-workers[54,110] observed an increase in the serum vasopressin level upon i.v. injection of small amounts of β-EP. This effect could not be blocked by naloxone, just as the similar action of morphine. Thus, the effect could not be mediated by opiate receptors. More difficult to explain are the data of Bisset et al.,[108] who found a marked antidiuretic action upon i.v. and i.c.v. treatments with β-EP, Leu-E, (D-Ala^2,D-Leu^5)-E in water-loaded rats, which was accompanied with increased vasopressin excretion. Since the plasma vasopressin concentration has not been measured in this study,[108] its increased renal excretion cannot be excluded as an alternative explanation. Another group,[112] however, observed a naltrexone-reversible decrease in the plasma vasopressin level. In a recent work[46] an increase in free water clearance, decrease in urinary vasopressin concentration, and prolactin release has been observed in humans upon i.v. injection of FK 33-824. Of these actions only the prolactin mobilization could be reversed by naloxone. Consequently, different receptors mediated the enkephalin analogue induced diuretic effect and prolactin release. Also Grossman et al.[113] recorded a distinct diuretic action (increase in urine flow and free water clearance) in water-deprived humans upon i.v. infusion of FK 33-824, the action being naloxone resistant. Giving naloxone alone, the vasopressin production was found inhibited also in humans.[114] In animal experiments naloxone promoted the isoprenaline-elicited vasopressin release.[115]

Working in vitro with isolated rat neural lobes, β-EP did not facilitate the vasopressin release even in the heroic concentration of 9×10^{-6} moles.[54] But in a recent study of Iversen et al.[66] the Ca^{++}-dependent liberation of vasopressin from nerve terminals of the neurohypophysis evoked by electrical stimulation of the rat pituitary stalk could be suppressed by morphine, β-EP, and (D-Ala^2,Leu^5)-E as well. The inhibitory action of the latter was reversed by naloxone.[66] Thus, in the more specific studies opioids proved to suppress the vasopressin release. However, contrary to this in vitro study,[66] the inhibition by FK 33-824 of vasopressin release could not be reversed by naloxone[46,113] just as the alleged vasopressin releasing action of β-EP.[110] It is tempting to suppose that the opiate receptors mediating the suppression of vasopressin liberation are of δ-type, therefore more naloxone is needed to antagonize this action than those elicited by μ-receptors. However, in the above-mentioned human studies[46,113] fairly high doses (4 to 8 mg) have been applied. Thus, the question requires further clarification.

Rossier et al.[116] reported that the Leu-E containing axons in the neurohypophysis (belonging to neurones in the supraoptic and paraventricular nuclei of hypothalamus) lost their opioid content 5 days after substitution with 2% saline of their drinking fluid. In these animals also the vasopressin stores were depleted.[116] It is tempting to speculate that this hypothalamo-pituitary pathway inhibits the vasopressin release. Upon lasting dehydration not only the vasopressin stores lose their content, but also the enkephalinerg fibers, which tonically counteract this effect.

Finally it should be mentioned that in rats with hereditary diabetes insipidus (Brattleboro strain), i.e., in animals unable to synthesize and release vasopressin, the opioid content of posterior pituitary is decreased.[109] In these animals the opiate receptor concentration is decreased in the neurohypophysis.[117] In the light of these data, the seemingly curious findings might be explained according to the drinking behavior induced by water deprivation or angiotensin injection can be inhibited by natural enkephalins and (D-Ala^2,D-Leu^5)-E.[118] As vasopressin might mediate the dipsogenic effect of water deprivation and angiotensin as well, these observations are in good agreement with the postulated vasopressin release inhibitory effect of enkephalins.

As for oxytocin, the only available in vitro study is in contradiction with the in vivo data. In vitro β-EP was found to facilitate the spontaneous oxytocin release from the isolated neurohypophysis.[58] However, inhibition by opiate agonists of oxytocin release is well established in vivo.[119] The contradiction of these[58,119] data are difficult to explain. Nevertheless, it is probably justified to assume that only those in vitro experiments might be relevant where enkephalins are examined and not endorphins, which are virtually absent from the neurohypophysis, and instead of spontaneous release that induced by natural stimuli, or by addition of Ca^{++} and K^+ is interesting from physiological point of view.

In a recent work[120] the in vivo oxytocin liberation induced by suckling, carbachol, or electrical stimulation of the neurohypophysis was inhibited by β-EP and (D-Ala^2,Met^5)-E. Naloxone not only reversed these actions but also facilitated the liberation if given alone. Since the same substances did not modify the spontaneous activity of oxytocin producing neurones, opioids might act also in this case presynaptically and the oxytocin liberation is inhibited by endogenous opioids tonically.

VI. OTHER HORMONAL EFFECTS

Though the endocrine effects of opioid peptides are primarily mediated by modulation of the hypothalamic and hypophyseal hormones, some direct peripheral actions must be considered.

Working with isolated perfused dog pancreas, the somatostatin release was found inhibited[70] just as in the case of mediobasal hypothalamic slices.[65] Thus, the action seems to be similar whether the hormone affected is located centrally or peripherally.

As for glucagon and insulin, their production in islet cell cultures was inhibited[62,67] by Met-E, Leu-E, and some synthetic analogues, e.g., (D-Ala^2,Leu^5)-E, the effect being naloxone sensitive. In the same study[67] morphine had opposite action and β-EP was reported to facilitate the liberation of insulin and glucagon in isolated perfused dog pancreas.[70]

The corresponding in vivo study of Ferri et al.[121] seems to confirm the former works,[62,67] namely in their work D-Met,Pro-EA reduced the insulin liberation and increased the hyperglycemia induced by glucose loading in rats.

Summarizing the data so far available on the effects of endogenous opioid peptides, some conclusions might be drawn.

1. The main site of action of endogenous opioid peptides is the hypothalamus including its distal extension, i.e., the posterior hypophysis.
2. Endogenous opioid peptides presynaptically inhibit the liberation of certain releasing and pituitary hormones such as somatostatin, LHRH, vasopressin, and oxytocin. The liberation of some others is enhanced, but this enhancement is due to the suppression of dopamine production (prolactin, GH).
3. Their inhibitory action is always presynaptically mediated.
4. Opioids also inhibit the production of some peripheral hormones such as pancreatic somatostatin, glucagon, and insulin.
5. In some cases their inhibitory action seems to be tonic (prolactin, oxytocin) in other cases mostly phasic.
6. In some cases the production of the same hormone is simultaneously facilitated and inhibited by different mechanisms all activated by opioids (ACTH, GH).

REFERENCES

1. **Meites, J., Bruni, J. F., van Vugt, D. A., and Smith, A. F.,** Relation of endogenous opioid peptides and morphine to neuroendocrine functions, *Life Sci.,* 24, 1325, 1979.
2. **Van Vugt, D. A. and Meites, J.,** Influence of endogenous opiates on anterior pituitary function, *Fed. Proc. Fed. Am. Soc. Exp. Biol.,* 39, 2533, 1980.
3. **Lien, E. L., Morrison, A., and Dvonch, Wm.,** The effects of partial opiate agonists on plasma prolactin and growth hormone levels in the rat, *Life Sci.,* 25, 1709, 1979.
4. **Lien, E. L., Fenichel, R. L., Garsky, V., Sarantakis, D., and Grant, N. H.,** Enkephalin-stimulated prolactin release, *Life Sci.,* 19, 837, 1976.
5. **Rivier, C., Vale, W., Ling, N., Brown, M., and Guillemin, R.,** Stimulation in vivo of the secretion of prolactin and growth hormone by β-endorphin, *Endocrinology,* 100, 238, 1977.
6. **Dupont, A., Cusan, L., Labrie, F., Coy, D. H., and Li, C. H.,** Stimulation of prolactin release in the rat by intraventricular injection of β-endorphin and methionine-enkephalin, *Biochem. Biophys. Res. Commun.,* 75, 76, 1977.
7. **Cocchi, D., Santagostino, A., Gil-Ad, I., Ferri, S., and Müller, E. E.,** Leu-enkephalin stimulated growth hormone and prolactin release in the rat: comparison with the effect of morphine, *Life Sci.,* 20, 2041, 1977.
8. **Shaar, C. J., Frederickson, R. C. A., Dininger, N. B., and Jackson, L.,** Enkephalin analogues and naloxone modulate the release of growth hormone and prolactin — evidence for regulation by an endogenous opioid peptide in brain, *Life Sci.,* 21, 853, 1977.
9. **Dupont, A., Cusan, L., Garon, M., Labrie, F., and Li, C. H.,** β-endorphin: stimulation of growth hormone release in vivo, *Proc. Natl. Acad. Sci. USA,* 74, 358, 1977.
10. **Grandison, L. and Guidotti, A.,** Regulation of prolactin release by endogenous opiates, *Nature (London),* 270, 357, 1977.
11. **Cusan, L., Dupont, A., Kledzik, G. S., Labrie, F., Coy, D. H., and Schally, A. V.,** Potent prolactin and growth hormone releasing activity of more analogues of Met-enkephalin, *Nature (London),* 268, 544, 1977.
12. **Van Vugt, D. A., Bruni, J. F., and Meites, J.,** Naloxone inhibition of stress-induced increase in prolactin secretion, *Life Sci.,* 22, 85, 1978.
13. **Bruni, J. F., van Vugt, D., Marshall, S., and Meites, J.,** Effects of naloxone, morphine and methionine enkephalin on serum prolactin, luteinizing hormone, follicle stimulating hormone, thyroid stimulating hormone and growth hormone, *Life Sci.,* 21, 461, 1977.
14. **Brown, B., Dettmar, P. W., Dobson, P. R., Lynn, A. G., Metcalf, G., and Morgan, B. A.,** Opiate analgesics: the effect of agonist-antagonist character on prolactin secretion, *J. Pharm. Pharmacol.,* 30, 644, 1978.
15. **Lien, E. L., Clark, D. E., and McGregor, W. H.,** Stimulation of growth hormone and prolactin release by a potent enkephalin analog, *FEBS Lett.,* 88, 208, 1978.
16. **Meltzer, H. Y., Miller, R. J., Fessler, R. G., Simonovic, M., and Fang, V. S.,** Effects of enkephalin analogues on prolactin release in the rat, *Life Sci.,* 22, 1931, 1978.

17. **Kato, Y., Iwasaki, Y., Abe, H., Ohgo, S., and Imura, H.**, Effects of endorphins on prolactin and growth hormone secretion in rats, *Proc. Soc. Exp. Biol. Med.*, 158, 431, 1978.
18. **Takahara, J., Kageyama, J., Yunoki, S., Yakushiji, W., Yamauchi, J., Kageyama, N., and Ofuji, T.**, Effects of 2-bromo-α-ergocryptine on β-endorphin-induced growth hormone, prolactin and luteinizing hormone release in urethane anesthetized rats, *Life Sci.*, 22, 2205, 1978.
19. **Hart, I. C. and Cowie, A. T.**, Effect of morphine, naloxone and an enkephalin analogue on plasma prolactin, growth hormone, insulin and thyroxine in goats, *J. Endocrinol.*, 77, 16P, 1978.
20. **Chihara, K., Arimura, A., Coy, D. H., and Schally, A. V.**, Studies on the interaction of endorphins, substance P, and endogenous somatostatin on growth hormone and prolactin release in rats, *Endocrinology*, 102, 281, 1978.
21. **Graffenried, von, B., del Pozo, E., Roubicek, J., Krebs, E., Poldinger, W., Burmeister, P., and Kerp, L.**, Effects of the synthetic enkephalin analogue FK 33-824 in man, *Nature (London)*, 272, 729, 1978.
22. **Stubbs, W. A., Jones, A., Edwards, C. R. W., Delitala, G., Jeffcoate, W. J., Ratter, S. J., Besser, G. M., Bloom, S. R., and Alberti, K. G. M. M.**, Hormonal and metabolic responses to an enkephalin analogue in normal man, *Lancet*, 2, 1225, 1978.
23. **Gold, M. S., Redmond, D. E., Jr., Donabedian, R. K., Goodwin, F. K., and Extein, I.**, Increase in serum prolactin by exogenous and endogenous opiates: evidence for antidopamine and antipsychotic effects, *Am. J. Psychiatry*, 135, 1415, 1978.
24. **Muraki, T. and Tokunaga, Y.**, Effects of morphine on the serum prolactin levels of morphine-tolerant and nontolerant male rats and on the in vitro release of pituitary prolactin, *Jpn. J. Pharmacol.*, 28, 803, 1978.
25. **Janowsky, D., Judd, L., Huey, L., Roitman, N., Parker, D., and Segal, D.**, Naloxone effects on manic symptoms and growth-hormone levels, *Lancet*, 2, 320, 1978.
26. **Locatelli, V., Cocchi, D., Bajusz, S., Spampinato, S., and Müller, E. E.**, Prolactin and growth hormone releasing activity of (D-Met^2,Pro^5)-enkephalinamide in the rat after systemic administration, *Experientia*, 34, 1560, 1978.
27. **Hermann, I., Borvendég, J., Bajusz S., Rónai, A. Z., and Székely, J. I.**, Effect of synthetic enkephalin analogues on prolactin and LH release in rats, in *Proc. XVIII Int. Cong. Physiol.*, Budapest, Hungary, 1980, Abstracts, 470.
28. **Van Vugt, D. A., Bruni, J. F., Sylvester, P. W., Chen, H. T., Ieiri, T., and Meites, J.**, Interaction between opiates and hypothalamic dopamine on prolactin release, *Life Sci.*, 24, 2361, 1979.
29. **Spampinato, S., Locatelli, V., Cocchi, D., Vicentini, L., Bajusz, S., Ferri, S., and Müller, E. E.**, Involvement of brain serotonin in the prolactin-releasing effect of opioid peptides, *Endocrinology*, 105, 163, 1979.
30. **Rubin, P., Swezey, S., and Blaschke, T.**, Naloxone lowers plasma-prolactin in man, *Lancet*, 1, 1293, 1979.
31. **Gold, M. S., Redmond, D. E., Jr., and Donabedian, R. K.**, Prolactin secretion, a measurable central effect of opiate-receptor antagonists, *Lancet*, 1, 323, 1978.
32. **Gold, M. S., Redmond, D. E., Jr., and Donabedian, R. K.**, The effects of opiate agonist and antagonist on serum prolactin in primates: possible role for endorphins in prolactin regulation, *Endocrinology*, 105, 284, 1979.
33. **Koenig, J. I., Mayfield, M. A., McCann, S. M., and Krulich, L.**, Stimulation of prolactin secretion by morphine: role of the central serotonergic system, *Life Sci.*, 25, 853, 1979.
34. **Martin, J. B., Tolis, G., Woods, I., and Guyda, H.**, Failure of naloxone to influence physiological growth hormone and prolactin secretion, *Brain Res.*, 168, 210, 1979.
35. **Muraki, T., Tokunaga, Y., Nakadate, T., and Kato, R.**, Inhibition by cholinergic agonists of the prolactin release induced by morphine, *Naunyn-Schmiedeb. Arch. Pharmacol.*, 308, 249, 1979.
36. **Tache, Y., Brown, M., and Collu, R.**, Effects of neuropeptides on adenohypophyseal hormone response to acute stress in male rats, *Endocrinology*, 105, 220, 1979.
37. **Foley, K. M., Kourides, I. A., Inturrisi, C. E., Kaiko, R. F., Zaroulis, C. G., Posner, J. B., Houde, R. W., and Li, C. H.**, β-endorphin: analgesic and hormonal effects in humans, *Proc. Natl. Acad. Sci. USA*, 76, 5377, 1979.
38. **Gold, M. S., Pottash, A. L. C., Extein, I., Redmond, D. E., Jr., and Kleber, H. D.**, Decrease in serum prolactin by naloxone: evidence against antidopamine and antipsychotic effects, *Am. J. Psychiatry*, 136, 1339, 1979.
39. **Murray, G. M.**, The influence of morphine and other opioids on the secretion of anterior pituitary hormones: the CNS sites of action and the possible receptors mediating the hormonal responses, *Diss. Abstr. Intern.*, 39, 4300, 1979.
40. **Shaffer, Tannenbaum, G., Panerai, A. E., and Friesen, H. G.**, Failure of β-endorphin antiserum, naloxone, and naltrexone to alter physiologic growth hormone and insulin secretion, *Life Sci.*, 25, 1983, 1979.

41. **Casanueva, F., Betti, R., Vellani, D., Cocchi, D., and Müller, E. E.,** Enkephalin-induced growth hormone release in the dog, *Pharm. Res. Commun.,* 12, 461, 1980.
42. **Casanueva, F., Betti, R., Frigerio, C., Cocchi, D., Mantegazza, P., and Müller, E. E.,** Growth hormone-releasing effect of an enkephalin analog in the dog: evidence for cholinergic mediation, *Endocrinology,* 106, 1239, 1980.
43. **Rossier, J., French, E., Rivier, C., Shibasaki, T., Guillemin, R., and Bloom, F. E.,** Stress-induced release of prolactin: blockade by dexamethasone and naloxone may indicate β-endorphin mediation, *Proc. Natl. Acad. Sci. USA,* 77, 666, 1980.
44. **Morley, J. E.,** Endocrine effects of naloxone-induced opiate receptor blockade, *J. Clin. Endocrinol. Metab.,* 50, 251, 1980.
45. **Evans, G. and Groom, G. V.,** The interrelationship between hormones, psychotropic drugs and the opiate receptor — a pilot study, *Postgraduate Med. J.,* 56, 65, 1980.
46. **Brownell, J., del Pozo, E., and Donatsch, P.,** Inhibition of vasopressin secretion by a met-enkephalin (FK 33-824) in humans, *Acta Endocrinol.,* 94, 304, 1980.
47. **Volavka, J., Bauman, J., Pevnick, J., Reker, D., James, B., and Cho, D.,** Short-term hormonal effects of naloxone in man, *Psychoneuroendocrinology,* 5, 225, 1980.
48. **Blank, M. S., Panerai, A. E., and Friesen, H. G.,** Effects of naloxone on luteinizing hormone and prolactin in serum of rats, *J. Endocrinol.,* 85, 307, 1980.
49. **Van Loon, R., de Souza, E. B., and Shin, S. H.,** Dopaminergic mediation of β-endorphin-induced prolactin secretion, *Clin. Res.,* 27, 261A, 1979.
50. **Stubbs, G. M.,** The metabolic and hormonal effects of an enkephalin analogue in normal, GH deficient and acromegalic man, *Diabetologia,* 15, 273, 1978.
51. **Höllt, V. and Teschemacher, Hj.,** Hydrophobic interactions responsible for unspecific binding of morphine-like drugs, *Naunyn-Schmiedeb. Arch. Pharmacol.,* 288, 163, 1975.
52. **Rossier, J., French, E. D., Rivier, C., Ling, N., Guillemin, R., and Bloom, F. E.,** Foot-shock induced stress increases β-endorphin levels in blood but not brain, *Nature (London),* 270, 618, 1977.
53. **Tseng, L.-F., Loh, H. H., and Li, C. H.,** β-endorphin as a potent analgesic by intravenous injection, *Nature (London),* 263, 239, 1976.
54. **Weitzman, R. E., Fischer, D. A., Minick, S., Ling, N., and Guillemin, R.,** β-endorphin stimulates secretion of arginine vasopressin in vivo, *Endocrinology,* 101, 1643, 1977.
55. **Enjalbert, A., Ruberg, M., Arancibia, S., Priam, M., and Kordon, C.,** Endogenous opiates block dopamine inhibition of prolactin secretion in vitro, *Nature (London),* 280, 595, 1979.
56. **Hollander, C. S., Hirooka, Y., Prasad, J., Richardson, S., Ferinand, P., and Thaw, C.,** Human beta endorphin modulates hormonal release from organ cultures of rat hypothalamus, *Clin. Res.,* 26, 491A, 1978.
57. **Rotsztejn, W. H., Drouva, S. V., Pattou, E., and Kordon, C.,** Met-enkephalin inhibits in vitro dopamine-induced LHRH release from mediobasal hypothalamus of male rats, *Nature (London),* 274, 281, 1978.
58. **Vizi, E. S. and Volbekas, V.,** Modulatory role of dopamine and β-endorphin in the release of oxytocin from neural lobe of pituitary, in *Proc. 7th Int. Cong. Pharm.,* Paris, 1978, Abstracts, 562.
59. **Enjalbert, A., Ruberg, M., Fiore, L., Arancibia, S., Priam, M., and Kordon, C.,** Effect of morphine on the dopamine inhibition of pituitary prolactin release in vitro, *Eur. J. Pharmacol.,* 53, 211, 1979.
60. **Login, I. S. and MacLeod, R. M.,** Failure of opiates to reverse dopamine inhibition of prolactin secretion in vitro, *Eur. J. Pharmacol.,* 60, 253, 1979.
61. **Melmed, S., Morley, J. E., Kurtzman, G., Carlson, H. E., Briggs, J. E., and Hershman, J. M.,** Opiates and dopamine enhance prolactin release from cultured pituitary tumor cells, *Clin. Res.,* 27, 257A, 1979.
62. **Kanter, R. A., Fujimoto, W. Y., and Ensinck, J. W.,** Opiates modulate insulin and glucagon secretion, *Diabetes,* 28, 395S, 1979.
63. **Shanker, G. and Sharma, R. K.,** β-endorphin stimulates corticosterone synthesis in isolated rat adrenal cells, *Biochem. Biophys. Res. Commun.,* 86, 1, 1979.
64. **Gibson, A., Ginsburg, M., Hall, M., and Hart, S. L.,** The effects of opioid drugs and of lithium on steroidogenesis in rat adrenal cell suspensions, *Br. J. Pharmacol.,* 65, 671, 1979.
65. **Drouva, S. V., Epelbaum, J., Tapia-Arancibia, L., Laplante, E., and Kordon, C.,** Met-enkephalin inhibition of K^+-induced LHRH and SRIF release from rat mediobasal hypothalamic slices, *Eur. J. Pharmacol.,* 61, 411, 1980.
66. **Iversen, L. L., Iversen, S. D., and Bloom, F. E.,** Opiate receptors influence vasopressin release from nerve terminals in rat neurohypophysis, *Nature (London),* 284, 350, 1980.
67. **Kanter, R. A., Ensinck, J. W., and Fujimoto, W. Y.,** Disparate effects of enkephalin and morphine upon insulin and glucagon secretion by islet cell cultures, *Diabetes,* 29, 84, 1980.
68. **Matsuoka, H., Mulrow, P. J., and Li, C. H.,** β-lipotropin: a new aldosterone-stimulating factor, *Science,* 209, 307, 1980.

69. **Gudelsky, G. A. and Porter, J. C.**, Morphine- and opioid peptide-induced inhibition of the release of dopamine from tuberoinfundibular neurons, *Life Sci.*, 25, 1697, 1979.
70. **Ipp, E., Dobbs, R., and Unger, R. H.**, Morphine and β-endorphin influence the secretion of the endocrine pancreas, *Nature (London)*, 276, 190, 1978.
71. **Ferland, L., Fuxe, K., Eneroth, P., Gustafsson, J.-A., and Skett, P.**, Effects of methionine-enkephalin on prolactin release and catecholamine levels and turnover in the median eminence, *Eur. J. Pharmacol.*, 43, 89, 1977.
72. **Muraki, T., Tokunaga, Y., Matsumoto, S., Yasuyama, M., and Hosoya, E.**, Effect of morphine on serum prolactin levels in male rats, *Jpn. J. Pharmacol.*, 27, 51P, 1977.
73. **Okajima, T., Motomatsu, T., Kato, K., and Ibayashi, H.**, The stimulatory effect of beta-endorphin on the plasma prolactin levels was diminished in the rats treated with 6-hydroxydopamine, *Life Sci.*, 26, 699, 1980.
74. **Lal, H., Brown, W., Drawbaugh, R., Hynes, M., and Brown, M.**, Enhanced prolactin inhibition following chronic treatment with haloperidol and morphine, *Life Sci.*, 20, 101, 1977.
75. **Collu, R., Taché, Y., and Charpenet, G.**, Role of biogenic amines in opioid peptide-induced secretion of pituitary hormones, *Pharm. Res. Commun.*, 12, 403, 1980.
76. **Mendelson, J. H., Ellingboe, J., Kuehnle, J. C., and Mello, N. K.**, Heroin and naltrexone effects on pituitary-gonadal hormones in man: interaction of steroid feedback effects, tolerance and supersensitivity, *J. Pharmacol. Exp. Ther.*, 214, 503, 1980.
77. **Cicero, T. J., Wilcox, C. E., Bell, R. D., and Meyer, E. R.**, Acute reductions in serum testosterone levels by narcotics in the male rat: stereospecificity, blockade by naloxone and tolerance, *J. Pharmacol. Exp. Ther.*, 198, 340, 1976.
78. **Cicero, T. J., Meyer, E. R., Wiest, W. G., Olney, J. W., and Bell, R. D.**, Effects of chronic morphine administration on the reproductive system of the male rat, *J. Pharmacol. Exp. Ther.*, 192, 542, 1975.
79. **Packman, P. M. and Rothchild, J. A.**, Morphine inhibition of ovulation: reversal by naloxone, *Endocrinology*, 99, 7, 1976.
80. **Barraclough, C. A. and Sawyer, C. H.**, Inhibition of the release of pituitary ovulatory hormone in the rat by morphine, *Endocrinology*, 57, 329, 1955.
81. **Pang, C. N., Zimmermann, E., and Sawyer, C. H.**, Morphine inhibition of the preovulatory surges of plasma luteinizing hormone and follicle stimulating hormone in the rat, *Endocrinology*, 101, 1726, 1977.
82. **Rotsztejn, W. H., Drouva, S. V., Pattou, E., and Kordon, C.**, Effect of morphine on the basal and the dopamine-induced release of LHRH from mediobasal hypothalamic fragments in vitro, *Eur. J. Pharmacol.*, 50, 285, 1978.
83. **Kinoshita, F., Nakai, Y., Katakami, H., Kato, Y., Yajima, H., and Imura, H.**, Effect of β-endorphin on pulsatile luteinizing hormone release in conscious castrated rats, *Life Sci.*, 27, 843, 1980.
84. **Cicero, T. J., Wilcox, C. E., Bell, R. D., and Meyer, E. R.**, Naloxone-induced increases in serum leuteinizing hormone in the male: mechanisms of action, *J. Pharmacol. Exp. Ther.*, 212, 573, 1980.
85. **Ieiri, T., Chen, H. T., and Meites, J.**, Naloxone stimulation of luteinizing hormone release in prepubertal female rats; role of serotoninergic system, *Life Sci.*, 26, 1269, 1980.
86. **Steger, R. W., Sonntag, W. E., van Vugt, D. A., Forman, L. J., and Meites, J.**, Reduced ability of naloxone to stimulate LH and testosterone release in aging male rats; possible relation to increase in hypothalamic Met^5-enkephalin, *Life Sci.*, 27, 747, 1980.
87. **Cicero, T. J., Meyer, E. R., Gabriel, S. M., Bell, R. D., and Wilcox, C. E.**, Morphine exerts testosterone-like effects in the hypothalamus of the castrated male rat, *Brain Res.*, 202, 151, 1980.
88. **Blank, M. S., Panerai, A. E., and Friesen, H. G.**, Opioid peptides modulate luteinizing hormone secretion during sexual maturation, *Science*, 203, 1129, 1979.
89. **LaBella, F. S., Song Kim, R., and Templeton, J.**, Opiate receptor binding activity of 17-alpha estrogenic steroids, *Life Sci.*, 23, 1797, 1978.
90. **Sheridan, P. J. and Buchanan, J. M.**, The effects of opiates on androgen binding in the forebrain of the rat, *Intern. J. Fert.*, 25, 36, 1980.
91. **Hahn, E. F. and Fishman, J.**, Changes in rat brain opiate receptor content upon castration and testosterone replacement, *Biochem. Biophys. Res. Commun.*, 90, 819, 1979.
92. **Maughan, G., Johnson, J. H., and Rosecrans, J. A.**, Effects of microinjection of morphine into the midbrain raphes on ovulation in rats, *Pharmacologist*, 19, 156, 1977.
93. **Parvizi, N. and Ellendorff, F.**, β-endorphin alters luteinizing hormone secretion via amygdala but not the hypothalamus, *Nature (London)*, 286, 812, 1980.
94. **Mumford, L. and Kumar, R.**, Sexual behavior of morphine-dependent and abstinent male rats, *Psychopharmacology*, 65, 179, 1979.
95. **Meyerson, B. J. and Terenius, L.**, β-endorphin and male sexual behavior, *Eur. J. Pharmacol.*, 42, 191, 1977.

96. **Gessa, G. L., Paglietti, E., and Quarantotti, B. P.**, Induction of copulatory behavior in sexually inactive rats by naloxone, *Science*, 204, 203, 1979.
97. **Marko, M.**, Inhibition of sexual behaviour and luteinizing hormone secretion in female rats by an enkephalin analogoe and morphine, *Acta Endocrinol.*, 91 (Suppl. 225), 246, 1979.
98. **Hetta, J.**, Effects of morphine and naltrexone on sexual behavior of the male rat, *Acta Pharmacol. Toxicol.*, 41, Suppl. 4, 53, 1977.
99. **Pellegrini-Quarantotti, B., Paglietti, E., Bonanni, A., Petta, M., and Gessa, G. L.**, Naloxone shortens ejaculation latency in male rats, *Experientia*, 35, 524, 1979.
100. **Ho, W. K. K., Lam, S., Leung, K. C., Au, K. K., Wong, H. K., Tsang, Y. F., and Wen, H. L.**, Effect of naloxone on morphine-induced changes in ACTH, corticosterone, and cyclic nucleotides, *Neuropharmacology*, 17, 397, 1978.
101. **Eisenberg, R. M. and Sparber, S. B.**, Effects of naloxone following a single administration of levorphanol tartrate in the naive rat, *Pharmacologist*, 19, 141, 1977.
102. **Gold, M. S., Pottash, A. L. C., Extein, I., and Kleber, H. D.**, Anti-endorphin effects of methadone, *Lancet*, 2, 972, 1980.
103. **Eisenberg, R. M.**, Effects of naloxone on plasma corticosterone in the opiate-naive rat, *Life Sci.*, 26, 935, 1980.
104. **Gibson, A., Ginsburg, M., Hall, M., and Hart, S. L.**, The effects of opiate receptor agonists and antagonists on the stress-induced secretion of corticosterone in mice, *Br. J. Pharmacol.*, 65, 139, 1979.
105. **Reina, R. A., Costa, G., Scoto, G. M., Spadaro, C., Spampinato, S., and Ferri, S.**, Increase of plasma ACTH in the rat following the administration of D-Met^2-Pro^5-enkephalinamide, *Pharm. Res. Commun.*, 12, 479, 1980.
106. **Smythe, G. A., McGinley, R., and Chisholm, D. J., Lazarus, L.**, Effect of leu-enkephalin on serum cortisol levels in man, *Clin. Exp. Pharmacol. Physiol.*, 7, 47, 1980.
107. **Marton, J., Molnâr, J., and Halâsz, B.**, Effects of (D-Met^2,Pro^5)-enkephalinamide on ACTH and GH release in rat, in *Proc. XVIII Int. Cong. Physiol.*, Budapest, Hungary, 1980, Abstracts, 569.
108. **Bisset, G. W., Chowdrey, H. S., and Feldberg, W.**, Release of vasopressin by enkephalin, *Br. J. Pharmacol.*, 62, 370, 1978.
109. **Mata, M. M., Gainer, H., and Klee, W. A.**, Effect of dehydration on the endogenous opiate content of the rat neuro-intermediate lobe, *Life Sci.*, 21, 1159, 1977.
110. **Firemark,, H. M. and Weitzman, R. E.**, Effects of β-endorphin, morphine and naloxone on arginine vasopressin secretion and the electroencephalogram, *Neuroscience*, 4, 1895, 1979.
111. **Huidobro, F. and Hidobro-Toro, J. P.**, Antidiuretic effect of morphine and vasopressin in morphine tolerant and non-tolerant rats, differential effects on urine composition, *Eur. J. Pharmacol.*, 59, 55, 1979.
112. **Van Wimersma Greidanus, T. B., Thody, T. J., Verspaget, H., de Rotte, G. A., Goedemans, H. J. H., Croiset, G., and Van Ree, J. M.**, Effects of morphine and β-endorphin on basal and elevated plasma levels of α-MSH and vasopressin, *Life Sci.*, 24, 579, 1979.
113. **Grossman, A., Besser, G. M., Milles, J. J., and Baylis, P. H.**, Inhibition of vasopressin release in man by an opiate peptide, *Lancet*, 2, 1108, 1980.
114. **Lightman, S. L. and Forsling, M. L.**, Evidence for endogenous opioid control of vasopressin release in man, *J. Clin. Endocrinol. Metab.*, 50, 569, 1980.
115. **Knepel, W., Nutto, D., Anhut, H., and Hertting, G.**, Naloxone promotes stimulus-evoked vasopressin release in vivo, *Eur. J. Pharmacol.*, 65, 449, 1980.
116. **Rossier, J., Battenberg, E., Pittman, Q., Bayon, A., Koda, L., Miller, R., Guillemin, R., and Bloom, F.**, Hypothalamic enkephalin neurones may regulate the neurohypophysis, *Nature (London)*, 277, 653, 1979.
117. **Rigter, H., Messing, R. B., Vasquez, B. J., Jensen, R. A., Martinez, J. L., Jr., Crabbe, J. C., Jr., and McGaugh, J. L.**, Regional analysis of brain opiate receptors in rats with hereditary hypothalamic diabetes insipidus, *Life Sci.*, 25, 1137, 1979.
118. **De Caro, G., Micossi, L. G., and Venturi, F.**, Drinking behavior induced by intracerebroventricular administration of enkephalins to rats, *Nature (London)*, 277, 51, 1979.
119. **Haldar, J. and Sawyer, W. H.**, Inhibition of oxytocin release by morphine and its analogs, *Proc. Soc. Exp. Biol. Med.*, 157, 476, 1978.
120. **Clarke, G., Wood, P., Merrick, L., and Lincoln, D. W.**, Opiate inhibition of peptide release from the neurohumoral terminals of hypothalamic neurones, *Nature (London)*, 282, 746, 1979.
121. **Ferri, S., Spampinato, S., and Costa, G.**, Changes induced by D-Met^2-Pro^5-enkephalinamide in plasma insulin response to glucose in the rat, *Biochem. Pharmacol.*, 28, 3346, 1979.

INDEX

A

A 23187, ionophore, 3
Acetylcholine
analgesic and behavioral effects studies, 116, 134
electrographic studies, 55, 60, 62, 65—68, 88—92
mode of action studies, 2, 6, 9, 11, 13, 15, 18—21
release, 2, 6, 9, 11, 18—19, 55, 88, 90—92, 161—162, 179—180, 191
turnover, 15, 18—21, 134
vegetative regulation studies, 161—162, 179—180, 191
Acetylcholinesterase, 67
N-Acetyl-DALA, 209
N-Acetyl-(D-Ala^2, Lys^3)-Met-E-amide, 49, 97—98
N-Acetyl-(D-Ala^2,Lys^3,Met^5)-E-amide, 52
N-Acetyl-Met-E, 51, 211
Acetylsalicylic acid, 191
Ach, see Acetylcholine
ACTH, 21, 121, 139, 146, 163, 186, 193—194
secretion, effects on, 212, 215, 220—221, 224
ACTH fragments, 36, 139
Action, opioids, see Mode of action, opioids
Action potentials, 90, 92, 94—96
Active avoidance responses, 138—139
Addiction, 141—142, 164
Addicts, 35, 37, 194, 217, 220
Adenine nucleotides, 9
Adenohypophysis
neuroendocrine function studies, 212—213
vegetative regulation studies, 186, 193
Adenosine, 88, 95
Adenosine monophosphate, cyclic, see Cyclic adenosine monophosphate
Adenosine nucleotides, 9
Adenosine triphosphatase, 91
Adenosine triphosphate, 4—5
Adenylate cyclase, 2—5, 20, 54—55, 95, 122, 180
ADH, 146
Adrenal chromaffin cells, cultured, vegetative regulation studies, 59
Adrenal glands, vegetative regulation studies, 158, 160, 162, 164, 186, 195, 198
Adrenaline, 15, 18, 43, 84, 197—198, 221
Adrenal medulla, vegetative regulation studies, 195
Adrenergic neurones, 88
Adrenergic receptors, 42
α-Adrenergic receptors, 62—63, 191
α_2-Adrenergic receptors, 3
α-Adrenergic stimulants, 6
A δ Afferent sural fibers, electrographic studies, 50
A-fibers, 69—70, 84—85
Agonist, opiate and opioid
analgesic and behavioral effects studies, 115, 122—123, 126, 138, 140—141, 143, 145—146
electrographic studies, 35, 38, 42, 44, 54—55, 57, 61, 63, 65—66, 68—69, 75, 79—82, 84, 89, 91
mixed agonists — antagonists, see Mixed agonists — antagonists
mode of action studies, 2—4, 7—9
neuroendocrine function studies, 208, 211, 216, 223
vegetative regulation studies, 156, 161, 170, 174, 191
Akinesia, 131
(D-Ala^2, D-Leu^5)-E
analgesic and behavioral effects studies, 122, 133, 138, 146
ED_{50}, 168
electrographic studies, 48—49, 81, 94, 96, 98
mode of action studies, 19
neuroendocrine function studies, 222—223
vegetative regulation studies, 156—157, 165—166, 168, 196
(D-Ala^2, D-Leu^5)-E-amide, 209, 214
(D-Ala^2, D-Leu^5)-enkephalin, 61—62
(D-Ala^2, D-Leu^5-NH-ethyl)-E, 211
(D-Ala^2, D-Met^5)-E, 211, 213, 223
(D-Ala^2, D-Nle^5)-E-amide, 208, 210
(D-Ala^2)-γ-endorphin, 137
(D-Ala^2)-Leu-E, 209, 215
(D-Ala^2, Leu^5)-E, 10, 12, 13, 15, 19, 122
(D-Ala^2)-Leu-E-amide, 15
(D-Ala^1)-Met-E, 94
(D-Ala^2)-Met-E, 10, 49, 51, 57, 94, 97—98
(D-Ala^2, Met^5)-E, 122, 134—135, 156—157, 165, 176, 196
(D-Ala^2)-N-Me-Met^2-E-amide, 15, 19
(D-Ala^2, Pro^5)-E-amide, 51, 84
Aldosterone, 215, 221
Alertness, increased, 137
N-Allyl-norphenazocine, 143
γ-Aminobutyric acid
analgesic and behavioral effects studies, 134, 146
electrographic studies, 57, 67, 69, 73, 77, 96
metabolism, 19—20
mode of action studies, 9, 12, 14, 19—21
release, 9, 20
turnover, 12, 19, 21
γ-Aminobutyric acid transaminase, 14
Amphetamine, 40, 134—135, 140, 146
Amygdala
analgesic and behavioral effects studies, 133—134, 138
electrographic studies, 36, 38—39, 41, 43
mode of action studies, 14
neuroendocrine function studies, 219
Analgesic agents, 120
Analgesic effects
behavioral effects and, 110—147

depressant potencies compared with, 93—94
electrographic studies, 34—35, 40—42, 44, 53, 59, 62—63, 66, 72—74, 87—88, 93
mode of action studies, 2, 4, 8, 10, 18, 20—21
naloxone antagonizing, 142—145
neuroendocrine function studies, 208, 211—212
pain threshold, opiate antagonists affecting, 115—116, 121, 123—130
vegetative regulation studies, 157, 160, 164, 169—171, 174—175, 178, 182—186, 190—191, 193—194, 196—198
Androgens, 219
Angiotensin, 146, 167, 223
Anorexigenic effects, 146
Antagonists, opiate and opioid
analgesic and behavioral effects studies, 114, 122—130, 137, 138—140, 142—147
cardiovascular effects in vivo, 164—168
electrographic studies, 39—40, 43—52, 57, 60—61, 63, 66—69, 73—75, 77—81, 89, 93, 95
food and water intake modulated by, 144—147
mixed agonists — antagonists, see Mixed agonists — antagonists
mode of action studies, 6—7, 11—15
neuroendocrine function studies, 205—211, 213, 216—221
pain threshold affected by, 115—116, 121, 123—130
respiratory effects in vivo, 172—173
thermoregulation, effects on, 188—189
vegetative regulation studies, 156, 158—160, 163—168, 174—175, 179—180, 182, 186—194, 197
Anterior hypophysis, neuroendocrine function studies, 212
Anterior hypothalamic nucleus, mode of action studies, 14
Anterior hypothalamus, vegetative regulation studies, 188, 191, 193
Anterior pituitary
analgesic and behavioral effects studies, 121
neuroendocrine function studies, 214—215
Anticholinergic drugs, 134
Antidiarrheal action, 175—178, 180
Antidipsogenic action, 145—146
Antidiuretic action, 9, 146, 157—158, 160, 163, 166—168, 194, 222
Antidopamine actions, 78, 213—216
Antiepileptic agents, 40, 43
Antinociceptive action
analgesic and behavioral effects studies, 110, 114—116, 121—123, 130—132, 135—139, 141, 143, 146
electrographic studies, 38, 63, 69—70, 73—74, 76, 78, 80—81
mode of action studies, 19
neuroendocrine function studies, 212
vegetative regulation studies, 156, 171, 174—175, 177, 193
Antiperistaltic action, 180
Antipropulsive effects, 174—176, 178, 181
Antiprostaglandin activity, 180
Antipsychotic agents, 52—53, 134, 140
Anxiolytic effects, 144
Apomorphine, 40, 54, 78, 134, 140, 191, 216
Appetitive behavior, see Food intake
Arcuate nucleus, electrographic studies, 62
Area preoptica, electrographic studies, 56
Area supraoptica, electrographic studies, 56
Areflexia, 132
Aromatic amino acid decarboxylase, 16
Arousal, see Excitation (arousal, stimulation)
Ascending pain-inhibitory pathways, electrographic studies, 59, 63, 82
ATP, see Adenosine triphosphate
Atropine, 77, 134, 162, 178—180, 191, 217
Attention reaction, 37
Autonomic ganglia, vegetative regulation studies, 163
Autonomic reactions, 142—143, 174—175, 190, 194, 196
Autoreceptor, 54
Aversive character, pain stimulus, 128—129
Avoidance responses
active, 138—139
passive, 139

B

Barium ion, 4
Baroreceptor, 160—161, 164
Basal secretion, effects on, 176—177, 180—181
Behavioral effects studies, 115—116, 121, 123—147, 194
central administration and, see Central administration
conditioned reactions, effects on, 137—140
depression, see Depression (inhibition)
euphorigenic effects and stimulus properties, 138, 140—142, 147
excitation, see Excitation (arousal, stimulation)
food and water intake, modulation of, 144—147
hypermotility and stereotypy, 135—136, 140, 186
naloxone antagonizing, 142—145
pain threshold, opiate antagonists affecting, 115—116, 121, 123—130
peripheral administration and, see Peripheral administration
stupor, see Stupor
Benzodiazephine, 53
Bicuculline, 57, 146
Biphasic effects, 32, 58, 136, 145, 156, 181, 187
Blood-brain barrier
analgesic and behavioral effects studies, 121, 137, 141—143
electrographic studies, 39
neuroendocrine function studies, 211—212, 221
vegetative regulation studies, 157, 163, 174—175, 196
Blood pressure, fall in, 158—163

Body temperature, effects on, see Hyperthermia; Hypothermia; Thermoregulatory modulation studies
2-Br-α-ergocryptine, 216
Bradycardia, 156, 160—166, 168, 194
Bradykinin, 69—70, 129—130
Brain hemispheres, mode of action studies, 14
Brain slice preparation, electrographic studies, 92
Brain stem
  analgesic and behavioral effects studies, 110, 114—115
  electrographic studies, 44, 46—48, 58—63, 68, 72, 80, 86
  mode of action studies, 10, 13—14
  vegetative regulation studies, 156, 160, 169
Brain stem neurones, 58—59, 68
Brain synaptosomes, mode of action studies, 13
(+)-Butaclamol, 53

C

Calcium ion
  electrographic studies, 90—93, 95
  mode of action studies, 3—5, 7—8, 13, 20
  neuroendocrine function studies, 223
  uptake, 4—5, 13, 20
  vegetative regulation studies, 159
cAMP, see Cyclic adenosine monophosphate
Capsaicin, 74, 86—88
Carbachol, 146
Carbogen, 170
Carbon dioxide tension, arterial, 170, 172—173
Cardiodepressant action, 156—157, 161
Cardiovascular effects, 156—168
Cardiovascular system, physiological regulation and pathology of, 159—164
Carrageenan, 122
Casein hydrolysate, 180
Catalepsy, 20, 39, 77, 131—135, 140, 143—144, 164, 186, 190, 193, 219
Cataleptogenic action, 20, 76, 132—134, 136, 138, 143—145, 186
Cataractogenic action, 197—198
Catatonia, 131—132, 134
Catecholamine
  analgesic and behavioral effects studies, 134
  electrographic studies, 35, 52—53, 74, 78
  mode of action studies, 6, 12, 16—18
  neuroendocrine function studies, 216
  release, 6, 159—160, 167
  synthesis, 12
  turnover, 12, 16—18
  uptake, 6
  vegetative regulation studies, 158—160, 162, 164, 195—198
Cat nictitating membrane
  mode of action studies, 7
  relaxation of, 196
Caudate neurones, 18
Caudate nucleus, see Nucleus caudata
Caudate units, 54—56
Ceiling effect, naloxone, 123
Cellular level, mode of action at, 2—5
Central administration, effects of, 131—135, 139, 143—144, 175, 194—198, 211—212
Centrally mediated effects, 156—157
Central nervous system depressant action, 132—133, 135
Cerebellum, mode of action studies, 13, 20
Cerebral cortex
  electrographic studies, 36, 44—53, 56, 88
  mode of action studies, 4, 6, 11—13, 16, 18—19
Cerebral ganglia, mode of action studies, 14
Cerium ion, 4
C-fibers, 69—70, 78, 80, 82, 84—85, 88
cGMP, see Cyclic GMP
Chicken dorsal root ganglia, electrographic studies, 95—96
Chick rectum, contractile effects, 179
Chlordiozepoxide, 53
Chlorine ion, 91, 157, 167, 222
p-Chlorophenylalanine, 35, 56, 74, 122, 134, 191, 195, 217
Cholecystokinin, 177, 180
Choline, 19, 216—217
Cholinergic agonists, 42
Cholinergic neurones, 67, 92
Cholinergic receptor, 3
Cholinoceptive neurones, 67
Cinanserine, 56, 179
Clomipramine, 122, 134
Clonidine, 63, 161, 167—168
Codeine, 36, 170
Conditioned reactions, effects on, 137—140
Conditional reflex inhibitory action, 143—145
Conduction velocity, 90
Constipating actions, 171, 174—175, 177—180
Contractile activity, gastrointestinal tract, effects on, 175, 177—179
Convulsive activity, see Seizure activity
Core temperature, effects on, 181—189
Cortex, electrographic studies, 36—38, 45—47, 61, 88, 95
Corticosteroids, 146, 163, 215, 220
Corticosterone, 220—221
Cortisol, 220—221
Creatinine, 157, 222
Cross-tolerance, 45, 69
Aα-Cutaneous afferents, 69
Cyclazocine, 35, 75—76, 78, 126
Cyclic adenosine monophosphate, 2—5, 20, 95, 122—123
Cyclic GMP, 13—14, 20
Cyproheptadine, 162, 179

D

DA, see Dopamine
DALA
  analgesic and behavioral effects studies, 133, 135, 137—138

electrographic studies, 33, 41—42, 48—49, 52, 77, 86, 92, 94—97
mode of action studies, 10—13, 15—16, 18—20
neuroendocrine function studies, 208—210, 214—216, 219
vegetative regulation studies, 156—157, 161, 166—167, 169, 172—173, 178—179, 187, 189
Dependence, opiate and opioid, 3, 53, 57, 66—68, 142
Depolarization, 50, 85, 91—92, 95—96
Depression (inhibition)
analgesic and behavioral effects studies, 132—133, 135—140
depressant potencies compared to analgesic effects, 93—94
electrographic studies, 34—36, 40, 44—45, 53, 56—58, 60—62, 64—67, 69, 71—72, 74—77, 79, 82, 84, 90, 93—95
vegetative regulation studies, 156—157, 159—175, 181
Descending pain inhibitory pathways, 59—60, 70, 72—74
Descending reticulospinal pathways, electrographic studies, 38
Des-OH-Met-E, 47
Des-Tyr[1]-γ-endorphin, 17, 140
Desynchronization, see also Synchronization, 34—40, 43—44
Dextrorphan, 57, 65—66, 76, 89, 93
Diarrhea, action against, see Antidiarrheal action
Diazepam, 40, 53
Dibutyryl-cAMP, 95, 122—123
Diencephalon
analgesic and behavioral effects studies, 121
mode of action studies, 14
Dihydrotestosterone, 219
Dihydroxyphenylacetic acid, 16, 18, 134
Dihydroxytryptamine, 122, 217
Diprenorphine, 93, 125
Diuresis, 165
Diurnal fluctuation, opioid content of brain, 130
Domperidone, 55, 217
l-DOPA, 16—17, 54, 134, 137
DOPAC, see Dihydroxyphenylactic acid
Dopamine
analgesic and behavioral effects studies, 134—136, 140, 143
electrographic studies, 54—55
metabolism, 17
mode of action studies, 3, 6—9, 11—18, 20—21
neuroendocrine function studies, 213—216, 218, 221, 224
release, 7—8, 16—18, 54—55, 218
turnover, 11—18, 20—23, 54, 213—216
vegetative regulation studies, 158—159, 191
Dopamine-β-hydroxylase, 16
Dopamine receptor, 18, 20, 158—159, 216
Dorsal horn, electrographic studies, 46, 48—51, 69—75, 82—84, 86, 93, 96
Dorsal root, electrographic studies, 51
Dorsal root ganglia, electrographic studies, 92—93, 95—96, 98
Dorsal root potential, 51, 84—85
Dorsomedial nucleus, mode of action studies, 13
Dorsomedial thalamus, electrographic studies, 41
Drinking behavior, see Water intake
Dynorphin, 136

E

Eating behavior, see Food intake
EGTA, 4
Electroencephalographic activity, spontaneous, see Spontaneous electroencephalographic activity
Electrographic analysis, opioid mechanisms, 34—98
Gate Control System, 76, 81—88
general discussion, 34, 95
neuronal activity in vitro, 88—98
single brain neurones, 44—75
spinal mono- and polysynaptic reflexes, 75—81
spontaneous electroencephalographic activity
opiate effects on, 34—39
opioid peptide effects on, 39—44
Endocrine functions, effects on, 208—224
Endogenous antipsychotic action, 141
Endogenous opioids
mobilization of, opiates acting through, 121—122
peripheral action, 122—123, 137
vegetative regulation, role in, see Vegetative regulation, role of endogenous opioids in
Endorphin
analgesic and behavioral effects studies, 121, 123, 129—131, 133, 142, 144
electrographic studies, 61, 76, 82
neuroendocrine function studies, 208, 213, 220, 223
vegetative regulation studies, 160—161, 163, 170—171, 187—188, 192
α-Endorphin
analgesic and behavioral effects studies, 116, 137, 139—140
electrographic studies, 47, 52, 93—94, 96
mode of action studies, 13, 17—19
neuroendocrine function studies, 209—211, 214
vegetative regulation studies, 156, 177—179
β-Endorphin
analgesic and behavioral effects studies, 110—111, 117—119
ED_{50}, 117—119
electrographic studies, 39, 45, 52, 73, 81
vegetative regulation studies, 156
γ-Endorphin
analgesic and behavioral effects studies, 116, 133, 139—140
neuroendocrine function studies, 209—211
vegetative regulation studies, 179, 187—188
Endotoxin, 163, 166
Enkephalins
analgesic and behavioral effects studies, 110—116, 120—123, 130—132, 134, 136—139, 141—143, 145—147

analogues, see also Synthetic enkephalin analogues
analgesic and behavioral effects studies, 110, 115, 122, 133—134, 136—138, 141—143
electrographic studies, 76, 81, 84—85, 93—94
mode of action studies, 6—7, 9—10, 16—17, 19
neuroendocrine function studies, 208—209, 211—212, 219—222
stabilized, 52, 115, 156, 169, 211
vegetative regulation studies, 156, 158, 169, 171, 177, 182, 186—187, 194—197
electrographic studies, 39—44, 52—58, 60—62, 64, 66—69, 71, 73—75, 80—95
mode of action studies, 6—10, 16—20
natural, see Natural enkephalins
neuroendocrine function studies, 208—209, 211—213, 219—223
spinal modulation of pain reflexes (Gate Control System), 76, 81—88
vegetative regulation studies, 156—163, 169, 174—175, 177—180, 182, 186—187, 190, 194—197
β-EP
analgesic and behavioral effects studies, 111, 116, 120—122, 131—137, 134—144, 146—147
ED_{50}, 120, 132—133, 164, 168, 219
electrographic studies, 34, 39—44, 47, 49, 51—53, 57—58, 60—61, 86, 89, 91, 93—94, 96
mode of action studies, 3, 6—7, 9—21
neuroendocrine function studies, 208—219, 221—223
pA_2, 143—144
vegetative regulation studies, 156—160, 162—173, 175—180, 182, 184, 187—198
Epilepsy, 34, 41, 43
Epileptiform activity, 34, 36—43, 57—58
Ergometrine, 42, 77
Esophageal sphincter contraction, effects on, 177—178, 181
Estradiol, 218
Estrone derivatives, 219
Ethylketocyclazocine, 38—39, 80,171
Etonitazene, 134, 141
Etorphine
analgesic and behavioral effects studies, 133
electrographic studies, 45, 61—62, 64—67, 93—94
mode of action studies, 8, 19
neuroendocrine function studies, 211
vegetative regulation studies, 196
Euphorigenic effects, 138, 140—142, 147
Excitation (arousal, stimulation)
analgesic and behavioral effects studies, 132, 135—136
electrographic studies, 34—35, 37, 39—40, 43—45, 53, 55—61, 63, 65—67, 69, 74, 77, 79, 82, 85, 90—91
vegetative regulation studies, 181, 186
Excitatory agents, 79
Excitatory junction potentials, 92, 96
Excitatory postsynaptic potentials, 80, 90, 93—94, 97, 159
Exophthalmos, 132, 196—198
Expectation, pain, 128—129
Exploratory behavior, effects on, 144
Explosive hyperactivity syndrome, 135
Extensor α-motoneurones, 76—79
Eye effects studies, 193—198
Eyelid retraction, effects on, 196—198

F

Feline mania, 36, 137
Female sexual behavior, inhibition of, 219—220
Fentanyl, 38, 141, 161—162
Fibers
large diameter, 82, 88
small diameter, 82, 88
PK 33—824
analgesic and behavioral effects studies, 141—142
ED_{50}, 168
electrographic studies, 40, 50—51, 77, 84—85, 92—94, 96—98
mode of action studies, 10, 15—16, 19
neuroendocrine function studies, 208, 210, 217—222
vegetative regulation studies, 159, 161—162, 168—169, 172—173, 176—180, 186, 188, 195
Flexor α-motoneurones, 76—78
Fluoxetine, 162
α-Flupenthixol, 52
Food intake, modulation of, 146—147
Forebrain
analgesic and behavioral effects studies, 115
electrographic studies, 38, 41, 44, 57
mode of action studies, 11
Fourth ventricle, see Ventricles, fourth
Frog isolated spinal cord, electrographic studies, 97
Frog sympathetic ganglion, electrographic studies, 92, 97
FSH, 217—220

G

GABA, see γ-Aminobutyric acid
Gamma-aminobutyric acid, see γ-Aminobutyric acid
Gangliosides, 5
Gastric secretions, effects on, 174, 177, 180
Gastrointestinal effects studies, 171—181
Gate control system, 76, 81—88
schematic diagram, 83
GH, see Growth hormone
GHRH, 214
Globus pallidus, mode of action studies, 9, 11—12, 18—20

Glomerular filtration, 222
Glucagon, 215, 223—224
Glucosensitive neurones, 146
Glutamate, 94, 96
Glycine, 66—68
Glycoproteins, 5
Glycosyl transferase, 5
Gonadal hormones, 219—220
Grooming, 136, 139
Growth hormone, 17, 208—218, 221—222, 224
GTP, 3
Guanine nucleotide, 3
Guinea pig ileum
analgesic and behavioral effects studies, 121—122
mode of action studies, 4, 6—9
myenteric plexus neurones, see Myenteric plexus neurones
vegetative regulaton studies, 171, 174, 179—180
Guinea pig vas deferens, mode of action studies, 7

H

HA, see Histamine
Haloperidol
analgesic and behavioral effects studies, 134, 139—141
electrographic studies, 40, 42, 54—55, 77
neuroendocrine function studies, 216—218
vegetative regulation studies, 159, 195
Halothane, 167
Heroin, 35
Hexamethonium, 90, 159
5-HIAA, see 5-Hydroxyindoleacetic acid
Hippocampal epilepsy, role of opioid peptides in, 41
Hippocampal slices, electrographic studies, 94, 97—98
Hippocampus
electrographic studies, 35—39, 41—43, 47, 49—51, 56—58, 72, 95
mode of action studies, 4, 6, 9—11, 13, 15, 18—19
Histamine
release, 8, 158, 164, 180
vegetative regulation studies, 158, 164, 176, 180
Histamine receptor blocking agents, 158
Homocysteate, 65—66, 68
Homovanillic acid, 16, 134
Hormonal effects, 208—224
5-HT, see Serotonin
5-HTP, 175—177
Hunger center, 56—57
HVA, see Homovanillic acid
3-Hydroxybenzylhydrazine, 16
6-Hydroxydopamine, 6, 216
5-Hydroxyindoleacetic acid, 10, 12
5-Hydroxytryptamine, 191
5-Hydroxytrytophan, 8, 10, 134, 180, 218
Hyperalgesic effects, 112, 115, 122—130
Hypermotility, 135—136, 186
Hyperphagia, 147
Hyperpolarization, 84—86, 90—92, 95—98
Hypertension, 156, 160—162, 166—167
Hyperthermia, 181—193
stress-induced, 192—193
Hypnotics, 141
Hypophysis
neuroendocrine function studies, 212—213, 217, 223
vegetative regulation studies, 186
Hypotension, 156—157, 160—168, 170, 194
Hypothalamic slices
electrographic studies, 92, 98
neuroendocrine function studies
Hypothalamus
analgesic and behavioral effects studies, 121, 146—147
electrographic studies, 36, 46, 56—58, 62
mode of action studies, 4, 8, 10, 12—13, 15, 17
neuroendocrine function studies, 208, 212—214, 216, 218—219, 221, 223—224
vegetative regulation studies, 162, 187—191, 193
Hypothermia, 181—193
Hypovolemia, 163—164, 167

I

Ileocecal sphincter contraction, effects on, 177—178, 181
Immobility, 131, 137
Impulse conduction, morphine congeners affecting, 2
Indomethacin, 191
Inhibition, see Depression (inhibition)
Inhibitory postsynaptic potentials, 92, 94, 97
Insulin, 147, 214, 223—224
Interneurones, electrographic studies, 49, 57—58, 64, 68—69, 71, 75, 77, 80, 82, 85, 94—95, 98
Intestinal fluid volume, effects on, 176, 180
Intestinal secretions, effects on, 174—177, 180—181
Intestinal tone, 171—174, 178—179
Intracisternal injection
analgesic and behavioral effects studies, 113, 115, 118, 121
vegetative regulation studies, 156, 160, 162—163, 165—166, 170, 172, 189, 195
Intranigral injection, analgesic and behavioral effects studies, 135
Intrathecal injection
analgesic and behavioral effects studies, 113, 115, 118, 121—123
vegetative regulation studies, 188—190
Ionophores, 3—4
Isoprenaline, 122, 222
Isopropylnoradrenaline, 42, 197

J

Jerking movements, 136

K

Kainic acid, 18, 20, 55, 58
Ketocyclazocine, 171
Kidney, mode of action studies, 14
Kyotorphin, 51, 69

L

Laminae, spinal cord dorsal horn, electrographic studies, 46, 48—51, 69—73, 80, 82, 84
Lanthanum ion, 4
Large diameter fibers, electrographic studies, 82, 88
Latency time, 110—111, 126—127, 129—130
Lateral hypothalamic area, electrographic studies, 56—57
Lateral ventricle
neuroendocrine function studies, 216
vegetative regulation studies, 160, 187
Lenticular opacity, induction of, 197—198
Leu-E
analgesic and behavioral effects studies, 110, 112—115, 121—122, 138—139, 141, 146—147
ED_{50}, 110, 112—113, 115, 168
electrographic studies, 39—43, 46—52, 57, 60, 64, 67, 71, 80, 84, 89—90, 93—94, 96
mode of action studies, 9, 11, 14—15, 18—19
neuroendocrine function studies, 208—209, 211, 214—215, 221—223
vegetative regulation studies, 156—157, 159, 162, 167—168, 175—179, 187—189, 194
Levallorphan, 66, 75, 78, 80, 124—125
Levorphanol
electrographic studies, 36—38, 57, 60, 65—67, 71—72, 76, 82, 89, 93
neuroendocrine function studies, 220
vegetative regulation studies, 197—198
LH, 215—220
LHRH, 214—215, 217—220, 224
Limbic brain, neuroendocrine function studies, 219
Limbic forebrain, mode of action studies, 10, 12, 16—17
Limbic structures
electrographic studies, 41, 43—44
mode of action studies, 11, 16—17
Limbic system, neuroendocrine function studies, 221
Lipid solubility, 211
β-Lipotropin
analgesic and behavioral effects studies, 116—121, 137, 139—141
ED_{50}, 116
mode of action studies, 13—14
neuroendocrine function studies, 215, 221
vegetative regulation studies, 159
β-Lipotropin fragments
analgesic and behavioral effects studies, 116—121, 137, 139—141
mode of action studies, 17
neuroendocrine function studies, 211
vegetative regulation studies, 179, 188
Local anesthetic action, 85, 90
Locomotor depression, 35, 131—132, 136, 144, 181, 193
Locomotor stimulation, 35, 40, 136—137, 140, 144, 181—182, 186
Locus coeruleus
electrographic studies, 48—50, 62—64, 92, 98
mode of action studies, 7
vegetative regulation studies, 161
Loperamide, 174, 180
Lordosis, 219—220
LPH, see β-Lipoprotein

M

Magnesium ion, 4, 159
Male sexual behavior, inhibition of, 219
Mamillary nuclei, vegetative regulation studies, 190
Mecamylamine, 77, 217
Mechanisms, opioids, see Mode of action, opioids
Medial septal nucleus, mode of action studies, 13
Medial thalamus, electrographic studies, 38, 40—43, 46, 53
Median eminence
mode of action studies, 11, 17
neuroendocrine function studies, 216, 218
Mediobasal hypothalamus, neuroendocrine function studies, 213, 219, 223
Medulla
analgesic and behavioral effects studies, 115
electrographic studies, 38, 46—50, 64—65, 73
vegetative regulation studies, 156, 160, 195
Medulla oblongata
electrographic studies, 50
mode of action studies, 13—14
Medullary respiratory centers, 171
Membrane conductance, 92, 94—96, 98
Membrane input potential, modification of, 84
Membrane potential, 50—51, 84, 91, 97
resting, see Resting membrane potential
Membrane resistance, 91, 96
Memory formation, 138
Meperidine, 76, 84, 90, 171
Mepyramine, 158
Mesencephalic area, electrographic studies, 38, 51, 58—59
Mesolimbic structures, analgesic and behavioral effects studies, 136
Metabolic effects, opioid peptides, in vivo studies, 9—21
Metacholine, 180

Met-E
analgesic and behavioral effects studies, 110—116, 122, 136—139, 141
ED_{50} 110, 112—115, 168, 176
electrographic studies, 39—42, 45—52, 56—62, 64, 66—67, 69, 80, 82, 84—85, 89—90, 92, 94—97
mode of action studies, 3, 6—20
neuroendocrine function studies, 209—211, 213—219, 221, 223
vegetative regulation studies, 156—159, 165, 168—169, 171—173, 175—180, 186—189, 191, 194
Met-E-amide, 46, 48—49, 71, 177, 209
Met-E-ethylamide, 209
Met-E, Leu-E, (D-Ala^2)-Met-E, 214—215
Met-E-methyl ester, 221
Metergoline, 217
Methadone
analgesic and behavioral effects studies, 141
electrographic studies, 35—37, 57, 76, 84, 90
mode of action studies, 7—8
neuroendocrine function studies, 220
vegetative regulation studies, 170—171
Methergoline, 162
Methoxytyramine, 16, 18
α-Methyldopa, 161, 167
α-Methyl-*p*-tyrosine, 35, 78, 134
Methysergide, 42, 73—74, 133—134, 162, 179, 191, 217
D-Met, Pro-EA
analgesic and behavioral effects studies, 111, 114, 132—136, 141—145
ED_{50}, 132—133, 175—178, 197—198, 212
electrographic studies, 81, 93—94, 96
lenticular opacity-producing activity, 198
mydriatic effect, 194—195
neuroendocrine function studies, 210, 212, 216—217, 221, 223
pA_2, 114, 143—144
vegetative regulation studies, 156, 162, 166, 169—172, 175—178, 182, 185, 194—198
Mianserine, 133, 162
Midbrain
analgesic and behavioral effects studies, 121
electrographic studies, 36, 38, 43—44, 47, 60, 62, 65
mode of action studies, 11, 13—14
Miosis, 193, 195—196, 198
Mixed agonists-antagonists, 35, 66, 78, 123
Mode of action, opioids, 2—21
cellular level, 2—5
electrographic analysis, see Electrographic analysis, opioid mechanisms
metabolic effects, in vivo studies, 9—21
neural regulation, in vito studies, 2, 5—9
Monoamines, 191
Monosynaptic reflexes, electrographic studies, 75—81
postsynaptic inhibition of, 29
post-tetanic potentiation of, 80
Morphine
analgesic and behavioral effects studies, 110—111, 114, 116, 120—122, 130—139, 141—146
ED_{50}, 110, 116, 132—133, 143—144, 175—177, 193, 197—198
electrographic studies, 34—43, 45, 52—57, 59—77, 79—82, 84—86, 88—94
lenticular opacity-producing activity, 197—198
mode of action studies, 2, 4, 6—9, 16, 20
mydriatic effect, 194—195
neuroendocrine function studies, 208, 211—214, 216—223
pA_2, 81, 114, 143—144, 170—171, 191, 197
vegetative regulation studies, 157—159, 161, 170—171, 174—183, 186—187, 190—198
Morphine congeners
analgesic and behavioral effects studies, 121, 132, 135, 137, 143
electrographic studies, 61—62, 66—67, 81
mode of action studies, 2—3, 6, 17, 20
vegetative regulation studies, 156, 158—159, 170—174, 181, 193, 196
Morphine derivatives, 110, 132, 135, 137, 142—143, 175, 181
Motoneurones, electrographic studies, 50, 65, 68—69, 75—80
α-Motoneurones, electrographic studies, 68—69, 76—79
extensor, see Extensor α-motoneurones
flexor, see Flexor α-motoneurones
γ-Motoneurones, electrographic studies, 76
Mouse cultured spinal neurones, electrographic studies, 92—94, 96
Mouse vas deferens
analgesic and behavioral effects studies, 122
electrographic studies, 92—93, 96
mode of action studies, 4, 7
β-MSH, 136, 139
Muscimol, 20, 146
Muscle relaxation effects, 178—180
Muscular activity, see Locomotor depression; Locomotor stimulation
Muscular rigidity, 71, 131—135, 164, 219
Muscular tone, 131—132, 136
Mydriasis, 132, 193—196, 198
Myenteric plexus neurones, electrographic studies, 88—93, 95—96
Myoclonic twitches, 39—40, 42, 77
Myogenic action, 179

N

Nalorphine
analgesic and behavioral effects studies, 126, 144
electrographic studies, 35, 61, 75—76, 78—79, 84
mode of action studies, 8
pA_2, 197
vegetative regulation studies, 197
Naloxone

analgesic and behavioal effects studies, 110, 114, 122—130, 133—140, 142—147
antagonism, 142—145
electrographic studies, 39—40, 43—46, 48—51, 53—54, 56—57, 59—69, 71—82, 84—86, 88—93, 96—98
mode of action studies, 4, 6—13, 16—20
neuroendocrine function studies, 208—223
neuronal activity in vitro, effects on, 96—98
pA_2, 81, 114, 143, 170—171, 175, 191, 197
pain threshold affected by, 123—130
quaternary, 174
tertiary, 174
vegetative regulation studies, 156—163, 165—168, 170—182, 186—197
^{3}H-Naloxone, 219
Naltrexone
analgesic and behavioral effects studies, 125, 130, 134, 139, 144
electrographic studies, 44, 47—48, 52, 78—79
mode of action studies, 6, 11, 17, 19—20
neuroendocrine function studies, 210, 213—215, 219, 222
vegetative regulation studies, 160, 163, 165, 168, 188—191
^{3}H-Naltrexone, 219
Narcotic analgesics, 78
Natural enkephalins
analgesic and behavioral effects studies, 110—116, 120—121, 133, 136—139, 146
electrographic studies, 39—40, 42—43, 52, 56, 64—69, 71, 81, 89, 93
mode of action studies, 16—17
neuroendocrine function studies, 208, 211—213, 223
vegetative regulation studies, 156—157, 169, 175, 178, 180, 187, 194
NE, see Norepinephrine
Neural lobe, neuroendocrine function studies, 214—215
Neural membranes, composition, changes in, 5
Neural regulation, opioids in, in vitro studies, 2, 5—9
Neuroblastoma x glioma hybrid cell NG 108—15, see NG 108—15
Neuroendocrine functions, effects on, 208—224
ACTH and TSH secretion, 212, 215, 220—221, 224
LHRH, LH, and FSH secretion, 214—220, 224
prolactin and growth hormone secretion, 208—218, 221—222, 224
vasopressin and oxytocin secretion, 214—215, 221—224
Neurohypophysis, neuroendocrine function studies, 222—223
Neuroleptic agents, 17, 131—132, 134, 138, 140—141, 144
Neuromodulators, 5, 86, 95, 142
Neuronal activity, in vitro, electrographic studies, 88—98
Neurotransmitters
analgesic and behavioral effects studies, 133, 142
electrographic studies, 64, 67, 73—74, 86, 94—95
mode of action studies, 2, 5—9, 20—21
vegetative regulation studies, 159, 162, 187, 191
NG 108—15, cell line, mode of action studies, 3—4
Nicotine, 65—66, 89—90, 159
Nictitating membrane
mode of action studies, 7
relaxaion of, 194, 196, 198
Nigro-striatal system
analgesic and behavioral effects studies, 134, 136
neuroendocrine function studies, 216
Nociceptive neurones, 70—75, 80
Nociceptive responses
analgesic and behavioral effects studies, 115, 123, 126—129
antinociceptive action, see Antinociceptive action
electrographic studies, 38, 45, 51, 53, 63—65, 69—73, 75, 77—80, 82, 86, 88
enhancement by naloxone, 78
vegetative regulation studies, 190
Non-nociceptive stimulation, 53, 65, 71—72, 82, 85—86
Nonspecific activating system, 38
Noradrenaline
analgesic and behavioral effects studies, 116
electrographic studies, 42, 60, 62, 73, 86, 92
mode of action studies, 11—17
release, 158—159, 196
turnover, 11—14, 17
vegetative regulation studies, 158—159, 191, 196—197
Noradrenergic nerve elements, mode of action studies, 6—7
Noradrenergic neurones, 161
Noradrenergic pathways, 74
Noradrenergic system, electrographic studies, 63
Norepinephrine, 6—7, 89
release, 6—7
Normorphine, 45, 52, 57, 89—90, 92, 220
NSD 1015, 16
Nucleus accumbens
analgesic and behavioral effects studies, 134—136
electrographic studies, 47, 49, 56
mode of action studies, 11, 18
Nucleus ambiguous, vegetative regulation studies, 161
Nucleus arcuatus of hypothalamus, analgesic and behavioral effects studies, 121
Nucleus caudata
analgesic and behavioral effects studies, 134
electrographic studies, 46—47, 54, 56
mode of action studies, 9, 11—13, 18—20
Nucleus interstitialis, mode of action studies, 13
Nucleus raphe magnus, see also Raphe nuclei
analgesic and behavioral effects studies, 141
electrographic studies, 59, 64, 72, 82
vegetative regulation studies, 161
Nucleus reticularis

electrographic studies, 72
mode of action studies, 13—14
Nucleus reticularis gigantocellularis
analgesic and behavioral effects studies, 113
electrographic studies, 46, 50, 64, 70, 72—74
vegetative regulation studies, 161
Nucleus reticularis paragigantocellularis, electrographic studies, 64—65, 73—74
Nucleus superopticus, vegetative regulation studies, 157
Nucleus tractus solitarius
mode of action studies, 13—14
vegetative regulation studies, 160—161
Nucleus ventromedialis, electrographic studies, 51

O

Oculomotor nuclear complex, visceral part, vegetative regulation studies, 195
Oddi sphincter contraction, effects on, 180
Oligaria, 222
Opiates
mobilization of endogenous opioids, acting through, 121—122
single brain neurones affected by, electrographic stucies, 44—75
spontaneous electroencephalographic activity affected by, 34—39
thermoregulation modulation, 181—189
Ornithine decarboxylase, 14, 21
Ouabain, 6—7, 9, 91
Ovulation, inhibition of, 217—218
Oxymorphone, 77
Oxytocin, release of, 9, 214, 223—224

P

Pain afferents, 74
Pain expectation, 128—129
Pain inhibitory mechanisms, 75
Pain-inhibitory pathways
ascending, see Ascending pain-inhibitory pathways
descending, see Descending pain inhibitory pathways
Pain-receptors, 69—70
Pain reflexes, spinal modulation by enkephalins (Gate Control System), 76, 81—88
Pain threshold, 115—116, 121, 123—130, 162, 170
Pain transmission, 64—65, 71, 75, 80, 82, 115, 121
Pain transmission-inhibiting substances, 80
Palpebral fissure, widening of, 196—197
Pancreas, neuroendocrine function studies, 213, 223
Pancreatic secretions, effects on, 174—175, 177, 180—181
Paradox effect, 164—169
Parasympathetic centers, vegetative regulation studies, 156
Parasympathetic tone, 174, 198
Paraventricular neurones, 92, 98
Paraventricular nucleus
mode of action studies, 13—14
neuroendocrine function studies, 223
Parietal cortex, mode of action studies, 15, 19
Passive avoidance responses, 139
Passive immobility, 137
Pathological waves, EEG activity, 36—37
Pentazocine, 35, 126
Pepsin, 176, 180
Periqueductal gray matter
analgesic and behavioral effects studies, 112—116, 125, 131, 133—136
electrographic studies, 38, 41—43, 46—48, 58—65, 72—74
mode of action studies, 8
Peripheral action, endogenous opioids, 122—123, 137
Peripheral administration, effects of, 137—139, 142—144, 157, 169, 174—175, 194—198, 211—212
Peripheral autonomic ganglions, vegetative regulation studies, 159
Peripherally mediated vascular effects, 158—159
Peristaltic reflex, effects on, 176, 179, 181
Pethidine, 8
PGE, see Prostaglandin E
PGF, see Prostaglandin F
(D-Phe^4)-Leu-E, 215
(D-Phe^4, Met^5)-E, 138
Phenoxybenzamine, 74, 159, 195—196
Phentoin, 40
Phentolamine, 74, 179, 191, 195
Phe-Phe-Gly-Leu-Met-NH_2 fragment, 8
Phosphatidylinositol, 20
1-Phosphatidylinosito—3,4-biphosphate, 20
Phosphoinositides, 14
Phospholipid, metabolism, 21
Phosphorus, 214—215
Phosphorylation, 5
Physostigmine, 191
Pilocarpine, 217
Piloerection, 132
Pinozide, 54—55, 191, 218
Piperoxane, 63
Pituitary-adrenal system, neuroendocrine function studies, 220—221
Pituitary gland
neuroendocrine function studies, 213—215, 220—221, 223—224
vegetative regulation studies, 162, 193
Plasma, mode of action studies, 15
Polysynaptic reflexes, electrographic studies, 50, 75—81, 97
Pons
electrographic studies, 58, 64—65
vegetative regulation studies, 160
Pontin cells, electrographic studies, 59
Pontin respiratory centers, 171
Pontin slices, electrographic studies, 98
Population spike response, 58
Posterior pituitary, analgesic and behavioral

effects studies, 147
Postganglionic nerve terminals, vegetative regulation studies, 159
Postsynaptic mechanisms, spasmogenic action, 178—179, 181
Postsynaptic potentials
excitatory, see Excitatory postsynaptic potentials
inhibitory, see Inhibitory postsynaptic potentials
Postural reflexes, 131—134
Potassium ion, 4—9, 91, 157, 167, 213, 215, 222—223
Preoptic area, hypothalamus, vegetative regulation studies, 187—189, 191
Pressor effects, 156—157, 159—160, 162, 169, 194—195
Presynaptic inhibition, 85, 94—95, 158—159, 178—179, 181, 196, 216, 218, 223—224
Primary afferents
depolarization of, 50, 85
electrographic studies, 50, 84—87, 93, 95, 97
hyperpolarization of, 84—86, 97
small diameter, 84, 86—87
Primary sensory pathways, pain transmission in, 37
Prolactin, 7, 17, 208—218, 221—222, 224
stress-induced release, 208—211, 217
Proopiocortin, 163
Propanolol, 74, 195
Propulsion, gastrointestinal, action against, see Antipropulsive effects
Prostacycline, 122
Prostaglandin, 95, 122—123, 180—181, 191
Prostaglandin E_1, 3, 176, 180
Prostaglandin E_2, 95, 122, 176, 179
Prostaglandin F_2, 159
Prostaglandin $F_{2\alpha}$, 176, 180
Protein secretion, effects on, 175, 177, 180
Psychoactive drugs, 197
Psychostimulants, 141
Psychotropic effects, 115
Pupillary size, effect on, 193—196
Purine nucleotides, release of, 9
Pylorus contraction, effects on, 174—175, 178, 181
Pyramidal cells, electrographic studies, 49, 51, 57—58, 72, 97

Q

Quaternary naloxone, 174
Quipazine, 134

R

Rabbit ear artery
mode of action studies, 7
vegetative regulation studies, 158
Rabbit heart, mode of action studies, 7
Rabbit portal vein, mode of action studies, 7
Rabbit pulmonary artery, mode of action studies, 7
Rabbit vas deferens, mode of action studies, 7
Raphe nuclei, see also Nucleus raphe magnus
analgesic and behavioral effects studies, 125
electrographic studies, 58—65, 72—73
neuroendocrine function studies, 219
Rat vas deferens, mode of action studies, 7—8
Receptor binding assay, neuroendocrine function, 208, 212, 219
Receptors, opiate and opioid
analgesic and behavioral effects studies, 110, 114—115, 120, 122—123, 130, 133—139, 141—143, 145—146
electrographic studies, 38—40, 42—45, 52, 54, 56—57, 61—63, 67—71, 73—74, 80—82, 84, 86, 92—93, 95
mode of action studies, 2—5, 7, 10, 17, 19, 21
neuroendocrine function studies, 208, 211—213, 216—223
population (subclass), difference or sameness, effects of, 19, 42, 61—62, 114, 122, 136, 142—144, 170—171, 174—175, 178, 181, 191, 193, 197, 211, 219
supersensitivity, 55, 80, 216—217
vegetative regulation studies, 156—161, 163—164, 170—171, 174—175, 178—182, 191—198
α-Receptors, 159
α_2-Receptors, 161
δ-Receptors
analgesic and behavioral effects studies, 122, 143, 146
antagonist to, 43
electrographic studies, 42—43, 62, 68, 73, 80—81, 89, 92—93, 95
vegetative regulation studies, 156, 175, 178
ε-Receptors, analgesic and behavioral effects studies, 143
μ-Receptors
analgesic and behavioral effects studies, 122, 143
electrographic studies, 42—44, 62, 80—81, 89, 92—93, 95
neuroendocrine function studies, 222
vegetative regulation studies, 156, 171, 175, 198
χ-Receptors
electrographic studies, 38, 80
vegetative regulation studies, 171
Region A, mode of action studies, 13—14
Reinforcing properties, 141
Renal opiate receptors, vegetative regulation studies, 157—158
Renshaw cells, electrographic studies, 46, 48, 65—69, 72, 75, 77, 80
Reproductive functions, effects on, 217—219
Reserpine, 10, 78, 122
Respiratory depression, 174—175
Respiratory effects studies, 164—173
depression, 164—173
Respiratory frequency, decrease in, 169, 172—173

Resting membrane potential, 90, 92, 95—96
Reticular formation
electrographic studies, 37—38, 43—44, 47—49, 51, 58—65
mode of action studies, 19
Reticulospinal pathways, descending, see Descending reticulospinal pathways
Righting reflex, 131—132
Rigidity, muscular, see Muscular rigidity
Running fit, 36, 136

S

Satiety center, 56—57, 146—147
Scopolamine, 40, 62
Secretin, 177, 180
Secretion
gastrointestinal, 174—177, 180—181
neuroendocrine, 208—224
Sedative effects, 131, 135, 140, 186, 194
Seizure activity, 36, 38—43, 95, 136
Self-administration, 140—141
Self-stimulation, 142
Septum, mode of action studies, 13—14, 19
(D-Ser[2], Met[5])-E-methyl-ester, 49
Serotonin
analgesic and behavioral effects studies, 122, 133—134, 137
electrographic studies, 35, 56, 60, 63, 73—74, 84, 89—90
metabolism, 8, 10—17
mode of action studies, 8, 10—17, 21
neuroendocrine function studies, 216—218
release, 8, 122, 179
turnover, 10—16, 21
uptake, 8
vegetative regulation studies, 162, 179, 181, 191
Serotoninergic neurones, 88
Serotonin receptor blocking drugs, 134
Sexual functions, effects on, 217—220
Shock, 163—164, 167—168, 170
Sialoglycosphingolipids, 5
Small diameter fibers, electrographic studies, 82, 88
Small diameter primary afferents, 84, 86—87
Sodium ion, 84, 91, 157, 167, 222
Somatostatin, 213, 216, 223—224
SP, see Substance P
Spasmogenic action, 174, 178—181
Spinal cord
dorsal horn, see Dorsal horn
dorsal root, see Dorsal root
Renshaw cells, see Renshaw cells
vegetative regulation studies, 162, 190, 193
ventral root, see Ventral root
Spinal modulation, pain reflexes, by enkephalins (Gate Control System), 76, 81—88
Spinal mono- and polysynaptic reflexes, electrographic studies, 75—81
Spinal space, vegetative regulation studies, 187
Spine, electrographic studies, 38
Spiroperidol, 52
Spontaneous electroencephalographic activity
opiate effects on, 34—39, 135, 181
opioid peptide effects on, 39—44, 181
Spontaneous peristalsis, effects on, 176, 179, 181
Spontaneous respiration, effects on, 164—173
SRIF, 214—215
Stabilized enkephalin analogues, 52, 115, 156, 169, 211
Stereotypy, 135—136, 140
Steroidogenesis, 215, 221
Steroids, 215, 219—220
synthesis, 215
Stimulation, see Excitation (arousal, stimulation)
Stimulus properties, 140—142
Straub phenomenon, 132, 196
Stress, effects of
analgesic and behavioral effects studies, 129, 144, 146
neuroendocrine function studies, 208—210, 212, 217—218, 220
vegetative regulation studies, 158—159, 163, 170, 176, 180, 186, 192—193
Stress-induced hyperthermia, 192—193
Stress-induced prolactin release, 208—211
Stressor effects, 162, 164, 186, 220—221
Striatum
analgesic and behavioral effects studies, 134, 136
electrographic studies, 47, 53—56, 88, 95
mode of action studies, 3—4, 6—8, 10—13, 16—20
Strontium ion, 4
Strychnine, 61
Stupor, 35—36, 135, 181
Subiculum, mode of action studies, 13
Substance P
electrographic studies, 62, 67, 74, 86—88, 95
Gate Control System, role in, 86—88
mode of action studies, 8—9
release, 8—9, 86—88
inhibition of, 86—88
vegetative regulation studies, 159
Substantia gelatinosa, electrographic studies, 71, 73—74, 80, 82—87
neurone population in, schematic diagram, 83
Substantia nigra
analgesic and behavioral effects studies, 134, 136
electrographic studies, 53—56
mode of action studies, 9, 19—20
Suckling behavior, antagonism to, 146
Sulpiride, 158
Superior cervical ganglion
electrographic studies, 47
vegetative regulation studies, 162
Supersensitivity, receptor, 55, 80, 216—217
Supraoptic nucleus, hypothalamus, neuroendocrine function studies, 223
Sympathetic tone, 159, 174
Synchronization, see also Desynchronization, 34—36, 38—39, 42—43, 53, 135

Synthetic enkephalin analogues, see also Enkephalins, analogues
analgesic and behavioral effects studies, 136—137
electrographic studies, 39—40, 56
mode of action studies, 6, 9—10, 19
neuroendocrine function studies, 208, 211, 213
vegetative regulation studies, 175, 187

T

Tachycardia, 156, 160—161, 165, 167
Tachyphylaxis, 43, 45, 61—62, 80, 89
Tachyphoea, 164—169, 172
Temperature regulatory effects, see Hyperthermia; Hypothermia; Thermoregulatory modulation studies
Tertiary naloxone, 174
Testosterone, 217, 219
Tetrabenazine, 74, 78
Tetrapeptides, 141
Tetrodotoxin, 178—179
Thalamus, electrographic studies, 36, 38, 41—42, 46—47, 49, 53
Thermoregulation modulation studies, 181—193
mechanism of alteration, 187—191
structure-activity and dose-response relationships, 181—189
Thermoregulatory centers, 190
Third ventricle, see Ventricles, third
Tidal volume, decrease in, 169, 172—173
Tolerance
analgesic and behavioral effects studies, 132, 135—136, 141—142, 145—147
cross-, see Cross-tolerance
electrographic studies, 34—35, 37, 43, 45, 57, 61—63, 69, 76, 89—90, 92—93
mode of action studies, 2—4, 20
vegetative regulation studies, 156, 160—161, 165, 167, 181—187, 193—194, 197
Tranquilizers, 141
Transpyloric flow, effects on, 175, 177—178
TRH, 192, 214
Tricyclic antidepressants, 137
Trigeminal ganglion, electrographic studies, 50
Trigeminal nuclei, electrographic studies, 72—73
Trigeminal subnucleus caudalis, electrographic studies, 72
Tryptophan, 8, 10, 12, 15
l-Trytophan, 134
Trytophan hydroxylase, 8, 10
TSH, 192, 221
Tuberculum olfactorium, mode of action studies, 11
Tubero-infundibular dopaminergic system, 216
Turtle isolated olfactory bulb, electrographic studies, 94, 97
Tyr-D-Ala-Gly-NH-C_2H_4-C_6H_5, 209
Tyr-D-Ala-Gly-Phe-Met-NH-C_3-H_2, 209
Tyr-Arg, 51, 69
Tyr-Gly-Gly-Phe, 141
(D-Tyr[1])-Met-E, 209, 211
Tyrosine, 12, 15—16
Tyrosine hydroxylase, 6, 16—18
Tyr-D-Ser-Gly-Phe-Met-*O*-methyl, 209

V

Vagal tone, 162, 164
Vagus nerve, vegetative regulation studies, 161
Vascular tone, 159—160
Vas deferens preparation, neuroendocrine function studies, 211
Vasoconstriction, 161, 194
Vasodepression, 156, 160, 162, 169
Vasodilatation, 156, 158—159, 161, 180
Vasopressin, release of, 9, 92, 157, 167, 214—215, 220—221, 224
Vasoregulatory centers, 160
Vegetative regulation, role of endogenous opioids in, 156—198
cardiovascular effects, 156—168
eye effects, 193—198
gastrointestinal effects, 171—181
respiratory effects, 164—173
thermoregulation modification, 181—193
Ventral root, electrographic studies, 48, 51—52, 65—69
Ventral tegmental area, analgesic and behavioral effects studies, 140
Ventral thalamus, analgesic and behavioral effects studies, 135
Ventricles
analgesic and behavioral effects studies, 110, 114, 123, 125
fourth
electrographic studies, 38
vegetative regulatio studies, 167
third
analgesic and behavioral effects studies, 114, 123, 125
electrographic studies, 38
vegetative regulation studies, 156, 160, 164, 186
vegetative regulation studies, 156, 160, 164, 167, 186—187, 190
Ventromedial hypothalamus, analgesic and behavioral effects studies, 146—147
Ventromedial nucleus of hypothalamus, electrographic studies, 56—57
Ventromedial thalamus, mode of action studies, 20
Veratridine, 9

W

Water intake, modulation of, 144—146
Waxy flexibility, 131
Wet dog shakes, 39, 41, 133—134, 136
Win 35, 197-2, 38
Withdrawal, 53, 57, 63, 79, 161

Z

Zona incerta, mode of action studies, 13—14